MULTIFUNCTIONAL PROTEINS

Multifunctional Proteins

Edited by

Hans Bisswanger and Eva Schmincke-Ott
Physiologisch-chemisches Institut
der Universität Tübingen, Germany

A WILEY-INTERSCIENCE PUBLICATION

JOHN WILEY & SONS, New York • Chichester • Brisbane • Toronto

Library of Congress Cataloging in Publication:

Main entry under title:
Multifunctional proteins.

"A Wiley-Interscience publication."
Includes index.
1. Proteins. 2. Enzymes. 3. Structure-activity relationship (Pharmacology) I. Bisswanger, Hans. II. Schmincke-Ott, Eva. [DNLM: 1. Proteins—Physiology. QU55.3 M961]

QP551.M84 574.1'9245 79-16055
ISBN 0-471-04270-6

Printed in the United States of America

10 9 8 7 6 5 4 3 2 1

CONTRIBUTORS

Hans Bisswanger, Physiologisch-chemisches Institut der Universität Tübingen, Tübingen, West Germany

Dietmar G. Braun, Basel Institute for Immunology, Basel, Switzerland

Georges N. Cohen, Unité de Biochimie Cellulaire, Département de Biochimie et Génétique Microbienne, Institut Pasteur, Paris, France

R. John Collier, Department of Microbiology, College of Letters and Science and The Molecular Biology Institute, University of California, Los Angeles, California

Irving P. Crawford, Department of Microbiology, University of Iowa, Iowa City, Iowa

Norman P. Curthoys, Department of Biochemistry, University of Pittsburgh School of Medicine, Pittsburgh, Pennsylvania

Alice Dautry-Varsat, Unité de Biochimie Cellulaire, Département de Biochimie et Génétique Microbienne, Institut Pasteur, Paris, France

Rebecca P. Hughey, Department of Biochemistry, University of Pittsburgh School of Medicine, Pittsburgh, Pennsylvania

Jürgen Kania, Universitätskliniken Köln, Abteilung Experimentelle Innere Medizin, Köln, West Germany

Horst Kleinkauf, Max-Volmer-Institut für Physikalische Chemie und Molekularbiologie, Abteilung Biochemie, Technische Universität Berlin, Berlin, West Germany

Hans Klenow, Biokemisk Institut B, Panum Instituttet, University of Copenhagen, Denmark

Hans Koischwitz, Max-Volmer-Institut für Physikalische Chemie und Molekularbiologie, Abteilung Biochemie, Technische Universität Berlin, Berlin, West Germany

John J. Mekalanos, Department of Microbiology, College of Letters and Science, University of California, Los Angeles, California

Benno Müller-Hill, Institut für Genetik der Universität Köln, Köln, West Germany

Wolfgang Schalch, Basel Institute for Immunology, Basel, Switzerland

Eva Schmincke-Ott, Physiologisch-chemisches Institut der Universität Tübingen, Tübingen, West Germany

Eckhart Schweizer, Lehrstuhl für Biochemie Institut für Mikrobiologie und Biochemie Universität Erlangen-Nürnberg, Erlangen, West Germany

Howard Zalkin, Department of Biochemistry, Purdue University, West Lafayette, Indiana

PREFACE

A few years ago Kaspar Kirschner and I decided to write a review article for *Annual Reviews of Biochemistry* (K. Kirschner and H. Bisswanger, *Annual Reviews of Biochemistry*, 1977). Our interest was stimulated by the work on the multifunctional protein phosphoribosyl-anthranilate isomerase: indoleglycerol-phosphate synthase, an enzyme that carries two different catalytic functions on one polypeptide chain. We wondered whether proteins of this type originated only from random fusion of two adjacent structural genes and rather expected to find for our review a collection of peculiar proteins whose structure would merely reflect a whim of nature. As the search went on the number of examples grew, and we finally had to accept the fact that the structural principle of multifunctional proteins was that of a class of proteins comparable to multienzyme complexes. It proved to be impossible to cover all aspects in one article, and we had to be content with a general view of the field.

The favorable reaction to this review suggested that, in the meantime, multifunctional proteins have been recognized as an interesting and relevant class of proteins by many biochemists.

When I shifted my field of investigation, working now with Eva Schmincke-Ott on multienzyme complexes, the problem of multifunctionality remained of interest to us, and we decided to start this book in order to have a more thorough and expert description of multifunctional proteins than the previous review could afford.

It is our aim to demonstrate a series of individual systems and by way of their detailed characterization point to different typical aspects of multifunctional proteins. The introduction summarizes these aspects and dwells on some interesting questions raised by the evolution of multifunctional proteins. It also gives a compilation of the known and reasonably well investigated examples to date.

It was rather difficult to find an exact definition of multifunctional proteins since most proteins will comprise several functions depending on how one would conceive the term "function." Allosteric proteins for instance may unite a catalytic and a regulatory function on one polypeptide chain, proteinases and other proenzymes carry a covalently bound protective peptide, and a large number of proteins combine binding or structural functions with some biological activity.

On the other hand, there are indeed many obviously multifunctional proteins, and in the definition of multifunctional proteins we used we referred to proteins that carry on one polypeptide chain two or more virtually autonomous functions that in principle are measurable or occur independently of one another. (The latter is sometimes found when an organism other than the one producing the multifunctional protein is investigated.) Of course there are always examples that escape a clear-cut definition, therefore to some extent our selection is but a personal judgment, and the table of multifunctional proteins in the introduction must be read in a qualified sense, keeping in mind the difficulties that arise when defining independent functions or interpreting yet incomplete experimental evidence in order to achieve a fairly systematic compilation.

We are aware that this monograph represents only the momentary situation of research in this field. Considering the pace at which protein isolation methods are becoming more effective and the increased availability of protein characterizations, it is difficult to cover the continually growing number of multifunctional proteins. However, we hope to have collected enough examples to convey a useful concept of this field.

We are indebted to several people who have supported our work on this book. They are Dr. V. Braun, Dr. I. P. Crawford, Dr. U. Henning, Dr. K. Kirschner, Dr. E. Schweizer, and Dr. H. Zalkin, who have kindly read the introduction, Dr. R. Clark, who helped with the English phrasing, and Frau G. Knödler, who handled the mail and manuscripts.

Hans Bisswanger
Eva Schmincke-Ott

Tübingen, Germany
November 1979

CONTENTS

Chapter 1

INTRODUCTION

EVA SCHMINCKE-OTT AND HANS BISSWANGER
Physiologisch-chemisches Institut der Universität Tübingen
Tübingen, Germany

1.1 PERSPECTIVES

Former concepts of the relationship between the structure and function of proteins were based on the hypothesis that one polypeptide chain would be responsible for a single function, and at first the general experimental evidence supported this notion. This view has had to be altered, however, as more refined techniques, especially x-ray diffraction analysis, became available and allowed a more exact and detailed analysis of protein tertiary structure.

Largely as the result of comparative structural studies of dehydrogenases in the laboratory of M. G. Rossman (1, 2), as well as the elucidation of immunoglobulin structures by G. M. Edelman and colaborators (3, 4), it seems reasonable to consider a single polypeptide chain as subdivided into a number of globular regions called domains, each of which is responsible for a certain defined function. Domains then appear to constitute the structural and functional building blocks of proteins.

It seems to be a general principle in biology that complex organizational forms develop as the result of the repetition of simpler structures. An initial simple multiplication of a number of structural elements is followed by their variegation and specialization such that their common origin is no longer easily recognizable. This principle also appears to hold for the evolution of complex protein systems. Figure 1.1 gives an idea of the possible organizational schemes for complex proteins. In such proteins the combined operation of several structural and functional elements is the rule. The number of possible combinations of such elements is small. Either homogeneous or hetereogeneous elements may be combined, and these may either be covalently bound, or noncovalently associated.

A large number of proteins are composed of subunits that are identical. Most enzymes are constructed in this way (5), and of these some are subject to allosteric regulation. If the identical subunits are covalently connected, a protein containing repeated amino acid sequences results. Examples of this latter class are not rare, but in only a few cases is the original function maintained in all of the repeated sequences.

A smaller proportion of proteins [about 10%—roughly estimated from a compilation of oligomeric proteins by I. Klotz (5)] is made up of heterogeneous subunits. Where these are noncovalently associated they form a multienzyme complex in which autonomous functions are performed by separate polypeptide chains. If, on the other hand, the heterogeneous subunits are covalently connected, a multifunctional protein results in which autonomous functions are distributed among the domains of a polypeptide chain. The number of proteins made up of heterogeneous subunits appears to be about the same in both classes.

However, in comparison with the other classes of proteins multifunctional proteins *per se* have been given little attention until recently. This stems in

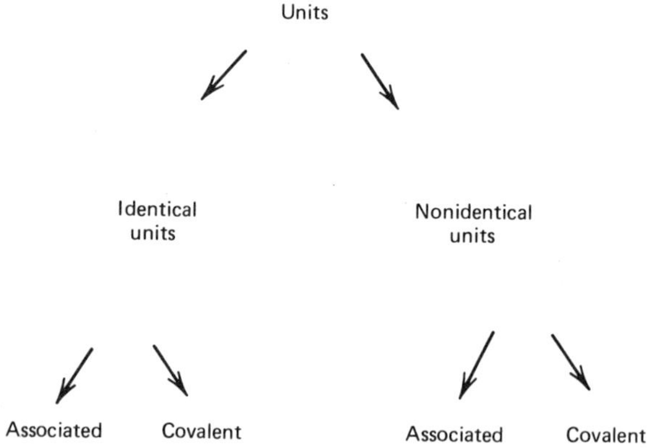

Fig. 1.1 Organizational scheme of complex proteins.

part from the fact that complex proteins with different functions used to be simply regarded as multienzyme complexes. Fatty acid synthetase, for example, used to be taken as an ideal example of a multienzyme complex (6, 7), but more critical investigations have revealed that it is in fact composed of multifunctional polypeptide chains (8–16).

Two simple techniques have led to the discovery of an increasing number of multifunctional proteins: improved purification procedures and polyacrylamide gel electrophoresis under protein denaturing conditions. Purification was improved by using gentle extraction and separation methods and performing them in the presence of proteolysis inhibitors. This is especially important since it has been found that many multifunctional proteins are particularly vulnerable to proteolytic attack in the area between two functional regions. This could very well lead to the destruction of the structural entity while the respective constituent activities remained unimpaired. For instance cytochrome b_5 was isolated, crystallized (17), and submitted to x-ray diffraction studies (18) before it was realized that a proteolytic artefact had been investigated (19). Polyacrylamide gel electrophoresis performed under denaturing conditions has provided a tool for estimating in a fast, simple, and fairly accurate fashion the minimal size of protein subunits.

The growing number of multifunctional proteins (20, 21, 22) shows that we are not simply dealing with a few cases of accidentally fused polypeptide chains, but with a typical organizational scheme for proteins comparable to that of multienzyme complexes. It would be interesting to find out whether the differences between the two organizational schemes are sufficiently significant to explain why in one organism multifunctional proteins occur and in the other multienzyme complexes, and why in some organisms the functions are completely separate.

1.2 CHARACTERIZATION OF MULTIFUNCTIONAL SYSTEMS

Definition of Multifunctional Proteins

Only proteins that combine several autonomous functions on one polypeptide chain should be accepted as multifunctional proteins. In its functional aspect this definition should describe autonomously measurable functions, not ruling out certain regulatory interdependencies between these functions. The definition is also meant to encompass the structural aspect of multifunctionality. Autonomy in this sense implies that each function is assigned to a distinct region, that is a domain, on the polypeptide chain.

Excluded are enzymes that can catalyze different reactions using the same reaction center, like asparaginase which can work as glutaminase (23), glucose-6-phosphatase which can catalyse two reactions (24), or phosphoglyceromutase which can catalyse three reactions always using the same reaction center (25, 26). In these cases we are only dealing with variations of one common function, so that the expression "multifunctional" is not appropriate. Allosteric enzymes and proenzymes will not be considered multifunctional either, although often they are proteins that carry a catalytic function and a regulatory or protective function on different regions of one polypeptide chain. However, these supplementary functions are not autonomous, they are only defined with respect to the catalytic function.

As constituents of multifunctional proteins we find mainly catalytical functions but also binding functions, as long as they are autonomous, like the membrane binding of amphipathic proteins, antigen binding of antibodies, or the DNA binding of lac repressor.

Review of Multifunctional Protein Systems

Table 1.1 includes a collection of the known examples of multifunctional proteins. For most of the systems, there is ample evidence showing them to be multifunctional proteins, but there are some systems listed that lack decisive evidence in this sense. Usually some details of their structure are not known. They are included for the sake of completeness however, and because they will be used as specific examples. The table shows that the class of multifunctional proteins is represented in all organisms from bacteria to mammals.

By far the largest group consists of enzymes that combine two or more functions. There are virtually no examples where the different functions are completely unrelated to each other. Usually they are catalytic functions in the same metabolic pathway, sometimes even of sequential steps. The non-catalytic functions in multifunctional proteins encompass transport and storage functions of yeast hemoglobin (134) and bovine serum albumin (119), the membrane binding of toxins (105) and amphipathic enzymes (94), as well as the binding functions of antibodies (127) and *lac* repressor (122, 123). The latter is an interesting and structurally well investigated example of cooperation between domains, the operator, and inducer binding sites. In myosin the structuring function of light meromyosin that builds up the strong filaments is combined with the function of motility executed by the S1 domain of heavy meromyosin and its ATPase activity (104).

Evidence for the Structure of Multifunctional Proteins

For a protein to qualify as multifunctional it must have had a thorough analysis of its structure. Two important criteria have to be fulfilled:

1 The existence of more than one function on one single polypeptide chain.

2 The autonomy of these functions, which may be demonstrated by the existence of distinct domains for the different functions on this polypeptide chain.

Genetic analysis may produce evidence for the existence of only one polypeptide chain when it is possible to obtain single point mutants deficient in more than one function. Pleiotropic effects must be excluded. If it is impossible to obtain complementation in mutants that are defective in alternative functions this fact may suggest multifunctionality in a protein as well.

The proper physical methods would be ultracentrifugation, gelfiltration, and polyacrylamide gel electrophoresis in the presence of and without denaturing agents. In the case of a multifunctional protein the number of autonomous functions must exceed the number of separable protein bands. One must be aware of the homogenity of the protein band, this may be achieved by endgroup analysis or the comparison of the theoretical number of tryptic peptides with the experimental amount.

Genetic analysis may also indicate autonomous domains if it is possible to isolate mutants that are defective in only one function. At the expense of time and means, very conclusive evidence may be derived from setting up an exact genetic map as was done for phosphoribosyl-anthranilate isomerase: indoleglycerolphosphate synthase from *E. coli* (37, 135). By a thorough analysis of a large collection of mutations, illustrative evidence was obtained about position and size of domains on a polypeptide chain in the case of an artificial multifunctional protein, *lac* repressor: β-galactosidase (122).

Comparison of the respective proteins from more or less related organisms may be helpful. Here one may find examples where the different functions are located on separate polypeptide chains whose molecular weights add up to about the molecular weight of the multifunctional chain. Distinct autonomous regions on a polypeptide chain usually show differing pH optima and temperature sensitivity with respect to their functions. It is possible to block one function selectively with competitive inhibitors or with amino acid specific agents without influencing the other function. Evidence along this line may come from enzyme kinetic studies or ligand binding studies. The most elegant and convincing method is to isolate and to characterize fragments that have retained their own function unimpaired. The N-terminal domains may be produced and isolated using polar mutants. Limited proteolysis is a more promising technique since in many systems the peptide bridges between domains are especially sensitive to proteolytic attack (136). Illustrative examples are DNA polymerase I from

Table 1.1 List of Multifunctional Proteins

System (source)	Structure[a] and Molecular Weights of the Subunits (daltons · 10^3)	Fused Functions	Size of the Domains (daltons · 10^3)	Cleavage in Active Fragments Possible (+/−)	References
Fused catalytic functions					
Two catalytic functions fused (bifunctional enzymes)					
Aspartokinase I: homoserine dehydrogenase I (*Escherichia coli*)	α_4 α: 86	A: aspartokinase B: homoserine dehydrogenase	A: 48 B: 55	A: + B: +	27, 29
Aspartokinase II: homoserine dehydrogenase II (*E. coli*)	α_2 α: 88	A: aspartokinase B: homoserine dehydrogenase	A: 35 B: 37	A: + B: +	30, 31
Anthranilate synthase (*Euglena gracilis*)	α α: 80	A: anthranilate synthase B: amidotransferase		−	32
Anthranilate synthase (*E. coli*, *Salmonella typhimurium*)	$\alpha_2\beta_2$ α: 62 β: 62	A: amidotransferase B: phosphoribosyltransferase	A: 20 B: 40	A: + B: −	33, 35
Indoleglycerol-phosphate synthase (*E. coli*)	α α: 48	A: phosphoribosyl-anthranilate isomerase B: indoleglycerol-phosphate synthase	A: 25 B: 25	A: (+) B: (+)	36, 37

Tryptophan synthase (*Neurospora crassa*)	α_2 α: 75	A: cleavage of indole-glycerol-phosphate B: combination of indole and serine	A: 29 B: 45	–	38
Chorismate mutase: prephenate dehydrogenase (*E. coli*, *S. typhimurium*, *Aerobacter aerogenes*)	α_2 α: 40	A: chorismate mutase B: prephenate dehydrogenase		A: (+) B: –	39–46
Chorismate mutase: prephenate dehydratase (*E. coli*, *S. typhimurium*, *A. aerogenes*)	α_2 α: 40	A: chorismate mutase B: prephenate dehydratase		A: (+) B: (+)	41, 42, 47–49
3-Deoxy-D-arabinoheptulosonate-7-P synthase: chorismate mutase (*Bacillus subtilis*)	α_4 α: 39	A: 3-deoxy-D-arabinoheptulosonate-7-phosphate synthase B: chorismate mutase	A: 29	A: + B: (+)	50–53
Formiminotransferase[b] cyclodeaminase (pig liver)	α_8 α: 64	A: formiminotransferase B: cyclodeaminase		A: – B: –	54, 55
Debranching enzyme (rabbit muscle)	α α: 160	A: amylo-1,6-glucosidase B: 4-α-glucanotransferase		A: – B: –	56–59
Flavocytochrome b_2 (*Saccharomyces cerevisiae*, *Hansenula anomala*)	α_4 α: 58,5	A: L-lactate-cytochrome c-oxidoreductase B: cytochrome b_2		–	60–63

Table 1.1 (*Continued*)

System (source)	Structure[a] and Molecular Weights of the Subunits (daltons · 10^3)	Fused Functions	Size of the Domains (daltons · 10^3)	Cleavage in Active Fragments Possible (+/−)	References
DNA polymerase[b] (bacteriophage T4)	α α: 115	A: DNA polymerase B: 3′-5′-exonuclease		–	64
More than two catalytic functions fused					
DNA polymerase I (*E. coli*)	α α: 109	A: DNA polymerase B: 5′-3′-exonuclease C: 3′-5′-exonuclease	A: 70 B: 35	A: + B: +	65–68
Anthranilate synthase (*N. crassa*)	$\alpha_2\beta_2$ α: 94 β: 70	A: glutaminase B: PRA isomerase[c] C: IGP synthase	A: 30 B + C: 80	A: + B + C: + B: – C: –	69, 70
Formyl-methenyl-methylenetetrahydro-folate synthetase (combined) (mammalian, *S. cerevisiae*)	α_2 α: 104	A: formyltetrahydro-folate synthetase B: methenyltetra-hydrofolate dehydrogenase C: methenyltetra-hydrofolate cyclohydrolase			71–73

Carbamylphosphate synthetase (ascites hepatoma cells)	α_4 α: 210	A: carbamylphosphate synthetase B: aspartate trans-carbamylase C: dihydroorotase	–	74
Arom-complex (*N. crassa*)	α_2 α: 150	A: DHQ synthetase[c] B: DHQase C: DHS reductase D: shikimate kinase E: EPSP synthetase	(+)	75–78
Gramicidin S-synthetase (*Bacillus brevis*)	α α: 280	specific joining of 5 different amino acids, cyclization reaction	–	79–81
Fatty acid synthetase (*S. cerevisiae*)	$\alpha_6\beta_6$ α: 185 β: 180	A: acetyl trans-acylase B: malonyl trans-acylase C. β-ketoacyl synthase D: β-ketoacyl reductase E: β-hydroxyacyl dehydrase F: enoyl reductase G: palmityl deacylase α: reactions B, E, F, G β: C, D and acyl-carrier protein	–	8, 9, 14, 82–84

Table 1.1 (*Continued*)

System (source)	Structure[a] and Molecular Weights of the Subunits (daltons · 10^3)	Fused Functions	Size of the Domains (daltons · 10^3)	Cleavage in Active Fragments Possible (+/−)	References
Fatty acid synthetase (*Mycobacterium smegmatis*)	$\alpha_{6\text{-}8}$ α: 290	reactions as the enzyme from *S. cerevisiae*		−	85
Fatty acid synthetase (avian, mammalian)	$\alpha\beta$ α: 220 β: 220	reactions as the enzyme from *S. cerevisiae* α: D, E, F, G β: A, B, E, G	G: 32	G: +	11–15 86–88
Fatty acid[b] synthetase (rabbit mammary gland)	α_3 α: 300-350	reactions as the enzyme from *S. cerevisiae*			16
Fatty acid[b] synthetase (goose uropygial gland)	α_2 α: 270	reactions as the enzyme from *S. cerevisiae*	G: 33	G: +	89, 90
Catalytic and Non-catalytic functions fused					
Hemoglobin: reductase (*Candida mycoderma*)	α α: 50	A: oxygen storage B: diaphorase		−	91, 92
Cytochrome b_5 (calf liver microsomes)	α α: 17	A: cytochrome b_5 B: anchor function	A: 97 amino acids B: 44 amino acids	A: + B: +	19, 93–96

NADH-Cytochrome b_5-reductase (calf liver microsomes)	α α: 43	A: NADH-cytochrome b_5-reductase B: anchor function	A: 292 amino acids B: 99 amino acids	A: + B: +	94, 97
Myosin (mammalian)	α_2 α: 250	A: ATPase, contractile element B: structural function (thick filaments)	A: 115 (heavy meromyosin) B: 150 (light meromyosin	A: + B: +	98–104
Diphtheria toxin (*Corynebacterium diphtheriae*)	α α: 62	A: inactivation of the elongation factor EF2 B: transport of the catalytic active protein across the membrane	A: 24 B: 38	A: + B: +	105–107
RNA ligase[b] (bacteriophage T4)	α α: 42	A: RNA ligase B: tail fiber attachment		–	108
Pyruvate carboxylase (*Pseudomonas citronellolis*)	$\alpha_2\beta_2$ α: 65 β: 54	A: biotin carboxylase B: biotin-carboxyl-carrier-protein (BCCP)	–		109
Pyruvate carboxylase (avian, mammalian, yeast, *Arthrobacter globiformis*, *Bacillus stearothermophilus*)	α_4 α: 130	A: biotin carboxylase B: BCCP C: carboxy transferase	B: 19	A: O: – B: +	109–111

Table 1.1 (*Continued*)

System (source)	Structure[a] and Molecular Weights of the Subunits (daltons · 10^3)	Fused Functions	Size of the Domains (daltons · 10^3)	Cleavage in Active Fragments Possible (+/−)	References
Acetyl-CoA carboxylase (*S. cerevisiae*, avian, mammalian)	α_4 α: 190 (*S. cerevisiae*) 125 (mammalian)	A: biotin carboxylase B: BCCP C: carboxy transferase		−	110
Acetyl-CoA carboxylase (*Candida lipolytica*)	α α: 230	A: biotin carboxylase B: BCCP C: carboxy transferase		−	112
β-Methylcrotonyl-CoA carboxylase (*Achromobacter*)	$\alpha_4\beta_4$ α: 96 β: 78	A: biotin carboxylase B: BPPC		−	113–115
Noncatalytic functions fused					
Serum albumin (human, bovine)	α_2 α: 65	independent binding and transport functions of tryptophan, bilirubin and long chain fatty acids	190 amino acids (3 homologous domains)	+	116–121

Lac repressor (*E. coli*)	α_4 α: 38	independent binding and regulatory functions of inducer (A) and operator (B)	A: about 80 amino acids B: about 90 amino acids	A: + B: +	122–125
Immunoglobulins (mammalian)	$\alpha_2\beta_2$ α: 53 (heavy chain) β: 22 (light chain)	antigen binding complement binding	12	+	4, 126–131
Ribosomal[b] protein S1 (*E. coli*)	α α: 65	A: mRNA-binding B: 16S rRNA-binding		–	132, 133

[a] The structure of the whole protein molecules is indicated. In the case of different polypeptide chains the multifunctional one is underlined.

[b] Less well documented examples.

[c] Abbreviations: PRA, *N*-(5′-phosphoribosyl)-anthranilate; IGP, indole-3-glycerol-phosphate; DHQ dehydroquinate; DHS, dehydroshikimate; EPSP, 3-enolpyruvylshikimate.

E. coli (66, 67, 137, 138), aspartokinase: homoserine dehydrogenase from *E. coli* (139, 140), anthranilate synthase: phosphoribosyl transferase from *S. typhimurium* (33, 34), and immunoglobulins (4, 128). By complementation experiments with fragments one may obtain valuable information about interdomain interactions.

1.3 BIOLOGICAL ASPECTS

As already mentioned, the single functions of multifunctional proteins always show some degree of relatedness to each other. This makes acceptable the statement that we are not just dealing with products of accidentally fused genes, but rather we may expect that the covalent connection between functional units has more often than not some selective advantage, or at least no disadvantage. The question becomes more direct when these selective advantages are considered in relation to those advantages that the noncovalent association found in a multienzyme complex can afford.

In some cases it becomes quite obvious that such connected functional domains draw an advantage from the rigid topology of their active centers. In the model of fatty acid synthesis as proposed by F. Lynen, the growing acyl chain moves as a substrate along six active centers during one reaction sequence. All this time it stays covalently bound to the acyl-carrier protein (ACP) by means of a 4-phosphopantethein spacer. This ACP-arm has to be able to reach all six reaction centers optimally. Very slight conformational deformations as are induced by changes in pH or salt concentrations, which might afflict the position of only one center, would suffice to block the whole reaction sequence. The rigid arrangement of all six centers on only two polypeptide chains (8) might thus guarantee optimal efficiency.

A similar model was suggested by L. J. Reed for the pyruvate dehydrogenase complex from *E. coli*, an enzyme system of associated monofunctional polypeptide chains (141). The protein bound lipoyl moiety was believed to oscillate between three different catalytic centers. More recent investigations, including the use of affinity markers (142, 143) and evaluation of molecular distances by energy transfer measurements (144) have shown that it is not necessary for one lipoyl moiety to react with these three active centers in given positions, but that the substrate could just as well be moved across several lipoyl moieties to various active centers, there being always 16 of a kind (145). The rigid arrangement which the multienzyme complex can not provide is thus seen to be avoided.

Another interesting example is the gramicidin S-synthetase from *B. subtilis*. Here five different amino acids have to be recognized, put into proper sequence, and finally connected to form a pentapeptide (146). One would also expect that, as in the case of fatty acid synthetase, the covalent linkage of reaction centers would prove more efficient for the total reaction than mere association of separate polypeptide chains (80, 81).

A different topological problem seems to be solved by the connection of functional units in the case of DNA polymerase I from *E. coli*, and in the glycogen debranching enzyme from yeast. In each of these cases two different reactions occur at spacially separate sites of the macromolecular substrate. DNA polymerase I is able to split off incorrect nucleotides with its 5′-3′-exonuclease activity producing single stranded DNA and at the same pace insert correct complementary nucleotides with its polymerase activity (68, 147). The debranching enzyme transfers the branches of a glycogen chain to the main chain, leaving a α-1,6 bound glucose residue at the branching point. This is hydrolysed by an amylo-1,6-glucosidase activity. During this reaction sequence the substrate macromolecule does not have to dissociate from the enzyme, although it changes its shape considerably (59).

Covalent connections between reaction centers not only may be superior under such topological apects but also by the mere strength of the connection. This kind of bond stays intact independent of any repellent forces, for instance, those between hydrophilic and hydrophobic compounds. Constructions like the amphipathic proteins become feasible: a markedly hydrophilic enzyme is anchored to the membrane by a strongly lipophilic region of the protein (93, 148). Myosin is another example where the strength of the bond between functional units is critical. The thick muscle filament is connected to the thin actin filament by a covalent peptide bridge. Conformational changes, probably induced by the adenosine triphosphatase activity of the S1 domain, shift the two filaments against each other. The resulting movement of the filaments is the basis for muscle contraction (149). In a similar fashion antibodies are constructed with several flexible regions between the different domains of the molecule. The so called "switch peptide" is actually the covalent connection between two functional units of the antibody molecule, the variable and the constant region. The flexibility of the two parts is well documented by fluorescent polarisation measurements (150), small angle x-ray scattering (151), and various other observations. It seems to play a part in the relationship of antigen binding and effector functions (3).

So far we have only mentioned examples where the covalent bond is clearly superior to the association of separate polypeptide chains. In other cases the proximity of reaction centers provides an advantage not appreciably different from the noncovalent association. By linking two sequential reaction steps the concentrations of intermediate reactants are elevated, and binding becomes more efficient. Here anthranilate synthase: phosphoribosyl transferase (33) and phosphoribosylanthranilate isomerase: indoleglycerol-phosphate synthase (36) are good examples. Tryptophan synthase (38) and chorismate mutase: prephenate dehydrogenase (43, 152) are multifunctional enzymes that show during their catalysis no detectable reaction intermediates. They are channeled from the first reaction center to the next one without being released from the protein because the proximity of reaction centers has led to the development of a compound active site. Coordination of function is also observed in the case of diphteria toxin and related toxins. The toxic enzymic domain is transported across the membrane into the interior of the cell by the membrane binding domain (105). Hemoglobin from yeast carries a fused reductase activity designed to protect the hemoglobin moiety from autoxidation (134).

The fusion of active centers may also serve to direct regulatory signals. This is the case with aspartokinase I: homoserine dehydrogenase I from *E. coli*. The feedback inhibitor threonine exerts a positive cooperative effect on the aspartokinase I domain which is enforced upon the homoserine dehydrogenase I domain (27). As an isolated fragment the latter does not show this property (139). This common regulation appears to be of biological significance as aspartokinase I catalyzes the initial step of the threonine biosynthesis, and homoserine dehydrogenase is active at the branching point of methionine and threonine biosynthesis. A further interesting case of coordinated regulation is met with the arom complex. The first substrate 3-deoxy-D-arabino-heptulosonate-7-phosphate activates simultaneously four out of five catalytic activities by binding to only one allosteric binding site. Induced by this substrate binding there seems to be a change in conformation that seizes the whole complex. In favor of this interpretation is the observation that this first substrate is able to protect all five catalytic activities against proteolysis.

Two final, less spectacular, aspects of the advantage of multifunctional structures should be mentioned. The covalent association of two domains makes the association independent of protein concentration. This effect, combined with the necessarily concerted regulation of biosynthesis because of synchronous protein expression serves the economy of protein synthesis. There is some speculation that this kind of concerted regulation of biosynthesis may be a general principle in eucaryotic cells, as a substitute of the operon structure on procaryotic cells (22). However so far there is no experimental evidence to justify this generalization.

Considering the sum of examples cited in this introduction, it appears that the class of multifunctional proteins comprises a whole spectrum of molecules more or less well adapted to various biological functions. Examples of very strong coordination between domains are confronted with equally clear independence of the linked domains. For instance in the case of the fused enzymes of the tryptophan operon anthranilate synthase component II: phosphoribosyl transferase and phosphoribosyl-anthranilate isomerase:indoleglycerol-phosphate synthase no interaction between domains has been found as yet. In some other organisms of comparable levels of organization these enzymes are fused, in others not: anthranilate synthase component II is fused with phosphoribosyl transferase in *E. coli* (33); in *Euglena gracilis* anthranilat synthase component I and anthranilate synthase component II are fused, phosphoribosyl transferase activity is located on a separate chain in this case (32). The gap between minimal and optimal interaction between functional units is larger than in the related class of multienzyme complexes. This may be explained by the completely different evolution of multifunctional proteins and multienzyme complexes.

1.4 EVOLUTIONARY ASPECTS

What is the part multifunctional proteins play in the evolution of protein structures? Are they precursors of multienzyme complexes, or do the latter represent an intermediate in the development of multifunctional proteins? Based on structural information about bacterial and *N. crassa* tryptophan synthase, Bonner et al. (155) have postulated in 1965 the development of complex proteins starting from single monofunctional polypeptide chains, passing on to associated, partially independent components, and ending with fused components as the most highly developed organizational scheme. The increasing number of multifunctional systems, as well as the fact that our knowledge of genetic recombination has become more profound, requires a second look at this fairly simple theory.

A comparative review of multienzyme complexes and multifunctional proteins does not furnish any corroboration for Bonner's theory on the one hand. For instance in the family of biotin enzymes all three types of structural organization are present, but there is no obvious system in the distribution of organization forms from lower to higher organisms (109, 110, 113). Even considering the tryptophan pathway, there is the phosphoribosyl transferase in *E. coli* combined with the amidotransferase component of the anthranilate synthase complex, whereas in *N. crassa* it exists as an isolated polypeptide chain (33, 69). In Table 1.1 there are some more examples listed that illustrate the same point. On the other hand, Bonner's theory has a series of improbable implications: the noncovalently associated proteins to be fused must have neighboring genes, the ends of the polypeptide chains to be connected must have the right orientation, furthermore, these parts should be dispensable as they would have to serve only as a bridge between fused domains, and last but not least, the very fusion has to have some selective advantage beyond the simple spacial closeness.

To set up a more differentiated model of multifunctional protein evolution it is helpful to discuss the possible conditions that can lead to the formation of these structures. There are basically two different possibilities for the evolution of a second functional unit on a polypeptide chain.

1 The development of a new functional unit on an originally monofunctional polypeptide chain.
2 The fusion of two genes, whereby the added function may be conserved or modified to a new one later on.

Though the development of a completely new function on a preexisting polypeptide chain seems to be a very rare and quite protracted event, the so-called "silent" or "intronic" regions within eucaryotic genes may be a good tool for the organism to probe such new functions (156). The fusion of two different genes may proceed in three different ways.

1 Gene duplication.
2 Gene translocation including insertion and gene recombination
3 Gene deletion.

Apparently the mechanism of gene duplication is of general importance for the evolution of proteins. The development of protein families like protamines (183), globins, hemoproteins, cytochrome c, and b type proteins (184–187) can be followed back to short ancestral proteins. These have suffered repeated gene duplications, partial and complete, amino acid exchange by point mutations, and deletions to find the shape and size they present today.

It is the gene duplication that is observed most frequently under experimental conditions. In bacteria it occurs spontaneously or after induction with ultraviolet light with a rate of 10^{-5}. One to 10% of the chromosome may be affected by a gene duplication event, but the duplicated genes are usually unstable (157–160). Stable partial duplications followed by fusion lead to proteins with repeating sequences. A fine example is hemocyanin from mollusces carrying about eight oxygen binding sites which look like pearls on a necklace in electron microscopy (161), or the transacetylase component from *E. coli* pyruvate dehydrogenase complex (143, 145, 162, 163). The repeating sequences are usually detected by amino acid sequence analysis, finger print pattern, or x-ray diffraction analysis.

Upon gene duplication the organism can tolerate modification of the newly won genes by further partial duplications or mutations to such an extent that often the original element can only be identified by comparison with corresponding proteins in other organisms that have not duplicated these genes. This approach has been successful with the t-RNA synthetases (164) or with phosphofructokinase (165, 166). Proteins that have functional units sufficiently similar to suggest their having developed from a gene duplication are listed in Table 1.2. Of the multifunctional proteins we have already mentioned serum albumin as an illustrative example for gene duplications. It is composed of three homologous domains that show different and independent binding functions (116, 117). Further candidates are gramicidin S-synthetase (146) and the antibody molecules (127). Finally the fusion products within an operon should be mentioned.

The close relationship between multienzyme complexes and multifunctional proteins that we have already mentioned is not repeated in the relationship between oligomeric proteins with identical subunits and proteins with repeated sequences. Whereas proteins with identical subunits have but one structural gene, and any genetic change must be expressed equally in all subunits, the protein derived from gene duplication may always keep one copy of the original gene. Then the other gene may develop without any risk for the organisms. An interesting observation is that there are sometimes clusters of homologous genes that are not fused, but undergo so called coincidental evolution as is the case with globins on immunoglobulins. One gene is mutated, and by some unknown mechanism the change is enforced upon the other genes present (182). In this case the phenotype comes out to be just like proteins with identical subunits.

Gene duplication and the translocation of such duplicates might well explain the occurrence of strongly homologous domains in otherwise functionally unrelated proteins like α-lactalbumin and lysozyme (188), haptoglobulin and Bence-Jones proteins (189), immunoglobulins and superoxide dismutase (129), or β-galactosidase and dihydrofolate reductase (152). The great resemblance of nucleotide binding domains in different dehydrogenases, kinases, and in phosphoglyceromutase may also be attributed to a translocation event (186). The same may be true considering a distribution of the glutamine amidotransferase subunit to anthranilate synthase and formylglycine-amidoribonucleotide amidotransferase and possibly other glutamine utilizing enzymes (33, 190). The discovery of transposable elements in procaryotic and eucaryotic systems shows that all cells possess the capacity for such a mechanism (191).

Deletion is probably the least frequent means, due to the concomittant loss of genetic material. Nevertheless translocation of deletions are postu-

Table 1.2 List of Proteins with Gene Duplications

System (source)	Structure and Molecular Weights of the Subunits (daltons · 10^3)	Duplicated Regions: Number	Duplicated Regions: Size (daltons · 10^3)	References
Different tRNA-synthetases, i.e. val- and leu-tRNA-synthetases (*B. stearothermophilus*)	α α: 110	2	50	164
Phosphofructokinase (rabbit muscle)	α_4 α: 85	2	40	165, 166
Dihydrolipoamide transacetylase (*E. coli*)	α_{16} α: 80	2	36–40	143, 145, 162, 163
Rhodanese (bovine liver)	α α: 32	2	16	167
Ferredoxin (bacteria)	α α: 54 amino acids	2	26–28 amino acids	168, 169
Transferrin (human)	α α: 77	2		170
Haptoglobulin α^2 (mammalian)	α α: 142 amino acids	2	70–80 amino acids	171, 172
β-Galactosidase (*E. coli*)	α_4 α: 1012 amino acids	2	380 amino acids	152

L-Arabinose-binding protein	α	2		173
Immunoglobulins (mammalian)	$\alpha_2\beta_2$ α: 53 β: 22	α: 4 β: 2	12 for α and β	3, 129
Serum albumin (human, bovine)	α_2 α: 65	3 homologous domains, each consisting of 3 nearly homologous subdomains	190 amino acids	116, 117
Murein-lipoprotein (*E. coli*)	α α: 57 amino acids	2 3–4	14 amino acids 6 amino acids	174
Protein A (*Staphylococcus aureus*)	α α: 42	4	58–62 amino acids	175
α-Tropomyosin (mammalian)	α α: 284 amino acids	7 repetitive units of about 42 amino acids, each consisting of 6 repetitive sequences of 7 amino acids		176
Hemoglobin (*Arthropoda*, i.e. *Artemia salina*)	α α; 120	7	18	177, 178
Hemocyanin (*Mollusces*, i.e. *Helix pomatia*)	α_{20} α: 260	7–8	50	161, 179, 180
Erythrocruorin (*Mollusces*, i.e. *Planorbis corneus*)	α_6 α: 300	12	22.3	181

lated in the biosynthesis of antibodies. It is known that in the undifferentiated cell the genes for the variable and the constant region are located apart from each other on the DNA. After cellular differentiation the genes are found to be neighboring and transcribed on one messenger (192, 193).

These three genetic events described in the foregoing paragraphs serve to arrange direct contact between two different genes but do not necessarily lead to their fusion. This will have to take place in a second step, by eliminating the stop signals between the genes. This is a mechanism that has been utilized in artificial gene fusion as well (194).

It is important to stress that, independent of how many mutations have prepared the situation, the fusion itself is always a one step event with the character of an all or nothing decision. This is where the basic difference between covalently bound and the noncovalently associated systems comes in. In the latter case gene products that are not located next to each other may adapt to a common function, as is the case with the dihydroliponamide dehydrogenase component of the α-keto acid dehydrogenase complexes from *E. coli* (195). First, different loose associations at various contacting regions of the two polypeptide chains may have been tried out, and then an optimal spatial orientation may have been selected where the reaction centers are close enough for optimal interaction without steric hindrance. In this way a binding site may have developed stepwise.

In contrast, gene fusion produces proteins that will almost inevitably experience negative interactions. They are unlikely to posses affinity for one another *a priori*, nor are such binding domains produced by the fusion itself. Since the fusion will always connect the carboxyl end of the first component and the amino end of the second component there is no choice of orientation. A fusion product will only be stable when it shows immediate advantages over the separate chains or is at least indifferent. With as little flexibility given for adaptation, successfully accomodated fusions must remain rare. This can also be noticed with artifical fusions where the functions are usually deteriorated, as in the case of histidinol dehydrogenase: imidazolacetol-P-aminotransferase from *S. typhimurium*, if they are not immediately destroyed by proteolysis as is the case with the artifical fusion product of the D and G gene in the his operon, histidinol dehydrogenase: ATP phosphoribosyltransferase (194).

A positive interaction on grounds of the fusion event may virtually be excluded: it can only develop afterwards.

Thus at the fusion level there is no selective pressure for multifunctional proteins. In contrast to multienzyme complexes the unproductive fusion of polypeptide chains is lost. From this point of view we can understand the wide spectrum of none to intense cooperation between the domains of multifunctional proteins.

It becomes obvious that the multifunctional proteins have developed by a principially different evolutionary mechanism than the multienzyme complexes. One class of proteins is not the precursor of the others; rather they seem to be two alternative structures that were generated in the train of complex protein evolution. A transition from multienzyme complexes to multifunctional proteins is probably very rare. It seems though that the choice between one structure or the other is not altogether random but, as discussed in Section 1.3, to a certain extent adapted to various biological problems.

ACKNOWLEDGMENTS

The authors thank V. Braun, I. Crawford, U. Henning, K. Kirschner; and E. Schweizer for valuable comments.

REFERENCES

1 A. Liljas and M. G. Rossmann (1974), *Ann. Rev. Biochem.* **43,** 475–507.

2 M. G. Rossmann, D. Moras, and K. W. Olson (1974), *Nature* **250,** 194–199.

3 G. M. Edelman (1970), *Biochemistry* **9,** 3197–3205.

4 R. Huber, (1976), *TIBS* **1,** 174–178.

5 I. M. Klotz, D. W. Darnall, and N. R. Langerman (1975), in H. Neurath and R. L. Hill, eds. *The Proteins*, Vol. 1, 3rd ed., Academic Press, New York, pp. 293–411.

6 M. Sumper and F. Lynen (1972), in E. Helmreich and R. Jaenicke, eds., *Protein-Protein Interactions*, Springer, Berlin, pp. 365–394.

7 P. C. Yang, P. H. W. Butterworth, R. M. Bock, and J. W. Porter (1967), *J. Biol. Chem.* **242,** 3501–3507.

8 A. Knobling and E. Schweizer (1975), *Eur. J. Biochem.* **59,** 415–421.

9 A. Knobling, D. Schiffmann, H.-D. Sickinger, and E. Schweizer (1975), *Eur. J. Biochem.* **56,** 359–367.

10 J. Elovson (1975), *J. Bact.* **124,** 524–533.

11 J. K. Stoops, M. J. Arslanian, Y. H. Oh, K. C. Aune, T. C. Vanaman, and S. J. Wakil (1975), *Proc. Natl. Acad. Sci. USA* **72,** 1940–1944.

12 A. W. Alberts, A. W. Strauss, S. Hennessy, and P. R. Vagelos (1975), *Proc. Natl. Acad. Sci. USA* **72,** 3956–3960.

13 I.-C. Kim, C. J. Unkefer, and W. C. Deal (1977), *Arch. Biochem. Biophys.* **178,** 475–485.

14 A. A. Qureshi, F. A. Lornitzo, R. A. Jenik, and J. W. Porter (1977), *Arch. Biochem. Biophys.* **177,** 364–378.

15 F. A. Lornitzo, A. A. Qureshi, and J. W. Porter (1975), *J. Biol. Chem.* **250,** 4520–4529.

16 N. Paskin and R. J. Mayer (1976), *Biochem. J.* **159,** 181–184.

17 R. H. Kretsinger, B. Hagihara, and A. Tsugita (1970), *Biochim. Biophys. Acta* **200,** 421–422.

18 F. C. Mathews, P. Argos, and M. Levine (1971), *Cold Spring Harbor Symp. Quant. Biol.* **36,** 387–395.

19 A. Ito and R. Sato (1968), *J. Biol. Chem.* **243,** 4922–4923.

20 K. Kirschner and H. Bisswanger (1976), *Ann. Rev. Biochem.* **45,** 143–166.

21 J. D. Yourno (1972), *Brockhaven Symp. Biol.*, **23,** 95–120.

22 G. R. Stark (1977), *TIBS* **2,** 64–66.

23 L. Davidson, D. R. Brear, P. Wingard, J. Hawkins, and G. B. Kitto (1977), *J. Bact.* **129,** 1379–1386.

24 R. C. Nordlie (1974), *Curr. Top. Cell. Regul.* **8,** 33–117.

25 K. Ikura, R. Sasaki, H. Narita, E. Sugimoto, and H. Chiba (1976), *Eur. J. Biochem.* **66,** 515–522.

26 R. Sasaki, S. Utsumi, E. Sugimoto, and H. Chiba (1976), *Eur. J. Biochem.* **66,** 523–533.

27 G. N. Cohen (1969), *Curr. Top. Cell. Regul.* **1,** 184–231.

28 P. Truffa-Bachi, M. Veron, and G. N. Cohen (1974), *CRC Crit. Rev. Biochem.* **2,** 379–415.

29 K. Wright and M. Takahashi (1977), *Biochemistry* **16,** 1548–1554.

30 A. Dautry-Varsat, L. Sibilli-Weill, and G. N. Cohen (1977), *Eur. J. Biochem.* **76,** 1–6.

31 A. Dautry–Varsat and G. N. Cohen (1977), *J. Biol. Chem.* **252,** 7685–7689.

32 C. N. Hankins and S. E. Mills (1976), *J. Biol. Chem.* **251,** 7774–7778.

33 H. Zalkin (1973), *Adv. Enzymol.* **38,** 1–39.

34 M. Grieshaber and R. Bauerle (1972), *Nat. New Biol.* **236,** 232–235.

35 E. N. Jackson and C. Yanofsky (1974), *J. Bact.* **117,** 502–508.

36 T. E. Creigthon (1970), *Biochem. J.* **120,** 699–707.

37 C. Yanofsky, V. Horn, M. Bonner, and S. Stasiowski (1971), *Genetics* **69,** 409–433.

38 W. H. Matchett and J. A. DeMoss (1975), *J. Biol. Chem.* **250,** 2941–2946.

39 G. L. E. Koch, D. C. Shaw, and F. Gibson (1971), *Biochim. Biophys. Acta* **229,** 795–804.

40 G. L. E. Koch, D. C. Shaw, and F. Gibson (1971), *Biochim. Biophys. Acta* **229,** 805–812.

41 R. G. H. Cotton and F. Gibson (1965), *Biochim. Biophys. Acta* **100,** 76–88.

42 J. Pittard and B. J. Wallace (1966), *J. Bact.* **91,** 1494–1508.

43 G. L. E. Koch, D. C. Shaw, and F. Gibson (1972), *Biochim. Biophys. Acta* **258,** 719–730.

44 G. L. E. Koch, D. C. Shaw, and F. Gibson (1970), *Biochim. Biophys. Acta* **212,** 375–386.

45 G. L. E. Koch, D. C. Shaw, and F. Gibson (1970), *Biochim. Biophys. Acta* **212,** 387–395.

46 M. J. Gething and B. E. Davidson (1976), *Eur. J. Biochem.* **71,** 327–336.

47 B. E. Davidson, E. H. Blackburn, and T. A. A. Dopheide (1972), *J. Biol. Chem.* **247,** 4441–4446.

48 J. C. Schmit and H. Zalkin (1971), *J. Biol. Chem.* **246,** 6002–6010.

49 M. I. Gibson and F. Gibson (1964), *Biochem. J.* **90, 248—256.**

50 E. W. Nester, J. H. Lorence, and D. S. Nasser (1967), *Biochemistry* **6,** 1553–1563.

51 L. Huang, W. M. Nakatsukasa, and E. W. Nester (1974), *J. Biol. Chem.* **249,** 4467–4472.

52 W. M. Nakatsukasa and E. W. Nester (1972), *J. Biol. Chem.* **247,** 5972–5979.

53 L. Huang, A. L. Montoya, and E. W. Nester (1974), *J. Biol. Chem.* **249,** 4473–4479.

54 R. Beaudet and R. E. MacKenzie (1976), *Biochim. Biophys. Acta* **453,** 151–161.

55 R. Beaudet and R. E. MacKenzie (1975), *Biochim. Biophys. Acta* **410,** 252–261.

56 R. C. White and T. E. Nelson (1975), *Biochim. Biophys. Acta* **400,** 154–161.

57 R. C. White and T. E. Nelson (1974), *Biochim. Biophys. Acta* **365,** 274–280.

58 E. J. Bates, G. M. Heaton, C. Taylor, F. C. Kernohan, and P. Cohen (1975), *FEBS Lett.* **58,** 181–185.

59 B. K. Gillard and T. E. Nelson (1977), *Biochemistry* **16,** 3978–3987.

60 M. Mevel-Ninio, Y. Risler, and F. Labeyrie (1977), *Eur. J. Biochem.* **73,** 131–140.

61 F. Labeyrie and A. Baudras (1971), *Eur. J. Biochem.* **25,** 33–40.

62 A. Baudras and A. Spyridakis (1971), *Biochimie* **53,** 943–955.

63 M. Gervais, O. Groudinsky, Y. Risler, and F. Labeyrie (1977), *Biochem. Biophys. Res. Commun.* **77,** 1543–1551.

64 W. M. Huang and I. R. Lehman (1972), *J. Biol. Chem.* **247,** 3139–3146.

65 T. Kornberg and A. Kornberg (1974), *Enzymes* **10,** 119–144.

66 H. Klenow and K. Overgaard-Hansen (1970), *FEBS Lett.* **6,** 25–30.

67 H. Klenow and I. Henningsen (1970), *Proc. Natl. Acad. Sci. USA* **65,** 168–175.

68 D. Uyemura, R. Bambara, and I. R. Lehman (1975), *J. Biol. Chem.* **250,** 8577–8585.

69 A. Arroyo-Begovich and J. A. DeMoss (1973), *J. Biol. Chem.* **248,** 1262–1267.

70 F. M. Hulett and J. A. DeMoss (1975), *J. Biol. Chem.* **250,** 6648–6652.

71 L. U. L. Tan, E. J. Drury, and R. E. MacKenzie (1977), *J. Biol. Chem.* **252,** 1117–1122.

72 J. L. Paukert, G. R. Williams, and J. C. Rabinowitz (1977), *Biochem. Biophys. Res. Commun.* **77,** 147–154.

73 J. L. Paukert, L. D. Straus, and J. C. Rabinowitz (1976), *J. Biol. Chem.* **251,** 5104–5111.

74 M. Mori and M. Tatibana (1975), *J. Biochem.* **78,** 239–242.

75 N. W. Giles, M. E. Case, C. W. H. Partridge, and S. I. Ahmed (1967), *Proc. Natl. Acad. Sci. USA* **58,** 1453–1460.

76 F. H. Gaertner and K. W. Cole (1977), *Biochem. Biophys. Res. Commun.* **75,** 259–264.

77 J. Lumsden and J. R. Coggins (1977), *Biochem. J.* **161,** 599–607.

78 J. Lumsden and J. R. Coggins (1978), *Biochem. J.* **169,** 441–444.

79 K. Kurahashi (1974), *Ann. Rev. Biochem.* **43,** 445–459.

80 C. Christiansen, K. Aarstad, T.-L. Zimmer, and S. G. Laland (1977), *FEBS Lett.* **81,** 121–124.

81 H. Koischwitz and H. Kleinkauf (1976), *Biochim. Biophys. Acta* **429,** 1052–1061.

82 P. Tauro, U. Holzner, H. Castorph, F. Hill, and E. Schweizer (1974), *Mol. Gen. Genet.* **129,** 131–148.

83 E. Schweizer, B. Kniep, H. Castorph, and U. Holzner (1973), *Eur. J. Biochem.* **39,** 353–362.

84 J. K. Stoops, E. S. Awad, M. J. Arslanian, S. Gunsberg, S. J. Wakil, and R. M. Oliver (1978), *J. Biol. Chem.* **253,** 4464–4475.

85 W. I. Wood, D. O. Peterson, and K. Bloch (1978), *J. Biol. Chem.* **253,** 2650–2656.

86 E. Agradi, L. Libertini, and S. Smith (1976), *Biochem. Biophys. Res. Commun.* **68,** 894–900.

87 S. Smith, E. Agradi, L. Libertini, and K. N. Dileepen (1976), *Proc. Natl. Acad. Sci. USA* **73,** 1184–1188.

88 J. K. Stoops, M. J. Arslanian, K. C. Aune, and S. J. Wakil (1978), *Arch. Biochem. Biophys.* **188,** 348–359.

89 C. J. Bedord, P. E. Kolattukudy, and L. Rogers (1978), *Arch. Biochem. Biophys.* **186,** 139–151.

90 P. E. Kolattukudy, J. S. Buckner, and C. J. Bedord (1976), *Biochem. Biophys. Res. Commun.* **68,** 379–386.

91 R. Oshino, T. Asakura, M. Tamura, N. Oshino, and B. Chance (1972), *Biochem. Biophys. Res. Commun.* **46,** 1055–1060.

92 R. Oshino, T. Asakura, K. Takio, N. Oshino, B. Chance, and B. Hagihara (1973), *Eur. J. Biochem.* **39,** 581–590.

93 J. Ozols and C. Gerard (1977), *Proc. Natl. Acad. Sci. USA* **74,** 3725–3729.

94 F. S. Mathews and E. W. Czerwinski (1976), in A. Martonosi, ed., *The Enzymes of Biological Membranes*, Vol. 4, Wiley, New York, pp. 143–197.

95 J. Ozols, C. Gerard, and C. Stachelek (1977), *J. Biol. Chem.* **252,** 5986–5989.

96 M. C. Berman, C. M. Adnams, K. M. Ivanetich, and J. E. Kench (1976), *Biochem. J.* **157,** 237–246.

97 L. Spatz and P. Strittmatter (1973), *J. Biol. Chem.* **248,** 793–799.

98 E. Mihalyi and A. G. Szent-Györgyi (1953), *J. Biol. Chem.* **201,** 189–196.

99 E. Mihalyi and A. G. Szent-Györgyi (1953), *J. Biol. Chem.* **201,** 211–219.

100 H. Mueller and V. S. Perry (1961), *Biochim. Biophys. Acta* **50,** 599–601.

101 H. Mueller (1965), *J. Biol. Chem.* **240,** 3816–3818.

102 S. Lowey, H. S. Slayter, A. G. Weeds, and H. Baker (1969), *J. Mol. Biol.* **42,** 1–29.

103 S. Highsmith, K. M. Kretzschmar, C. T. O'Konski, and M. F. Morales (1977), *Proc. Natl. Acad. Sci. USA* **74,** 4986–4990.

104 H. E. Huxley (1976), in L. M. G. Heilmeyer, J. C. Rüegg, and T. Wieland, eds., *Molecular Basis of Motility*, Springer, Heidelberg, pp. 9–25.

105 R. J. Collier (1975), *Bacteriol. Rev.* **39,** 54–85.

106 A. M. Pappenheimer, (1977), *Ann. Rev. Biochem.* **46,** 69–94.

107 T. Uchida, M. Yamaizumi, and Y. Okada (1978), *Biochem. Biophys. Res. Commun.* **81,** 268–273.

108 T. J. Snopek, W. B. Wood, M P. Conley, P. Chen, and N. R. Cozzarelli (1977), *Proc. Natl. Acad. Sci. USA* **74,** 3355–3359.

109 M. F. Utter, R. E. Barden, and B. L. Taylor (1975), *Adv. Enzymol.* **42,** 1–72.

110 M. Obermayer and F. Lynen (1976), *TIBS* **1,** 169–171.

111 H. G. Wood and R. E. Barden (1977), *Ann. Rev. Biochem.* **46,** 385–413.

112 M. Mishina, T. Kamiryo, A. Tanaka, S. Fukui, and S. Numa (1976), *Eur. J. Biochem.* **71,** 295–300.

113 F. Lynen (1974), in D. Richter, ed., *Lipman Symposium. Energy, Regulation and Biosynthesis in Molecular Biology*, W. de Gruyter, Berlin, pp. 671–698.

114 R. Apitz-Castro, K. Rehn, and F. Lynen (1970), *Eur. J. Biochem.* **29,** 237–248.

115 U. Schiele, R. Niedermeier, M. Stürzer, and F. Lynen (1975), *Eur. J. Biochem.* **60,** 259–266.

116 A. D. McLachlan and J. E. Walker (1977), *J. Mol. Biol.* **112,** 543–558.

117 J. R. Brown (1976), *Fed. Proc.* **35,** 2141–2144.

118 A. F.S.A. Habeeb, and M. Z. Atassi (1976), *J. Biol. Chem.* **251,** 4616–4621.

119 R. G. Reed, R. C. Feldhoff, and T. Peters (1976), *Biochemistry* **15,** 5394–5398.

120 T. P. King (1973), *Arch. Biochem. Biophys.* **156,** 509–520.

121 B. J. Soltys and J. C. Hsia (1978), *J. Biol. Chem.* **253,** 3023–3028.

122 B. Müller-Hill (1975), *Prog. Biophys. Molec. Biol.* **30,** 227–252.

123 S. Bourgeois and M. Pfahl (1976), *Adv. Proteinchem.* **30,** 1–99.

124 J. G. Files, K. Weber, and J. H. Miller (1974), *Proc. Natl. Acad. Sci. USA* **71,** 667–670.

125 K. Beyreuther, K. Adler, N. Geisler, and A. Klemm (1973), *Proc. Natl. Acad. Sci. USA* **70,** 3576–3580.

126 A. R. Williamson (1976), *Ann. Rev. Biochem.* **45,** 467–500.

127 J. D. Capra and A. B. Edmundson (1977), *Scient. Am.* **236,** 50–59.

128 D. R. Davies, E. A. Padlan, and D. M. Segal (1975), *Ann. Rev. Biochem.* **44,** 639–667.

129 R. J. Poljak (1975), *Nature* **256,** 373–375.

130 M. Colomb and R. R. Porter (1975), *Biochem. J.* **145,** 177–183.

131 D. Yasmeen, J. R. Ellerson, K. J. Dorrington, and R. H. Painter (1976), *J. Immunol.* **116,** 518–526.

132 D. E. Draper, C. W. Pratt, and P. H. Hippel (1977), *Proc. Natl. Acad. Sci. USA* **74,** 4786–4790.

133 H. Inouye, Y. Pollack, and J. Petre (1974), *Eur. J. Biochem.* **45,** 109–117.

134 N. A. Hendrickson (1977), *TIBS* **2,** 108–111.

135 O. Smith (1967), *Genetics* **60,** 95–105.

136 C. B. Anfinsen and H. A. Scheraga (1978), *Adv. Proteinchem.* **29,** 205–300.

137 P. Setlow, D. Brutlag, and A. Kornberg (1972), *J. Biol. Chem.* **247,** 224–231.

138 P. Setlow and A. Kornberg (1972), *J. Biol. Chem.* **247,** 232–240.

139 M. Veron, F. Falcoz-Kelly, and G. N. Cohen (1972), *Eur. J. Biochem.* **28,** 520–527.

140 J. Janin, P. Truffa-Bachi, and G. N. Cohen (1967), *Biochem. Biophys. Res. Commun.* **26,** 429–434.

141 L. J. Reed and D. J. Cox (1966), *Ann. Rev. Biochem.* **35,** 57–84.

142 J. H. Collins and L. J. Reed (1977), *Proc. Natl. Acad. Sci. USA* **74,** 4223–4227.

143 D. L. Bates, M. J. Danson, G. Hale, E. A. Hooper, and R. N. Perham (1977), *Nature* **268,** 313–316.

144 G. B. Sheperd and G. G. Hammes (1977), *Biochemistry* **16,** 5234–5241.

145 U. Henning, O. Vogel, W. Busch, and J. E. Flatgaard (1972), in E. Helmreich and R. Jaenicke, eds., *Protein-Protein Interactions*, Springer, Berlin, pp. 343–364.

146 S. G. Laland and T.-L. Zimmer (1973), *Essays Biochem.* **9,** 31–57.

147 R. B. Kelly, N. R. Cozzarelli, M. P. Deutscher, I. R. Lehman, and A. Kornberg (1970), *J. Biol. Chem.* **245,** 39–45.

148 N. C. Robinson and C. Tanford (1975), *Biochemistry* **14,** 369–378.

149 A. F. Huxley and R. M. Simmons (1971), *Nature* **233,** 533–538.

150 R. F. Ashman and H. Metzger (1971), *Immunochemistry* **8,** 643–656.

151 I. Pilz, O. Kratky, A. Licht, and M. Sela (1973), *Biochemistry* **12,** 4998–5005.

152 E. Heyde and J. F. Morrison (1978), *Biochemistry* **17,** 1573–1580.

153 G. R. Welch and F. H. Gaertner (1976), *Arch. Biochem. Biophys.* **172,** 476–489.

154 A. Vitto and F. H. Gaertner (1978), *Biochem. Biophys. Res. Commun.* **82,** 977–981.

155 D. M. Bonner, J. A. DeMoss, and S. E. Mills (1965), in V. Bryson and H. J. Vogel, eds., *Evolving Genes and Proteins*, Academic Press, New York, pp. 305–318.

156 W. Gilbert (1978), *Nature* **271,** 501.

157 W. R. Folk and P. Berg (1971), *J. Mol. Biol.* **58,** 595–610.

158 C. W. Hill and G. Combriata (1973), *Molec. Gen. Genet.* **127,** 197–214.

159 D. S. Straus and G. R. Hoffmann (1975), *Genetics* **80,** 227–237.

160 R. P. Anderson, C. G. Miller, and J. R. Roth (1976), *J. Mol. Biol.* **105,** 201–218.

161 R. J. Siezen and E. F. J. Van Bruggen (1974), *J. Mol. Biol.* **90,** 77–89.

162 R. N. Perham and J. O. Thomas (1971), *FEBS Lett.* **15,** 8–12.

163 C. Gebhardt, D. Mecke, and H. Bisswanger (1978), *Biochem. Biophys. Res. Commun.* **84,** 508–514.

164 G. L. E. Koch, Y. Boulanger, and B. S. Hartley (1974), *Nature* **249,** 316–320.

165 I. D. Walker, J. I. Harris, M. J. Runswick, and P. Hudson (1976), *Eur. J. Biochem.* **68,** 255–269.

166 K. Emerk and C. Frieden (1974), *Arch. Biochem. Biophys.* **164,** 233–240.

167 J. Bergsma, W. G. J. Hol, J. N. Jansonius, K. H. Kalk, J. H. Ploegman, and J. D. G. Smit (1975), *J. Mol. Biol.* **98,** 637–643.

168 R. V. Eck and M. O. Dayhoff (1966), *Science* **152,** 363–366.

169 J. N. Tsunoda, K. T. Yasunobu, and H. R. Whiteley (1968), *J. Biol. Chem.* **243,** 6262–6272.

170 K. G. Mann, W. W. Fish, A. C. Cox, and C. Tanford (1970), *Biochemistry* **9.** 1348–1354.

171 J. A. Black and G. H. Dixon (1968), *Nature* **218,** 736–741.

172 O. Smithies, G. E. Connell, and G. H. Dixon (1962), *Nature* **196,** 232–236.

173 F. A. Quiocho, G. L. Gilliland, and G. N. Phillips (1977), *J. Biol. Chem.* **252,** 5142–5149.

174 V. Braun and V. Bosch (1972), *Proc. Natl. Acad. Sci. USA* **69,** 970–974.

175 J. Sjödahl (1977), *Eur. J. Biochem.* **73,** 343–351.

176 A. D. McLachlan, M. Stewart, and L. B. Smillie (1975), *J. Mol. Biol.* **98,** 281–292.

177 L. Moens and M. Kondo (1977), *Biochem. J.* **165,** 111–119.

178 L. Moens and M. Kondo (1978), *Eur. J. Biochem.* **82,** 65–72.

179 M. Brouwer, M. Wolters, and E. F. J. Van Bruggen (1976), *Biochemistry* **15,** 2618–2623.

180 R. Lontie, M. De Ley, H. Robberecht, and R. Witters (1973), *Nat. New Biol.* **242,** 180–181.

181 E. J. Wood and L. J. Mosby (1975), *Biochem. J.* **149,** 437–445.

182 L. Hood, J. H. Campbell, and S. C. R. Elgin (1975), *Ann. Rev. Genet.* **9,** 305–353.

183 J. H. Black and G. H. Dixon (1967), *Nature* **216,** 152–154.

184 C. R. Cantor and T. R. Jukers (1966), *Proc. Natl. Acad. Sci. USA* **56,** 177–184.

185 W. M. Fitch (1966), *J. Mol. Biol.* **16,** 17–27.

186 M. G. Rossmann and P. Argos (1976), *J. Mol. Biol.* **105,** 75–95.

187 M. G. Rossmann and P. Argos (1975), *J. Biol. Chem.* **250,** 7525–7532.

188 R. L. Hill and K. Brew (1975), *Adv. Enzymol.* **43,** 411–490.

189 R. M. Waterson and W. H. Konigsberg (1974), *Proc. Natl. Acad. Sci. USA* **71,** 376–380.

190 S. Ohnoki, B.-S. Hong, and J. M. Buchanan (1977), *Biochemistry* **16,** 1065–1069.

191 P. Nevers and H. Saedler (1977), *Nature*, **268,** 109–115.

192 Y. Burstein, R. Zemell, F. Kantor, and I. Schechter (1977), *Proc. Natl. Acad. Sci. USA* **74,** 3157–3161.

193 N. Hozumi and S. Tonegawa (1976), *Proc. Natl. Acad. Sci USA* **73,** 3628–3632.

194 J. Yourno, T. Kohno, and J. R. Roth (1970), *Nature* **228,** 820–824.

195 J. R. Guest (1974), *J. Gen. Microbiol.* **80,** 523–532.

Chapter 2

ARTIFICIAL PROTEIN CHIMERAS

JÜRGEN KANIA
Abteilung Experimentelle Innere Medizin
Universitätskliniken Köln
Köln, West Germany

BENNO MÜLLER-HILL
Institut für Genetik der Universität Köln
Köln, West Germany

ABBREVIATIONS IS-elements: insertion elements.
SDS: sodium dodecyl sulfate

2.1 THE CONSTRUCTION OF ARTIFICIAL CHIMERIC PROTEINS

A Look at Some of the Proteins Involved

Any protein that is large enough to be composed of several domains has a good chance of recognizing more than one small molecule or of having more than one activity. To illustrate the point we would like to discuss the three major proteins of the *lac* system of *E. coli* (1).

The first protein, *lac* repressor, is a tetramer. Each subunit is composed out of 360 amino acids (2, 3). It consists of two major parts: a *headpiece* that is 59 residues long and that binds to DNA and recognizes *lac* operator (4, Gilbert, personal communication) and a *core* that has sites for aggregation and binding of inducer (5).

The second protein, β-galactosidase, is also a tetramer but is much larger than *lac* repressor. Each subunit is composed of 1021 residues (6). A computer analysis of the protein sequence has shown that it contains a duplication of residues 1 to 380 running from residue 398 to 782 and a C-terminal part running from residue 818 to 932 that resembles folate reductase residues 26 to 137 (7). β-Galactosidase has one binding site for galactosides per chain (1). Missense mutations are found in many parts of the gene. Mutants that lack the *z*-(β-galactosidase)-gene completely grow well on a range of carbon sources. Thus if β-galactosidase contains any activities besides the capability to hydrolyze β-galactosides, they are nonessential (1).

Let us finally consider the third protein, *lac* permease: it is composed of about 360 residues (Beyreuther, Bielig, Betteridge, and Müller-Hill, in preparation). It pumps lactose under energy consumption into the cell. It contains at least one binding site for the inner membrane in which it is found and one for the galactosides that it transports.

When we consider the evolution of proteins like *lac* repressor or *lac* permease it seems not completely unlikely that they have been put together from pieces that recognize DNA, membrane, or galactosides. Mechanisms exist in eucaryotes that allow the moving around of prefabricated pieces of genes and their fusion (8). The techniques for isolating such fusions in special systems in higher cells are just beginning to emerge (9). In *E. coli*, however, no elaborate mechanisms seem to exist to create fusions at a high rate that will produce chimeras.

Bacteria however are much more easily manipulated. Since fusions between genes (which are not induced by IS-elements) seem to be random at first approximation, the only limitation in fusion experiments seems to be the number of bacteria that can be manipulated and analysed. When a system is well understood, that is, the DNA cloned and sequenced and the protein function understood, it is even easier to do *in vitro* fusions using defined DNA restriction fragments.

A Look at Some Artificial Chimeras

Before going into details we will mention some typical chimeras produced by gene fusions isolated by us or others (10). In every case β-galactosidase is one of the components of the chimeras. The reason for this is that β-galactosidase is an excellent starting material for chimeras. To avoid confusion let us state here that when we say protein A is fused to protein B we always imply that protein A will provide the N-terminus and protein B the C-terminus of the chimeras.

1 The DNA-binding part of *lac* repressor has been fused to active β-galactosidase (11).
2 The membrane binding part of two membrane proteins has been fused to active β-galactosidase (12, 13).
3 An N-terminal fragment of the regulatory *ara C* protein has been fused to active β-galactosidase (14).
4 Active *lac* repressor has been fused to active β-galactosidase (15).

All these chimeras are produced from deletions that have occurred *in vivo* probably randomly. Finally one example of a fusion produced *in vitro* with DNA pieces of known sequence may be mentioned:

5 Inactive β-galactosidase has been fused to somatostatin (16).

This list does not imply that fusions are not possible *in vivo* or *in vitro* with other genes. Next we describe the principle of the *in vivo* fusions involving β-galactosidase.

2.2 β-GALACTOSIDASE AS MATERIAL FOR THE CONSTRUCTION OF CHIMERAS

The Principle of the Fusions Obtained *in vivo*

β-Galactosidase is an excellent starting material for fusions for two reasons. The enzyme activity is easily tested with simple colorimetric tests in extracts and in colonies. It is a large enzyme (according to sequence a subunit is composed of 1021 residues) that can be easily recognized on SDS polyacrylamide gels. Thus it may miss the first or the last few residues and have them replaced by foreign polypeptides without loosing activity (Fig. 2.1).

For the isolation of chimera producing fusions *in vivo* at the N-terminus a strongly polar nonsense mutation of the *z* gene (*u118*), which has an ochre codon, replacing the codon for Glu 17 of β-galactosidase was used (17, 18). The first fusions producing chimeras were achieved with the closely linked *i*-(repressor)-gene (15). In order to have adequate expression of the chimera a promoter mutation of the *i*- gene (i^q) that enhances the transcription tenfold was used to construct the double mutant $i^q z^{-u118}$. If bacteria having this genotype are plated on lactose about every fifth spontaneous lac^+-revertant is a fusion producing a repressor-β-galactosidase chimera. Such fusions can easily be identified by the *lac* constitutive character of the colonies (blue color) on bromochloroindolyl-β-D-galactoside containing glucose plates (15).

The fusions must cut deeper than 17 codons into the *z*-(β-galactosidase)-gene. However, the fact that the analoguous $i^q z^{-M15}$ mutant does not give rise to fusions producing chimeras shows that fusions may not cut deeper than codon 41 without impairing the activity of β-galactosidase since the internal *z* deletion *M15* has been determined to reach from codon 11 to codon 41 of the *z* gene (19). The inactive β-galactosidase produced from a strain containing the *M15*-deletion (it can be tested by its cross-reaction with antibodies raised against β-galactosidase) is dimeric; this may be the reason for its inactivity (19).

For most fusions the N-terminus of β-galactosidase has been used. Recently, however, *in vitro* (16) and *in vivo* (Müller-Hill, unpublished) fusions have been constructed using the C-terminus of β-galactosidase. In the *in vitro* fusions the last 16 residues of β-galactosidase are missing. As in the case of the N-terminal fusions one might suspect that the added proteins

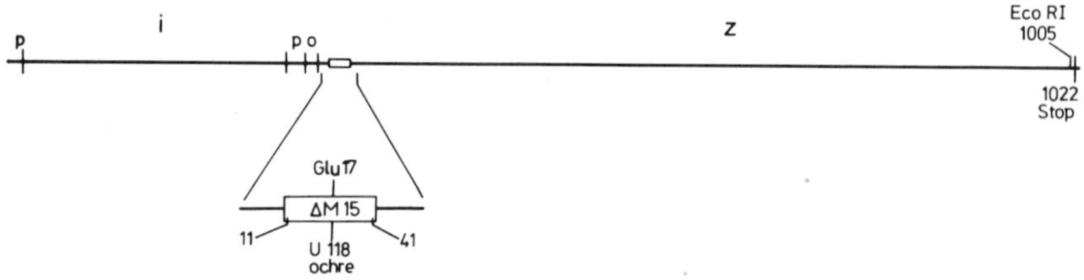

Fig. 2.1 Genetic map of the *i* gene, the promoter *p*, operator *o*, and the *z* gene (to scale). Both *i* gene (*lac* repressor) and *z* gene (β-galactosidase) are transcribed from left to right. *U 118* is an ochre mutation at position Glu17. The box within the *z* gene indicates the size of the *M 15* deletion, which has lost the codons 11–41.

are just hanging as tails or globules on otherwise normal tetrameric β-galactosidase.

A Method to Transpose the *z* Gene in the *E. coli* Chromosome

The possibility of using β-galactosidase for fusions *in vivo* with any possible gene product was severely limited by the fact that the *lac* operon could be moved only with difficulty to a few positions of the *E. coli* chromosome. This difficulty was overcome by Casadaban who developed a technique that makes it possible to move the *z* gene close enough to any other gene to obtain fusions at a reasonable frequency (20). For this purpose phage *mu* (containing the *Cts* mutation which leads to induction at 42°C), which can integrate in almost any sequence, is allowed to integrate in a gene close to the one whose product one wants to fuse to β-galactosidase. The bacteria used should carry a deletion of the *lac* operon. If the position and direction of integration are right (which is easily tested), a specially constructed derivative of phage *lambda* which lacks the *int* gene but carries a piece of *mu*-DNA and the *lac* operon with the *u118*-mutation is forced to integrate via the *mu*-homology. Lysogens for phage *lambda* are isolated. Such bacteria are plated and incubated on lactose plates at 42°C: Only bacteria that have lost *muCts* will survive, and only those that have acquired in addition a fusion producing a β-galactosidase chimera will grow on lactose. Since the phenotype of the chimeric β-galactosidase can be predicted, a biochemical analysis may prove the genetic analysis to be right.

2.3 VARIOUS CHIMERIC PROTEINS OBTAINED FROM *IN VIVO* FUSIONS

Fusions between Inactive *lac* Repressor and Active β-Galactosidase

More than a hundred fusions having parts of *lac* repressor grafted onto functioning β-galactosidase were isolated by screening *lac*$^{+}$-revertants from an *E. coli* strain of the $i^{q}z^{-u118}$ genotype (15). Most of the fusions were mapped against a set of deletions entering the *i* gene from its promoter proximal (*z* distal) end (21).

According to genetic analysis many of the fusions carried about 60 to 80 residues from the N-terminus of *lac* repressor grafted onto active β-galactosidase (for details see section 2.5). Two chimeric β-galactosidases of this type were purified from strains 71-68-4 and 71-68-16 (11). Limited tryptic digestion of the chimera from strain 71-68-4 released the first six N-terminal tryptic *lac* repressor peptides. This indicates that the 59 residues that make the headpiece were present. Both purified repressor-galactosidases bound to phosphocellulose almost as well as *lac* repressor, whereas wild type β-galactosidase does not bind at all. On SDS polyacrylamide gels the two chimeras were indistinguishable from β-galactosidase. We could have detected a 5% difference. This indicates that at the most 80 residues were added assuming that maximally 30 residues were lost from β-galactosidase. The sedimentation coefficient of both chimeras (16S) was the same as the one found for β-galactosidase. We conclude that these chimeras are tetrameric as is β-galactosidase.

There exists now good positive biochemical evidence that the N-terminal domains of *lac* repressor contain the DNA-binding site that interacts specifically with *lac* operator (4, Gilbert, personal communication). This domain, called the headpiece, is made up of 59 residues (4). The binding of the two chimeras to phosphocellulose indicated that the β-galactosidase had acquired the capability to bind to DNA backbone. We used the DNA-filter-binding assay to prove this. Both chimeras bound nonspecifically to DNA as well as *lac* repressor. Both were incapable of discriminating between λ and λ*plac* DNA. β-Galactosidase as a control did not bind at all to DNA. We speculate that the DNA-binding domains are grafted in such a way that they are not properly spaced to match accurately both sides of *lac* operator (11).

Fusions between Active *lac* Repressor and Active β-Galactosidase

We also purified a repressor-galactosidase which had active repressor covalently linked to active β-galactosidase. Antibody against *lac* repressor was capable of precipitating β-galactosidase activity and antibody against β-galactosidase was capable of precipitating *lac* repressor activity, whereas there is no cross-reaction between wild type *lac* repressor and β-galactosidase (15, 21). SDS polyacrylamide gel electrophoresis was used to prove the covalent linkage between the proteins. The position of the chimera on SDS gels suggested that at the most 60 residues were lost from both proteins by the fusion.

Recently a peptide that links *lac* repressor and β-galactosidase in this chimera has been isolated and sequenced: the chimera is missing the last four residues of *lac* repressor and the first 24 residues of β-galactosidase (Brake, Fowler, Zabin, Kania, and Müller-Hill, in preparation).

Fusions between Inactive *ara C* Protein and Active β-Galactosidase

The product of the *ara C* gene, the *ara C* protein, regulates initiation of transcription of the *ara* operon. This regulation is so complex that one would not be astonished if the transcription of the *ara C* gene was itself regulated. It is, however, impossible to measure accurately the concentration of *ara C* protein in cell extracts. Thus the *ara C* gene was fused in such a way with the *lac z* gene that an inactive N-terminal piece of it is fused to active β-galactosidase (14). This allowed one to study the effect of active *ara C* protein made by an intact *ara C* gene on an episome on the synthesis of the chimera on the chromosome. Less of the *ara C* protein β-galactosidase chimera is made when active *ara C* protein is present than when *ara C* protein is absent (14). In other words, the *ara C* protein represses its own synthesis severely.

Fusions between Inactive *mal F* Permease and Active β-Galactosidase

Membrane proteins are often made in small amounts and have laborious and often indirect assays. The technique of gene fusion is useful in circumventing problems with purification of such membrane bound proteins since β-galactosidase can be used as a *reporter* enzyme and since membrane proteins being part of β-galactosidase chimeras can easily be traced anywhere in the cell. This approach has been used by Silhavy et al. (12, 13) who identified two gene products of the *mal B* locus involved in maltose transport in the outer and inner membrane, respectively. Applying Casadabans technique (20), it was possible to bring the *mal F* gene close to the *lac* operon. Application of the selection procedure for chimera producing fusions allowed the isolation of strains producing chimeric proteins consisting of an inactive maltose transport protein attached to active β-galactosidase. After separating the inner and outer membranes on isopycnic sucrose gradients, β-galactosidase activity was found in the inner (cytoplasmic) membrane. Thus, using β-galactosidase as a tracer, it was possible to localize the so far unidentified maltose transport protein, the product of gene *mal F*. Moreover, it could be concluded that the N-terminal part of the maltose transport protein was essential and sufficient for integration into the inner cell membrane (12).

Fusions between Inactive Phage *lambda* Receptor and Active β-Galactosidase

The *lam B* gene product serves as receptor for phage *lambda* adsorption on the outer membrane of *E. coli*. It is also involved in the transport of certain maltodextrines. The above cited techniques have been used again to fuse the *lam B* gene to the *lac Z u118* gene in such a way that *lambda* receptor-β-galactosidase chimeras were obtained (13). Analoguous to the procedure described above (localization of the maltose permease-β-galactosidase chimera) here 50% of the *lambda* receptor-β-galactosidase chimera was found in the outer membrane of *E. coli*. Due to the high molecular weight of the polypeptide chain (MW $>$ 120,000 dalton), the chimera could be identified in crude extracts on SDS gels. Immunofluorescence techniques revealed the β-galactosidase part of the chimera facing outward from the outer membrane. The authors concluded from these results that the information that directs the *lambda* receptor to the outer membrane is located in the N-terminal part of the *lam B* gene product itself. Interestingly, another *lambda* receptor-β-galactosidase chimera, with a considerably smaller *lam B* part, occurred exclusively in the cytoplasm as does wild type β-galactosidase (13).

2.4 TRIPLE CHIMERAS

There is no reason that artificial chimeras should be composed only of two active proteins. Thus we tried to construct larger chimeras (22). Our starting material was a strain carrying the genotype of the fusions joining active *lac* repressor to active β-galactosidase, but with a normal i^+ promoter for the repressor producing *i* gene. This strain is essentially *lac*-negative. We isolated *ton B* deletions which had become at the same time phenotypically *lac*$^+$.

Our goal was actually to fuse active tryptophan synthase A to repressor core which itself was fused to β-galactosidase. We fell short of this goal by only being able to isolate a fusion in which inactive tryptophan synthase A protein was part of the fusion; however, the tryptophan synthase part was still capable of binding tryptophan synthase B subunits. Since the headpiece of *lac* repressor moves when inducer binds to the core (see Beyreuther ref. 1), one might potentially isolate enzymes fused to *lac*$^+$ repressor core that behave as allosteric enzymes with repressor core conferring the allosteric properties. This remains to be done.

2.5 PHYSICAL PROPERTIES OF THE CHIMERAS

Endpoints of the Fusions between *lac* Repressor and Active β-Galactosidase

The exact endpoints of the fusions may be determined by protein or DNA-sequencing methods. So far only one such endpoint has been determined (Brake, Fowler, Zabin, Kania and Müller-Hill, in preparation). Regions where fusions occur may be also determined genetically. In the case of the repressor-galactosidase fusions it became clear that the fusions may not cut much deeper than codon 41 of β-galactosidase (see Section 2.2). But where do the fusions end in *lac* repressor? Can almost any piece of *lac* repressor be fused to β-galactosidase, or are there strong restrictions? We have mapped about 100 repressor-galactosidase fusions using 15 deletions entering the *i* gene from its promoter side (21). This method can give only approximative results since the recombination frequency drops drastically the closer the deletions come together. It becomes clear from inspecting the data (Figure 2.2) that fusions occur with at least 14 endpoints in the *i* gene. However there seems to be a strong prevalence for fusions in which whole domains are fused to a β-galactosidase: The headpiece (26 out of 103 fusions) and the entire active repressor (12 out of 103 fusions) are among the pieces attached most frequently to β-galactosidase. In any case it is remarkable that such fusions fold correctly *in vivo*.

Aggregation of Chimeras

Since folding of a polypeptide chain is a rapid cooperative process starting at defined nucleation centers within the chain, the building of the final domain structure is finished just upon the release of the polypeptide chain from the ribosome.

If parts, that is, building blocks, of artificially fused proteins originally occur as identical multimeres themselves, the aggregation of such artificially created chimeras will be a more complex process than the aggregation of the different independently synthesized polypeptides. For naturally occurring bifunctional enzymes natural selection has solved the problem by reducing the number of possible forms of aggregation.

Yourno et al. (23) were the first to discuss the quaternary structure of an artifically created bifunctional enzyme (histidinol dehydrogenase: histidinol-phosphate aminotransferase) which, upon purification, eluted in three distinct peaks from DEAE-Sephadex. Since both wild type enzymes exist as dimers, a number of aggregation forms was predicted by the authors. Indeed, on Sephaex G-200 crude extracts showed dimers and oligomers up to a molecular weight of 10^6 daltons.

The quaternary structure and symmetry of repressor-galactosidases have been also investigated in detail. Figure 2.3 shows the separation of different aggregates of one identical polypeptide chain of the chimera from strain BMH 71-56-14 on phosphocellulose. This particular chimeric protein, in which fully active *lac* repressor is covalently linked to active β-galactosidase, was used as a system for the examination of models of the quaternary structure of *lac* repressor. We investigated the aggregation and symmetry of the material from the main peak in Figure 2.3. It contains homogeneous repressor-galactosidase tetramers, as determined by electron microscopy and sedimentation analysis, whereas the following minor peaks consist of

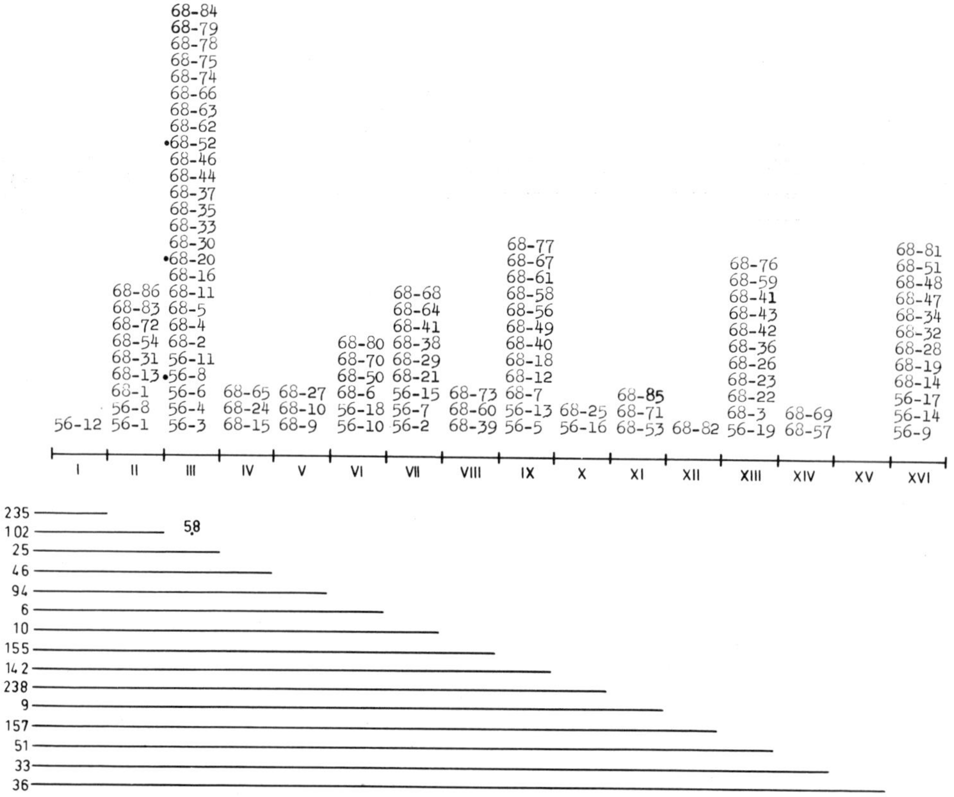

Fig. 2.2 Map of the i-z^+ fusions. The *F'lac pro* episomes were crossed into strains of the genotype $(tonB\ i^-)_\Delta\ z^+y^+\ (lac\ pro)_\Delta\ gal\ E^-SmR$. Numbers above the deletion groups are the strain numbers of the series BMH 71. Numbers such as 68-54 written above the second deletion group indicate that fusion BMH 71-68-84 did not recombine with deletion 102 but did recombine with deletion 235. We constructed for the selection of i^+ recombinants a double mutant carrying a *gal E* mutation and the i^{-d} point mutation BG 75 which has aspartic acid instead of glycine in position 58. All the fusions that did not recombine with deletion 25 but with 102 did recombine with the i^{-d} point mutation BG 58. Three fusions marked with a dot (68-52; 68-20; 56-8) did recombine with point mutation BG 78 but did not recombine even with deletion 102. According to H. J. Miller (personal communication) deletion 102 ends somewhere between codon 37 and 53, and deletion 25 ends somewhere between codon 62 and 78. The resolution of mapping is clearly not better and probably worse than 15 basepairs. For details see ref. 21. Reproduced from ref. 21 with permission of Elsevier/North-Holland Biomedical Press, Amsterdam.

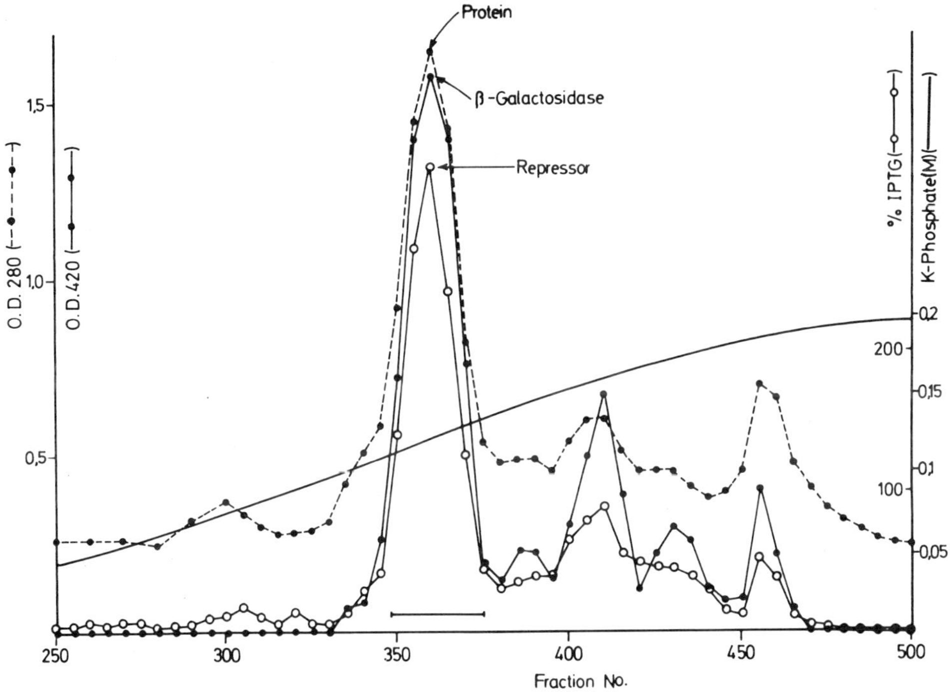

Fig. 2.3 Elution profile of repressor-galactosidase from strain 71-56-14 from phosphocellulose. In all fractions *lac* repressor activity cochromatographs with β-galactosidase activity. The main peak contains exclusively tetrameric repressor-galactosidase; the minor peaks consist of higher aggregates, as determined on glycerol gradients.

faster sedimenting aggregates. It is interesting to note that the separated aggregates were stable and not in a posttranslational state of equilibrium.

Also no activity could be restored after treatment of the chimera with 8M urea or 6M guanidine-HCl in contrast to both the wild type proteins. This may indicate that aggregation of this particular repressor-galactosidase is a sequential process in the cell starting with the repressor part, which is finished first, while the β-galactosidase part is still being synthesized.

Both *lac* repressor and β-galactosidase are stable tetrameric proteins, their subunits most probably arranged in D_2 symmetry, since the L1 deletion in the *i* gene and the M15 deletion in the *z* gene produce dimeric proteins, which excludes C_4 symmetry (24). As already indicated, the main fraction from chromatography on phosphocellulose of repressor-galactosidase contains exclusively tetramers, as visualized by electron microscopy. On the other hand, it is logically impossible that both the repressor and galactosidase parts of the chimera are tetrameric aggregates arranged in D_2

symmetry simultaneously. Therefore, one of the two protein parts must retain its biological activity in the repressor-galactosidase tetramer in the state of a dimer. We used the cross-linking agents dimethyl suberimidate, dimethyl adipinimidate, and iodine to determine which parts of the protein might become separated. All three agents do not cross-link native β-galactosidase under the conditions applied, whereas the diimidoesters cross-link *lac* repressor predominantly to dimers, trimers, and tetramers by way of N-terminal or C-terminal lysine residues. Iodine cross-links *lac* repressor predominantly to dimers at the aggregation-core sites of the molecule (25). Repressor-galactosidase at the same molar concentration as *lac* repressor, however, is cross-linked by all agents to dimers. Taking all results together, one may summarize the following conclusions: The β-galactosidase parts in the repressor-galactosidase are aggregated in a D_2 symmetry as in the wild type molecule, thus β-galactosidase separates one set of *lac* repressor dimers from the other (Figure 2.4). This led to the conclusion that two repressor subunits are sufficient for recognition of *lac* operator. It seems obvious that repressor-galactosidases, in which only the N-terminal domains of *lac*

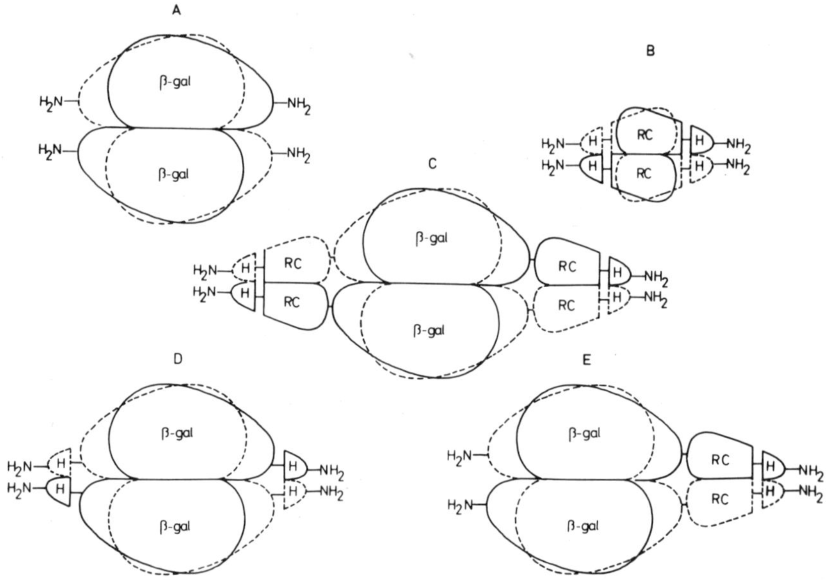

Fig. 2.4 Schematic drawing of β-galactosidase (*A*), *lac* repressor (*B*), repressor-galactosidase from strain BMH 71-56-14 (*C*), repressor-galactosidase from strain BMH 71-68-4 (*D*), hybrid built out of two repressor-galactosidase polypeptides from strain 71-56-14 and two β-galactosidase subunits (*E*). All models are drawn in D_2 symmetry. The aggregation of the repressor-galactosidase·β-galactosidase hybrid was proven by cross-linking experiments with dimethyl suberimidate. β-gal: β-galactosidase; RC: *lac* repressor core; H: *lac* repressor headpiece.

repressor are attached to β-galactosidase (11), are also aggregated by way of the β-galactosidase parts (Figure 2.4).

To prove these models an even more sophisticated system was chosen (26). We have prepared hybrid aggregated between subunits of the chimeric repressor-galactosidase (from strain BMH 71-56-14) and the enzyme β-galactosidase (Figure 2.4). As activity of repressor-galactosidase could not be restored after urea or guanidine-HCl denaturation, we constructed the desired hybrids *in vivo* assuming that a completed *lac* repressor polpeptide can fold independently and aggregate with another repressor part, even when it is fused to an unfinished β-galactosidase chain and attached to the ribosome (see above). From induced diploid cells with the genotype $i^- z^+/F'$ $pro^+ i^{q1} i^+ z^+_{71\text{-}56\text{-}14}$ in which approximately two β-galactosidase polypeptide chains are made per repressor-galactosidase chain, we were able to isolate and separate hybrid molecules with 1, 2, or 3 repressor-galactosidase polypeptides aggregated with 3, 2, or 1 β-galactosidase polypeptides, respectively.

We proved the aggregation between these different polypeptides by precipitation of the complexes with antiserum against *lac* repressor, which does not cross-react under several conditions with β-galactosidase. We interpreted the intensity of a band of a molecular weight of 120,000 on SDS gels as evidence that one or more β-galactosidase subunits were aggregated with the 155,000 polypeptide of repressor-galactosidase. The aggregation and symmetry of the isolated hybrids were again determined using electron microscopy and the cross-linking agent dimethyl suberimidate. The investigation of the structure and symmetry of the above hybrids between repressor-galactosidase chimeras and β-galactosidase led to an extension of the model for the *lac* repressor/*lac* operator interaction (26).

Stability of Chimeras; *in vivo* and *in vitro* Proteolysis

As already indicated, the retention of activity in the chimeras under consideration is most easily explained if it is assumed that folding of the polypeptide chain starts at discrete nucleation centers and cooperatively proceeds in both directions of the chain, thus independently forming protein globules (domains) connected by a flexible region of the polypeptide chain. This assumption should also be true for naturally occurring bifunctional enzymes or proteins that consist of more than one domain. Consequently one would expect the link between two domains to be more or less accessible to proteases, as has already been shown for a number of naturally occurring chimeric enzymes (27, 28).

Yourno et al. could demonstrate the release of one activity (histidinol dehydrogenase) from their artificially created chimera histidinol dehydrogenase: histidinolphosphate aminotransferase after mild proteolysis (23). The scission of the enzymatic globules was proven by SDS gel electrophoresis of the released fragments. The aminotransferase activity of the chimera was, as in the wild type enzyme, highly sensitive to proteolysis even under native conditions. Compared to the wild type activities, the aminotransferase of the chimera did not differ significantly from the wild type enzyme, whereas the dehydrogenase activity was slightly reduced and more heat labile with respect to the native enzyme. There is no evidence for proteolytic cleavage *in vivo* from the published data for this bifunctional enzyme.

Similar conclusions can be drawn for the repressor-galactosidase from strain BMH 71-56-14, in whic *lac* repressor and β-galactosidase activities are fully retained: No proteolysis could be detected *in vivo*. In contrast to this, if we used an extract from a strain, which carried the fusion on phage λ, after heat induction nearly all *lac* repressor activity was found separated from the β-galactosidase activity. It is not known, whether the protease was phage coded or activated by heat induction of the phage.

We also tested *in vitro* the accessibility of the junction between the repressor part and the galactosidase part. All β-galactosidase activity could be precipitated prior to incubation with trypsin by antiserum against *lac* repressor. After addition of trypsin most β-galactosidase activity remained in the supernatant (21). We interpreted this experiment proposing that repressor parts are connected with β-galactosidase parts by a (random coil?) region accessible to protease.

Proteolysis occurred probably *in vivo* before and *in vitro* during purification of the repressor-galactosidase from strain BMH 71-56-18. In this chimera the repressor part consists of about the N-terminal half of the *lac* repressor sequence. In contrast to the even shorter chimeras produced by strain BMH 71-68-4 and -16 (see Section 2.3), this chimera did not bind to phosphocellulose. We were not able to purify the protein due to the rapid loss of activity (21). Glycerol gradients revealed that this chimera did not exist in a stable aggregate form. Heat inactivation kinetics showed only a slightly higher, nonsignificant sensitivity to heat. The folding of the remaining half of the core of *lac* repressor is probably disturbed. This will enhance proteolysis. Here is the mechanistic reason for our genetically substantiated proposal that only readily folded domains can be covalently fused to chimeras without affecting the activities of both parts (see Section 2.5).

The analysis of even larger (triple) chimeras between tryptophan synthase A-subunit, *lac* repressor, and β-galactosidase supports this assumption, since in all strains examined partial degradation was detected. From the size of the fragments we inferred proteolysis at the joining points of domains (22).

The problem of proteolysis may become even larger when gentic material from other organisms is cloned in *E. coli* and a search for such proteins is made since these proteins have not evolved to be resistant against the proteases of *E. coli*. We discuss a typical example below.

2.6 CONSTRUCTION OF CHIMERIC PROTEINS BY *IN VITRO* TECHNIQUES

So far most of the chimeras we discussed were isolated as the result of more or less random deletions obtained *in vivo*. We have now to discuss the possibility of obtaining such fusions by *in vitro* techniques. Here we have two options: Either to try random fusions or to try fusions of defined and maybe sequenced restriction pieces of DNA. The former approach has to our knowledge not been tried so far for fusing proteins. Deletions, however, have been isolated by *in vitro* techniques in *SV40* (29). Potentially random *in vitro* fusion may become the ideal technique when properly developed. We have seen in the earlier sections that fusions can be isolated readily by *in vivo* techniques but to get a fusion with particular endpoints may require more plates or bacteria than one can easily handle, that is more than 10^{12} bacteria (see Section 2.4).

The other approach, which uses restriction fragments or synthetic DNA of known sequence, has been already used with success. The DNA-specifying somatostatin has been synthesized and introduced at two positions into the *z* gene of *E. coli* (16). When joined in such a way that the first 10 residues of β-galactosidase are followed by methionine to which somatostatin is fused no somatostatin activity could be found in extracts. The chimera was presumably degraded by proteases that occur in *E. coli* and specifically degrade proteins not evolved to be resistant against such an attack. It seems that this instability is probably due to slow folding into different semistable forms, that is the prevalence of random coils for prolonged times. When somatostatin, however, was fused in a similar way to methionine following residue 1005, measurable somatostatin activity was fused to inactive β-galactosidase. (The radioimmune assay for somatostatin activity was, however, performed after cyanogen bromide cleavage.) One does not need to be a prophet to predict that the technique of fusing restriction pieces of known sequences will become a standard technique of synthetic biochemists.

2.7 OUTLOOK

Fusing proteins by way of their genes is a game that is fun in itself almost irrespective of what is fused. The biologist is no longer doing analytical work but is beginning to do synthetic work. Unlike his colleague, the chemist, doing organic synthesis, who always has to worry about using up his synthesized material, the synthesizing biologist does not know these worries. DNA is stable and can be kept intact as such or inside bacteria in a deep freezer as long as energy is available on this planet. One molecule is enough to be amplified as much as one likes. It seems to us that the type of experiments rather superficially described in our article will become the major endeavor of the biochemist of the future.

ACKNOWLEDGMENT

Our work described in this article was supported by Deutsche Forschungsgemeinschaft through SFB 74.

REFERENCES

1 J. R. Beckwith and D. Zipser, eds. (1970), *The Lactose Operon*, Cold Spring Harbor Laboratory, New York; and J. Miller and W. Reznikoff *The Operon*, Cold Spring Harbor Laboratory, New York, in press.

2 K. Beyreuther, K. Adler, N. Geisler, and A. Klemm (1973), *Proc. Natl. Acad. Sci. USA* **70,** 3567–3580.

3 P. J. Farabaugh (1978), *Nature* **274,** 765–769; and K. Beyreuther (1978) *Nature* **274,** 767.

4 T. M. Jovin, N. Geisler, and K. Weber (1977), *Nature* **269,** 668–672.

5 J. G. Files and K. Weber (1976), *J. Biol. Chem.* **251,** 3386–3391.

6 A. V. Fowler and J. Zabin (1977). *Proc. Natl. Acad. Sci. USA* **74,** 1507–1510.

7 J. M. Hood, A. V. Fowler, and J. Zabin (1978), *Proc. Natl. Acad. Aci. USA* **75,** 113–116.

8 C. Brack and S. Tonegawa (1977), *Proc. Natl. Acad. Sci. USA* **74,** 5652–5656.

9 B. Liesegang, A. Radbruch, and K. Rajewsky *Proc. Natl. Acad. Sci. USA* in press.

10 P. Bassford, J. Beckwith, M. Berman, A. Brickman, M. Casadaban, L. Guarente, I. Saint-Girons, A. Sarthy, M. Schwartz, H. Shuman, and I. Silhavy in J. Miller and W. Reznikoff, eds., *The Operon*, Cold Spring Harbor Laboratory, New York, in press.

11 B. Müller-Hill, B. Gronenborn, J. Kania, M. Schlotmann, and K. Beyreuther (1977), in G. Vogel, ed., *Nucleic Acid-Protien Recognition*, Academic Press, New York, pp. 219–236.

12 T. J. Silhavy, M. J. Casadaban, J. A. Shuman, and J. R. Beckwith (1976), *Proc. Natl. Acad. Sci. USA* **73,** 3423–3427.

13 T. J. Silhavy, H. A. Shuman, J. Beckwith, and M. Schwartz (1977), *Proc. Natl. Acad. Sci. USA* **74,** 5411–5415.

14 M. J. Casadaban (1976), *J. Mol. Biol.* **104,** 557–566.

15 B. Müller-Hill and J. Kania (1974), *Nature* **249,** 561–563.

16 K. Itakura, T. Hirose, R. Crea, A. D. Riggs, H. L. Heynecker, F. Bolivar, and H. W. Boyer (1977), *Science* **198,** 1056–1063.

17 F. Jacob and J. Monod (1961), *Cold Spring Harbor Symposia on Quantitative Biology*, Vol. XXVI, Cold Spring Harbor, New York, pp. 193–211.

18 J. Zabin, A. V. Fowler, and J. R. Beckwith (1977), *J. Bacteriol.* **133,** 437–438.

19 K. E. Langley, M. R. Villarejo, A. V. Fowler, P. J. Zamenhof, and J. Zabin (1975), *Proc. Natl. Acad. Sci. USA* **72,** 1254–1257.

20 M. J. Casadaban (1976), *J. Mol. Biol.* **104,** 541–555.

21 B. Müller-Hill, G. Heidecker, and J. Kania (1976), in R. Markham and R. W. Horne, eds., *Proceedings of the Third John Innes Symposium* Elsevier/North-Holland, pp. 167–179.

22 G. Heidecker and B. Müller-Hill (1977), *Mol. Gen. Genet.* **155,** 301–307.

23 J. Yourno, T. Khono, and J. R. Roth (1970), *Nature* **228,** 820–824.

24 B. W. Matthews and S. A. Bernhard (1973), in L. J. Mullins, W. A. Hagins, and L. Stryer eds., *Ann Rev. of Biophysics and Bioengineering*, Vol. 2, Annual Reviews, Inc., Palo Alto, Calif., pp. 257–317.

25 J. Kania and D. T. Brown (1976), *Proc. Natl. Acad. Sci. USA* **73,** 3529–3533.

26 J. Kania and B. Müller-Hill (1977), *Eur. J. Biochem.* **79,** 381–386.

27 A. Kornberg (1972), *J. Biol. Chem.* **247,** 224–231.

28 M. Véron, F. Falcoz-Kelly, and G. N. Cohen (1972), *Eur. J. Biochem.* **28,** 520–527.

29 J. E. Mertz and P. Berg (1974), *Virology* **62,** 112–124.

Chapter 3

THE ASPARTOKINASES-HOMOSERINE DEHYDROGENASES OF *ESCHERICHIA COLI*

GEORGES N. COHEN and ALICE DAUTRY-VARSAT

Unité de Biochimie Cellulaire
Département de Biochimie
et Génétique Microbienne
Institut Pasteur
Paris, France

This chapter is dedicated to the memory of our dear friend Dean B. Cowie.

ABBREVIATIONS M_r: number average molecular weight; M_w: weight average molecular weight; M_z: z-average molecular weight.

3.1 INTRODUCTION

Historical

General In *E. coli* diaminopimelic acid and lysine, methionine, threonine, and isoleucine derive part or all of their carbon atoms from the four carbon atoms of aspartic acid. The corresponding pathway is summarized in Figure 3.1. The first reaction is catalyzed by three distinct aspartokinases (AK) which differ in the mode of regulation of their synthesis and activity. Aspartokinase I is inhibited by threonine, and its synthesis is repressed when the bacteria are grown in excess threonine *plus* isoleucine; the synthesis of aspartokinase II is repressed by methionine; lysine inhibits the activity and represses the synthesis of aspartokinase III. The reduction of aspartate semialdehyde is likewise catalyzed by two distinct homoserine dehydrogenases (HDH). Homoserine dehydrogenase I synthesis is repressed by threonine *plus* isoleucine and its activity is inhibited by threonine; whereas methionine represses the synthesis of homoserine dehydrogenase II.

It took some years to define these entities with respect one to another. Whereas the existence of AK I and AK III was recognized in 1961 (1), and that of a threonine-regulated homoserine dehydrogenase (HDH I) in 1963 (2), AK I and HDH I were treated as separate entities. AK II and HDH II, present in minute quantities in the extracts from wild type *E. coli* K12, were uncovered only in 1967 (3). The evidence that AK I and HDH I were carried by the same protein dates from 1965 (4, 5) and the definitive proof from 1968 (6). AK II and HDH II were shown to be carried by a single protein in 1969 (7). However, this was not sufficient to bestow upon AK I-HDH I and AK II-HDH II the dignity of multifunctional proteins, which

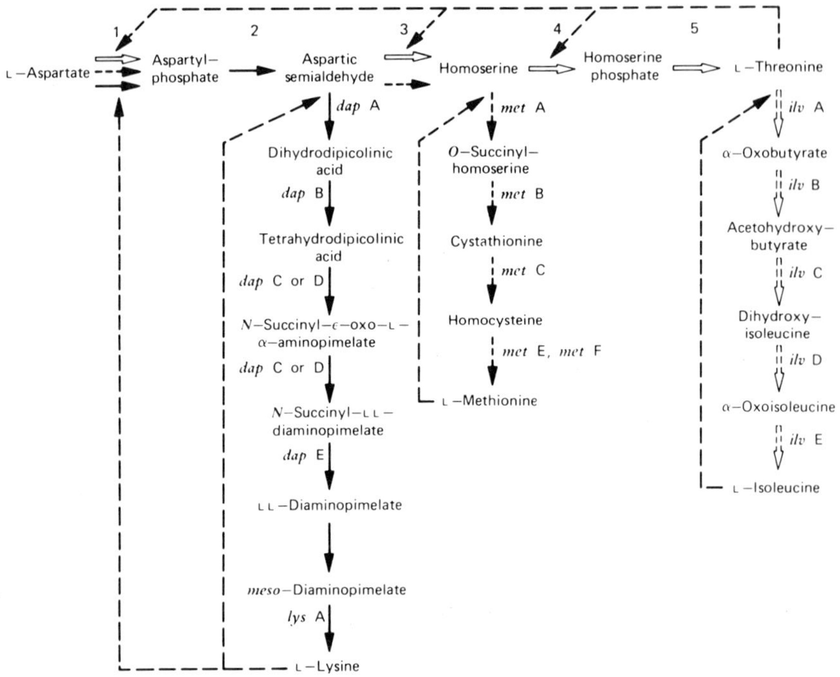

Fig. 3.1. Biosynthetic pathway for the amino acids deriving from aspartate. Symbols: →, enzymes repressed by lysine; ➞, enzymes repressed by threonine *plus* isoleucine; --➞, enzymes repressed by methionine; ⇒ enzymes repressed by isoleucine, leucine, and valine. The dashed lines point to the reactions subject to feedback inhibition. the underlined symbols represent the genes corresponding to each reaction. Aspartokinase II and homoserine dehydrogenase II are both coded by the single gene *metL*, instead of the previous *metLM* designation, since this article shows that both activities are carried by the same polypeptide. Reproduced from ref. 11 with permission of the American Society for Microbiology.

Reaction	Name of the Enzyme	Abreviation	Name of the Corresponding Genes	Regulation of Biosyn-thesis by:	Regulation of Activity by:
1	Aspartokinase I	AK I	*thr* A	thr and ile	thr
1	Aspartokinase II	AK II	*met* L	met	O
1	Aspartokinase III	AK III	*lys* C	lys	lys
2	Aspartic semialdehyde dehydrogenease	ASADH	*asd*	lys,thr,met	O
3	Homoserine dehydro-genase I	HDH I	*thr* A	thr and ile	thr
3	Homoserine dehydro-genase II	HDH II	*met* L	met	O
4	Homoserine kinase	HSK	*thr* B	thr and ileu	thr
5	Threonine synthetase	TS	*thr* C	thr and ileu	n.d.

pertains only to those proteins where the several functions are associated in the same polypeptide chain. An account follows describing how our laboratory arrived at the conclusion that in AK I-HDH I and AK II-HDH II, the two activities are carried by the same molecule.

Aspartokinase I-Homoserine Dehydrogenase I For the sake of clarity, we can for the moment dismiss the existence of AK II-HDH II which is present in such low amounts in *E. coli* K12 that it does not influence either the results or the rationale of what is exposed below.

It was interesting to obtain mutants that lacked one of the two main aspartokinases activities, AK I or AK III. In the wild type threonine and isoleucine repress the homoserine dehydrogenase (2, 8, 9); furthermore, the activity that remains is inhibited by threonine (2). This creates a methionine limitation; as a result, growth is severely impaired, unless exogenous methionine is added. These properties allowed us to isolate a mutant (Gif 54) lacking the lysine-sensitive aspartokinase (10). It had the phenotype of a prototroph growing perfectly on minimal medium, but it exhibited a conditional requirement for diaminopimelate in the presence of threonine, isoleucine, and methionine. The corresponding structural gene, *lysC*, was precisely localized at 80 min (11). Spontaneous revertants were selected on the basis of their capacity to grow on plates containing threonine, isoleucine, and methionine without added diaminopimelate. Some of the revertants were heavy threonine excretors and were studied further. They had not reacquired the lysine-sensitive aspartokinase, but their aspartokinase I was no longer inhibited by threonine, thus permitting a more rapid flow of aspartyl-phosphate and aspartate semialdehyde and a normal rate of synthesis of diaminopimelate and lysine. However, since they overproduce threonine, their levels of aspartokinase I and homoserine dehydrogenase I were at the repressed value, precluding the easy study of these proteins. Therefore, we transduced the part of their genome responsible for the excretion character into an isoleucine-requiring strain, in order to enable derepressed synthesis of these enzymes in a chemostat under isoleucine limitation (4). It was immediately apparent that the mutant under study possesses a modified threonine-sensitive aspartokinase. The threonine inhibition of homoserine dehydrogenase was also modified in a parallel manner in this mutant extract, as shown in Figures 3.2 and 3.3. (4). This prompted us to examine a mutant obtained several years before by resistance to α-amino-β-hydroxyvalerate, an entirely different selection method (12). We had found that its homoserine dehydrogenase was practically insensitive to L-threonine inhibition. As suspected, the allosteric inhibition of its threonine-sensitive aspartokinase was modified in a parallel way (5).

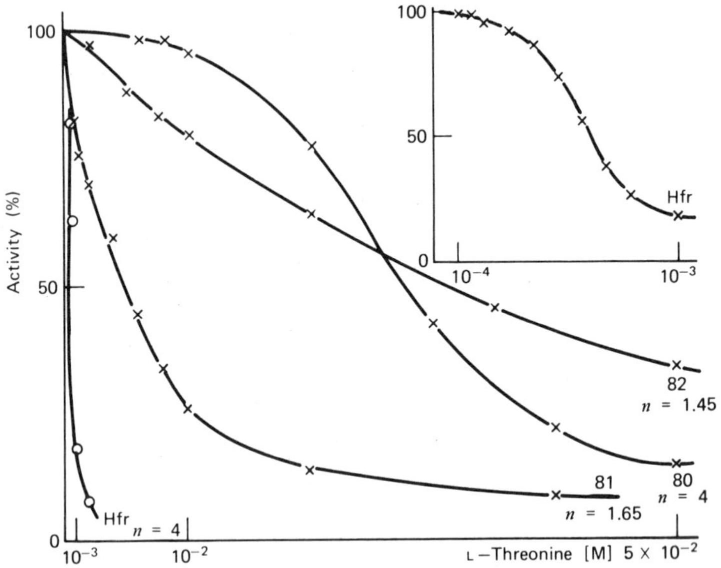

Fig. 3.2 Inhibition by L-threonine of threonine-sensitive aspartokinases. Hfr denotes the enzyme from the wild type. The numbers 80, 81, 82 refer to enzymes from mutants. The inset is an enlargement of the Hfr curve. The Hill coefficient is denoted by n. Reproduced from ref. 4 with permission of Academic Press.

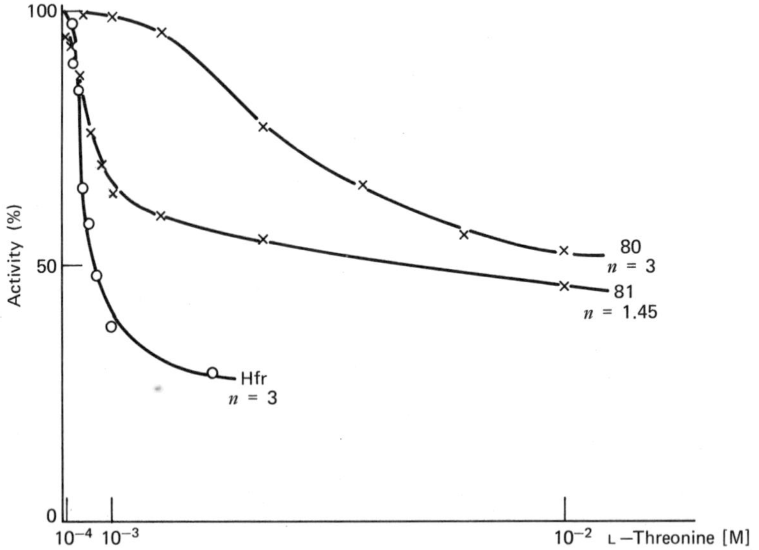

Fig. 3.3 Inhibition by L-threonine of homoserine dehydrogenases. See preceding figure for explanations (4).

The fact that both activities are affected in the same way by a single mutational event led us to the inescapable conclusion that the proteins carrying the two activities possess at least a common polypeptide chain, somehow involved in the allosteric inhibition by L-threonine.

In 1954 we had described auxotrophs of *E. coli* B that required homoserine or threonine *plus* methionine (13, 14). They turned out, as expected, to lack homoserine dehydrogenase (2). These mutants also lack the threonine-sensitive aspartokinase. All revertants to prototrophy examined recover both enzyme activities simultaneously. This surprising result led us to the attractive and nonconventional hypothesis (in 1966) that the two activities were carried by the same molecule. This was at the time received with considerable skepticism by many colleagues.

Indirect evidence that the same molecule was responsible for the two activities was adduced by the fact that the substrate of one reaction affected the course of the other: L-aspartate and ATP inhibit the homoserine dehydrogenase activity, and L-homoserine and NADP inhibit the aspartokinase activity. Another evidence is provided by the NADPH protection of both homoserine dehydrogenase *and* aspartokinase activities against thermal inactivation (5). At this point it appeared necessary to apply the tools of protein chemistry and to work with a well-defined molecular species rather than with more or less purified extracts. Preliminary results (5) showed that the two activities could not be separated upon purification by ammonium sulfate precipitation, DEAE cellulose chromatography, gel filtration, or hydroxyapatite chromatography. The ratio of the two activities remains constant throughout the purification (see for example Figure 3.4), which has to be carried in the presence of threonine in order to protect the enzymatic activities. Obtaining the protein in the homogeneous state (6) was greatly helped by the availability of the strain Tir 8 of *E. coli* (15), which is genetically derepressed for the synthesis of AK I-HDH I, as the result of three independent mutations (16).

Aspartokinase II-Homoserine Dehydrogenase II If the homoserine dehydrogenase reaction is carried by a single species, it should be possible to obtain mutants that, following the loss of this activity, would require either homoserine or threonine *plus* methionine. Such mutants have been described in *B. subtilis* (17), *Neurospora crassa* (18) and *S. cerevisiae* (19). Two such mutants (B 174 and B 184) had been described by us in *E. coli* B (13, 14); despite separate efforts, we have for a long time not been able to obtain a similar phenotype in *E. coli* K12, however. As a consequence, we have attempted to obtain a homoserine dehydrogenase deficient mutant of *E. coli* K12 by transduction of the corresponding chromosomal piece of B 184 with phage P1Kc. After the adequate selection, mutants were isolated

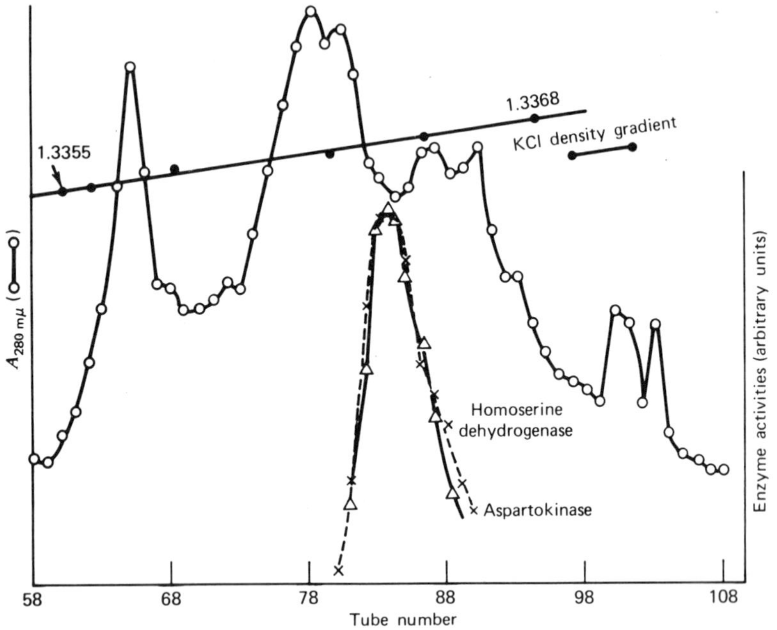

Fig. 3.4 Elution pattern of the new activities on a DEAE-cellulose column. Linear gradient: the buffer contains threonine. The two activities are normalized by equating the maxima of their peaks. Reproduced from ref. 5 with permission of Elsevier/North-Holland Biomedical Press, Amsterdam.

with the nutritional pattern indicated in Table 3.1. (3). It is immediately visible from inspection of the table that the transductants (Gif 88) do not have the same nutritional pattern as B 184; they can grow with threonine alone; in minimal medium they eventually grow in 48 h. They have no AK I-HDH I, however, weak aspartokinase and dehydrogenase activities are measurable. These activities are threonine-insensitive, and their level is greatly lowered by growth in the presence of methionine. Finally, methionine-requiring mutants have been prepared from Gif 88, which, when grown in a chemostat under methionine limitation, produce a high level of both threonine-insensitive activities (Table 3.2). We decided to call these two enzyme activities, aspartokinase II and homoserine dehydrogenase II (3); their synthesis is regulated by the level of the intracellular methionine pool (see Section 3.1).

During a systematic survey of spontaneous revertants of Gif 88, that is mutants that could grow rapidly without threonine, a mutant (Gif 881) was isolated that synthesized constitutively AK II and HDH II. The existence of this mutant allowed the purification of the two activities. They could not be

Table 3.1 Comparison of the Nutritional Requirements of Strains B 184 and Gif 88 (3)[a]

	Additions to the Culture Medium				
Strain	None	Threonine	Methionine	Threonine + Methionine	Homoserine
B 184	0	0	0	+++	++
Gif 88	+ (48 h)	++	0[b]	+++	++

[a] The amino acid concentrations used were 10^{-3} *M* for DL-threonine, and 3×10^{-4} *M* for DL-homoserine and DL-methionine. The growth was estimated by visual inspection of the turbidity of the culture medium after 16 h at 37°.

[b] The absence of growth of Gif 88 with added methionine is ascribable to the repression of aspartokinase II-homoserine dehydrogenase II: this repression in strains devoid of aspartokinase I-homoserine dehydrogenase I creates an absolute phenotypic requirement for threonine.

Table 3.2 Effect of Methionine on Repression and Derepression of Aspartokinase II and Homoserine Dehydrogenase II (3)[a]

		Specific Activities (nm/mg·min)	
Strain	Additions to the Culture Medium	Aspartokinase II	Homoserine Dehydrogenase II
Hfr H	None	Undetectable	Undetectable
Gif 95 (met^- mutant) of Gif 88	3×10^{-4} *M* DL-methionine	Undetectable	2.0
	Chemostat with 4×10^{-5} *M* (limiting) DL-methionine	8.2	71.0
Gif 881	None	22.0	80.0
	5×10^{-3} *M* DL-methionine	17.5	61.0

[a] When a limitation in the chemostat is reported, the bacteria were grown under limitation conditions for about three generations.

separated by the standard procedures of protein chemistry (3), and an homogeneous protein, aspartokinase II-homoserine dehydrogenase II was obtained in the pure state (7).

It is amusing in retrospect to notice that in *S. typhimurium* (20) and in other strains of *E. coli* (21), AK II and HDH II exist in such amounts that all the sophistication we applied to their detection in *E. coli* K12 would have been unnecessary with these organisms. But we would have missed all the fun.

The Reactions Catalyzed

They have been first described in yeast by Black and Wright in 1955 (22, 23).

Aspartokinase:

$$\underset{\text{L-aspartic acid}}{COOH{-}CH_2{-}CH(NH_2)COOH} + ATP \xrightleftharpoons{Mg^{2+}} \underset{\text{L-aspartyl-phosphate}}{COO(PO_3H_2){-}CH_2{-}CH(NH_2)COOH} + ADP$$

Homoserine dehydrogenase:

$$\underset{\text{L-aspartic semialdehyde}}{CHO{-}CH_2{-}CH(NH_2)COOH} + NAD(P)H + H^+ \rightleftharpoons \underset{\text{L-homoserine}}{CH_2OH{-}CH_2{-}CH(NH_2)COOH} + NAD(P^+)$$

The second reaction is specific for NADH in yeast, but in *E. coli* NADPH is a much better substrate (2).

The conditions of assay and the enzyme units have been repeatedly described (24, 25). Aspartokinase is best assayed, when pure, through a coupled assay using phosphoenolpyruvate, pyruvate kinase, lactate dehydrogenase, and NADH. This assay has, among others, the advantage of regenerating the ATP (26).

The Regulations at the Genetic Level

Regulation of the Synthesis of Aspartokinase I-Homoserine Dehydrogenase I
Let us call *thrA* the gene coding for aspartokinase I and homoserine dehydrogenase I, *thrB* the gene coding for homoserine kinase, and *thrC* the gene coding for threonine synthase. It has been shown in our laboratory that *thrA* is a complex gene composed of two cistrons, $thrA_1$ and $thrA_2$ (see Section 3.2)

and that *thrA*, *thrB*, and *thrC* constitute an operon transcribed in that order (27). The existence of this operon has been demonstrated by the partial polar effects of nonsense mutations in *thrA* on the expression of *thrB* and by the total polar effects caused by the insertion of bacteriophage Mu in the difference genes. Different classes of mutations have been obtained affecting the expression of the threonine operon.

OPERATOR CONSTITUTIVE MUTATIONS This first class affects the operator region. Mutants were selected by their resistance to α-amino-β-hydroxyvaleric acid (AVA), the threonine higher homologue. Mutants resistant to this analogue had already been isolated (9, 11). These were localized in *thrA* and affected feedback inhibition by threonine of both AK I and HDH I. The four activities involved in threonine biosynthesis (AK I, HDH I, homoserine kinase, and threonine synthase) are coordinately derepressed (four- to five-fold) in these mutants, and growth in the presence of threonine exerts a slight repression, if any. The mutations are localized on the *thrA* side of the *thrA-thrB-thrC* cluster and lead to derepression of the *thr* operon only when it is located in *cis* position, in merodiploids $thrO^+$ $thrO^c$ (28). Other mutations have been independently selected and isolated by Gardner and Smith (29) using an entirely different technique. They integrated bacteriophage lambda near the threonine operon. The presence of the prophage in the lysogenic bacteria causes a pleiotropic effect that leads to a lack of expression of the threonine operon: the bacteria become auxotrophic for threonine. Threonine-independent revertants have been obtained that synthesize the enzymes at the constitutive levels. These mutations map in the regulatory regions of the threonine operon and are similar to the *thrO* mutations.

MUTANTS OF THE THREONYL-T-RNA SYNTHETASE An *E. coli* K12 strain was constructed with two copies of the threonine operon. After mutagenesis, threonine-requiring mutants were sought and found. Such mutants must be located outside the operon. In one of these mutants, the enzymes of the threonine operon are not repressed by threonine *plus* isoleucine; its threonyl-t-RNA synthetase shows an apparent K_m for theonine 200-fold higher than that of the parental strain (30). The gene, *thrS*, coding for threonyl-t-RNA synthetase, was located at 37.5 min on the circular *E. coli* map (31). Regulatory properties of the mutant imply the involvement of charged threonyl-t-RNA or threonyl-t-RNA synthetase in the regulation of the threonine operon (30).

OTHER MUTATIONS AFFECTING THE REGULATION OF THE THREONINE OPERON The *lac* genes have been fused to the threonine operon by the general method described by Casadaban (32). In the strains obtained, the synthesis of β-galactosidase is under the control of the threonine operator. It is specifically repressed by threonine *plus* isoleucine. From these strains, it has been possible to select spontaneous mutants with a constitutive β-galactosidase synthesis. Some are *thrO*^{c}, analogous to the operator mutants already described. Some others carry a *thrX* mutation, distinct from *thrO* and *thrS*. They are resistant to α-amino-β-hydroxy valeric acid but do not excrete threonine. Since their exact location on the map is unknown, it is not possible to say whether the different *thrX* mutants are allelic. *ThrX* might code for a regulatory protein, or be the structural gene for one of the threonyl-t-RNA's, or affect the synthesis of an enzyme modifying threonyl-t-RNA (33).

Regulation of the Synthesis of Aspartokinase II-Homoserine Dehydrogenase II The addition of methionine to the growth medium causes repression of the synthesis of all the structural gene products of methionine biosynthesis and in particular of *metL* coding for aspartokinase II-homoserine dehydrogenase II (34, 35, 36, 3) (Table 3.2). These genes are found in several different regions of the *E. coli* chromosome (37), and although some of them are clustered, no two genes are known to be in a single operon. Two genes seem to be implied in the repression of those genes by methionine: *metK*, the structural gene for methionine adenosyltransferase; and *metJ*, whose function has not been established (the *metJ* product is thought to be a repressor protein, although this is not proved) (38). The effect of a *metK* mutation on the expression of the *metL* gene has not been investigated, but a *metJ* mutation leads to a pronounced derepression of aspartokinase II-homoserine dehydrogenase II (39) as well as of the other methionine biosynthetic enzymes (38).

3.2 ASPARTOKINASE I-HOMOSERINE DEHYDROGENASE I

Quaternary Structure of AK I-HDH I, Localization of the Two Functions on One Polypeptide Chain

Molecular Weight of the Native Protein The protein gives a single, sharp and symmetrical peak throughout the sedimentation process in the ultracentrifuge. Likewise, disc gel electrophoresis in the presence of 10 m*M* L-threonine (which protect against the alkaline denaturation of the enzyme) shows one main band of protein and of homoserine dehydrogenase activity (6) (Figure 3.5).

The molecular weight of AK I-HDH I has been determined by several methods. The value for the sedimentation coefficient obtained upon linear

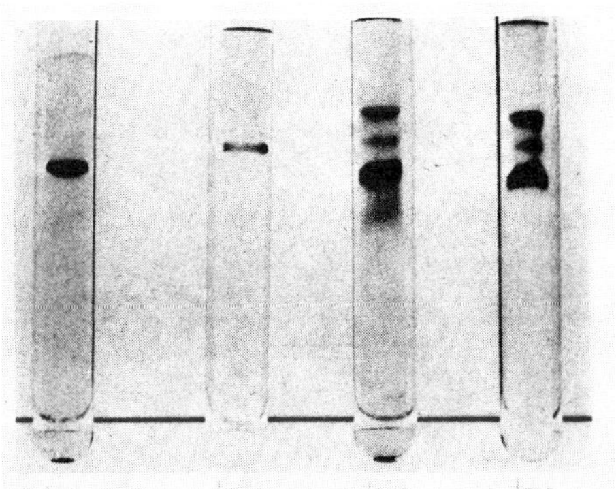

Fig. 3.5 Disc electrophoresis of the protein. From left to right: Electrophoresis of 50 μg of purified protein carried out in presence of 10 m*M* L-threonine; (I) protein stained with Amido Black, (II) revelation of homoserine dehydrogenase activity. Electrophoresis of 100 μg of the same protein sample carried out in absence of L-threonine; (III) protein staining; (IV) homoserine dehydrogenase activity. Reproduced from ref. 6 with permission of the European Journal of Biochemistry.

extrapolation to infinite dilution is $s^{\circ}_{20,w} = 11.5\ (\pm 0.2) \times 10^{-13}$ sec. This agrees very well with the results of sedimentation in a sucrose gradient using catalase and alcohol dehydrogenase as markers, where the peaks of both activities of AK I and HDH I have an *s* value of $11.0 \pm 0.5 \times 10^{-13}$ sec. Equilibrium sedimentation runs using interference optics were also done. At the completion of the photography of the Rayleigh pattern, schlieren pictures were also taken. This allowed to obtain values for M_w and M_z. The mean values, corresponding to the three protein concentrations used are: $M_w = 357{,}000 \pm 14{,}000$; $M_z = 350{,}000 \pm 30{,}000$ (6). Other workers have confirmed this value, also using hydrodynamic methods (40). An independent measurement using light scattering has yielded a value of $M_w = 358{,}000 \pm 35{,}000$ (41).

Determination of the Subunit Molecular Weight and Amino-Acid Composition The reduced protein in solution in 6 *M* guanidine chloride dissociates to subunits of equivalent molecular weight (42). A careful sedimentation equilibrium study taking into account the nonideal behavior of the protein under these conditions yields a value of approximately 80,000 for the molecular weight of the subunits (40).

Clark and Ogilvie (43), Starnes et al. (40), Wampler (44), and our own laboratory (41), using dodecylsulfate gel electrophoresis or gel filtration in 6 *M* guanidine chloride, found values varying between 80,000 and 90,000. Our own results indicate that mol wt = 86,000 ± 4000. If we assume a molecular weight of 86,000 for the subunit, then the native enzyme is a tetramer of mol wt = 344,000. Further support for the tetrameric structure is provided by cross-linking the subunits with dimethylsuberimidate followed by disc gel electrophoresis in dodecylsulfate. When the enzyme was cross-linked at pH 8.5 in the presence of 0.3 *M* KCl, only four bands were seen, with the mobilities expected for the monomer, dimer, trimer, and tetramer of the subunits (41).

The amino acid composition of aspartokinase I-homoserine dehydrogenase I has been determined several times (41, 42, 45). Since it has been redetermined many times in our work on the primary structure, we give in Table 3.3 what we think are the most reliable values.

Identity of Subunits As the protein is known to carry two enzymatic activities, it was important to determine whether or not the subunits were identical. We have already seen that they are equivalent in mass. Isoelectric

Table 3.3 Amino Acid Composition of Aspartokinase I-Homoserine Dehydrogenase I (63)

Amino Acid	Residues/Mole of Subunit
Lys	37
His	13
Arg	42
CMCys	12
Asp	77
Thr	34
Ser	57
Glu	91
Pro	36
Gly	61
Ala	84
Val	64
Met	20
Ile	42
Leu	83
Tyr	18
Phe	28
Trp	4
	803

focusing, in the presence of 6 *M* urea and 0.1% nonionic detergent, gives a single peak with no indication of charge heterogeneity in the subunits.

The results of chemical studies strongly support the conclusion that the subunits are identical (41):

1 The number of peptides that are detected by ninhydrin staining of the tryptic map of the soluble peptides is compatible with identical subunits (41).

2 The determination of the number of unique sequences around specific residues supplies the additional information that these sequences are present in all subunits.

Four unique sequences have been determined around tryptophan residues in distinct tryptic peptides:

T85	Ala-Asp-Ile-CMCys-Glu-Trp-Thr-Asp-Val-Asp-Gly-Val-Tyr-Thr-CMCys-Asp-Pro-Arg
T10′	Ala-Leu-Leu-Thr-Asx-Val-His-Gly-Leu-Asn-(Asn,Gln)-Leu-(Glu_3, Gly,Leu)-Trp-Ala-Gln-Ala-Lys
TB1	Gln-Gln-Ser-Trp-Leu-Lys
TB2	Thr-Leu-Ser-Trp-Lys

Twelve unique carboxymethylcysteine residues have been found (41, 45, 46):

T85	See above (2 CMCys residues)
T60	Tyr-Val-Gly-Asn-Ile-Asp-Glu-Asp-Gly-Val-CMCys-Arg
T30	Glx-Tyr-His-Leu-(CMCys,Asx_3,Thr,Ser_2,Glx_2,Pro,Ala_2,Val_3,Ile, Leu,Tyr)-Ala-Asp-Phe-Leu-Arg
T10	His-Val-Leu-His-Gly-Ile-Leu-Leu-Leu-Gly-Glu-CMCys-Pro-Ser-Asp-Ile-Asn-Ala-Ala-Leu-Ile-CMCys-Arg
TNa	Val-CMCys-Gly-Val-Ala-Asn-Ser-Lys
TNb	Thr-Ile-Thr-Pro-Ile-Ala-Gln-Phe-Gln-Ile-Pro-CMCys-Leu-Lys
Ch60*	CMCys-Val-Pro-Glu-(CMCys,Asx_2,Ser,Glx,Val_2)-Arg
TNc	Ala-Ala-CMCys-Leu-Arg
T40	Asx-(CMCys,$Asx_{1\text{-}2}$,Ser,Glx,Gly,$Ala_{2\text{-}3}$,Val,Leu_2,Tyr)-Arg

* Chymotryptic peptide from the tryptic core.

3 The amino terminal sequence of the protein was determined with an automatic sequenator:

NH_2-Met-Arg-Val-Leu-Lys-Phe-Gly-Gly-Pro- -Val-Ala- .....

At each turn of the sequenator, a single amino acid derivative was released, showing that all subunits share at least the same amino terminal sequence.

The C-terminal sequence, Leu-Gly-Val-COOH, was determined with carboxypeptidase A, with roughly 1 mole of each residue being released per mole of subunit.

Isolation of Functionally Active Fragments

ISOLATION OF A PROTEIN FROM A NONSENSE MUTANT THAT HAS ONLY THE ASPARTOKINASE ACTIVITY: THE AK I FRAGMENT In 1967 we reported (47) the isolation of a revertant of Gif 54 (see Section 3.1), which turned out to be entirely devoid of the threonine-sensitive HDH I activity but possessed a normal level of threonine-sensitive AK I activity when the extracts were made in the presence of high threonine concentration. The enzyme was threonine insensitive when this precaution was not taken. Its molecular weight, inferred from gel filtration studies, was estimated to be 180,000–200,000. In the absence of threonine, the enzyme dissociates to subunits of molecular weight estimated to be 45,000–50,000 by the same method (47). Later, SDS gel electrophoresis of the pure protein gave a molecular weight of 48,000 (48). These results show that its threonine-sensitive form is a tetramer made of subunits considerably shorter than those of the wild type protein. It has been since demonstrated that this mutation (Gif 101) is an *ochre* mutation (27). The corresponding enzyme has a methionine N-terminal residue as the wild type protein, and its C-terminal sequence is different from that of the wild type protein. Further evidence that the aspartokinase activity resides in the amino-terminal part of the wild type subunit is given below.

LIMITED PROTEOLYSIS OF AK I-HDH I: THE HDH I FRAGMENT Limited proteolysis of aspartokinase I-homoserine dehydrogenase I results in the production of a desensitized homoserine dehydrogenase fragment devoid of aspartokinase activity (48).

As shown in Figure 3.6, proteolysis by α-chymotrypsin leads to desensitization of the homoserine dehydrogenase activity, whereas the kinase activity is completely destroyed. At any given time before complete inactivation the remaining kinase activity is fully sensitive to threonine. The products of proteolysis were analyzed on dodecylsulfate gels, and we observed that the band corresponding to the native subunit of 86,000 mol wt disappeared while a faster band appeared. Upon calibration it was found that this band corresponds to a species of 55,000 mol wt. In the presence of 2 m*M* L-threonine, the two enzymatic activities are fully protected and retain their sensitivity toward threonine. Under these conditions, the patterns of the gels carried under denaturing or nondenaturing conditions are identical to those of the native enzyme and show that the allosteric inhibitor protects fully against proteolysis (48).

When trypsin, papain, or subtilisin were used as the proteolytic agent instead of α-chymotrypsin, the same effects were observed: the loss of kinase activity, the dehydrogenase desensitization, the characteristic band pattern upon electrophoresis, and the protection by threonine against

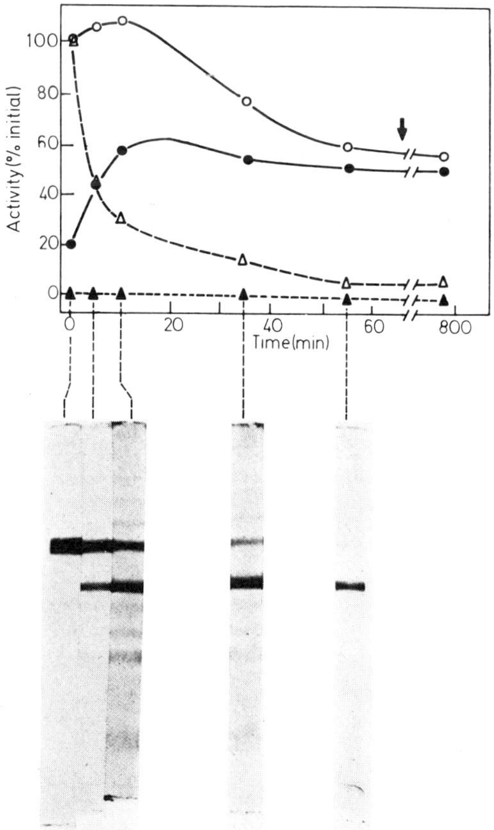

Fig. 3.6 Effect of proteolysis on AK I-HDH I. Upper part: enzymatic activities. Δ: aspartokinase activity; O: homoserine dehydrogenase activity. Full triangle and circles: same activities measured with 2 m*M* L-threonine. Lower part: electrophoresis on dodecylsulfate gels. Reproduced from ref. 48 with permission of the European Journal of Biochemistry.

proteolysis were obtained in all cases. With all proteolytic enzymes tested, the homoserine dehydrogenase fragment, as we shall now designate it, proved to be quite resistant to further degradation (48). This shows that, notwithstanding the specificity of the protease used, native AK I-HDH I possesses a sensitive region of limited length, since the fragments obtained with chymotrypsin and subtilisin, for example, are not distinguishable on denaturing gels. Whereas α-chymotrypsin yields a fragment with an heterogeneous N-terminal (Ile and Ala), subtilisin yields an homogeneous fragment whose N-terminal sequence has been determined (49).

After proteolysis, the homoserine dehydrogenase fragment has been purified by gel filtration. Its elution volume corresponds to a molecular weight of 110,000. This has also been determined by sedimentation equilibrium, and the same value has been found. By comparing with the value of 55,000 found by SDS gel electrophoresis, one concludes that the homoserine dehydrogenase fragment is a dimer of identical subunits (48).

The characterization of the C-terminal residues of the homoserine dehydrogenase fragment was made with carboxypeptidase A. After 1 h of reaction, roughly one equivalent of valine, leucine, and glycine were released per 55,000 (48). The C-terminal sequence of the entire enzyme being Leu-Gly-Val-COOH (41), this experiment thus proves that the homoserine dehydrogenase fragment corresponds to the C-terminal side of the polypeptide chain of aspartokinase I-homoserine dehydrogenase I.

The aspartokinase I fragment and the homoserine dehydrogenase I fragment overlap as evidenced by the sheer examination of numbers; the sum of the molecular weights of the two fragments (55,000 + 48,000 = 103,000) exceeds by 17,000 the molecular weight of the polypeptide subunit of aspartokinase I-homoserine dehydrogenase I (86,000).

Evidence for the Autonomy of Functions, Attributed to Discrete Domains of Aspartokinase I-Homoserine Dehydrogenase I

Even though the subunits of the homoserine dehydrogenase I and of the aspartokinase I fragment share a common sequence, the two catalytic activities are likely to be carried by two distinct regions of the native polypeptide chain, in the native enzyme. The aspartokinase activity resides in the amino-terminal and the homoserine dehydrogenase in the carboxyl-terminal section of the polypeptide chain.

The fact that modified protein can be obtained either by mutation or by limited proteolysis, gives some indications regarding the configuration of native aspartokinase I-homoserine dehydrogenase I. They both retain, almost unchanged, one or the other of the two catalytic activities and at least some of the association areas between subunits (the AK I fragment is tetrameric, and the HDH I fragment dimeric). Therefore, as has been proposed for β-galactosidase and its complemented derivative (50), their structure must be close to that of the corresponding region in the native protein. The subunit of aspartokinase I-homoserine dehydrogenase I appears thus in its native configuration to be composed of at least two rather independent regions, each being able to retain its own catalytic activity when the other is removed, either by mutation or by proteolytic cleavage.

Gene fusion has been proposed as an important mechanism in the evolution of complex protein (51) and aspartokinase I-homoserine dehydrogenase I may well have arisen by such a process; in which case, we might expect the subunit to be composed of at least two independently folding regions.

Selective Inactivation of Each of the Two Functions

BY A SUBSTRATE ANALOG 6-Mercapto-9-β-D-ribofuranosylpurine-5′-triphosphate (SHTP) is an adenosine triphosphate analog that can serve as a substrate of aspartokinase activity (52). Long incubation of SHTP in the absence of L-threonine leads to a modified protein of essentially the same molecular weight as the native protein that has specifically lost its aspartokinase activity; however, the associated homoserine dehydrogenase is active but desensitized to threonine. Reaction of the modified enzyme with mercaptoethanol releases four molecules of SHTP per protein molecule, this being associated with the regain of aspartokinase activity and the resensitization of homoserine dehydrogenase. Saturating levels of L-threonine protect the enzyme against SHTP.

ATP is not a good protector against SHTP inactivation, suggesting that either SHTP has a much greater affinity than ATP for the catalytic site or that it does not bind at this catalytic site but that, when covalently bound, it somehow interferes with catalysis. Whatever the case, the point that interests us here is that this reagent inactivates the activity associated with one domain, while the other stays functionally intact.

BY AN AFFINITY LABEL We have been able to obtain the selective inactivation of the dehydrogenase by using 2-amino-4-oxo-5-chloropentanoic acid (AOC) (53), an analog of aspartate semialdehyde:

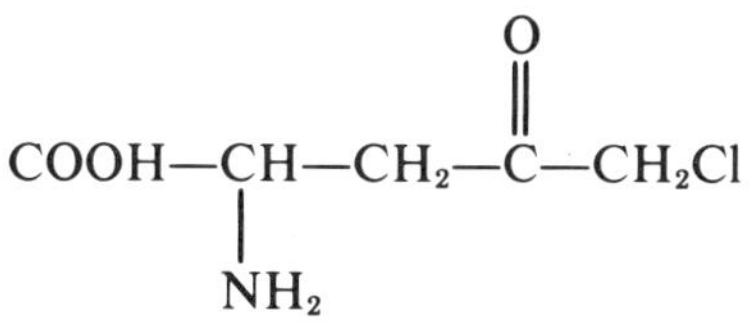

AOC

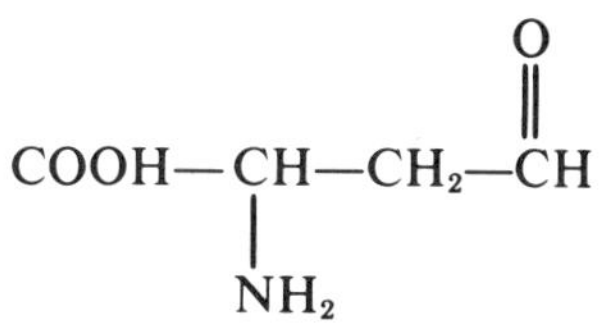

Aspartate semialdehyde

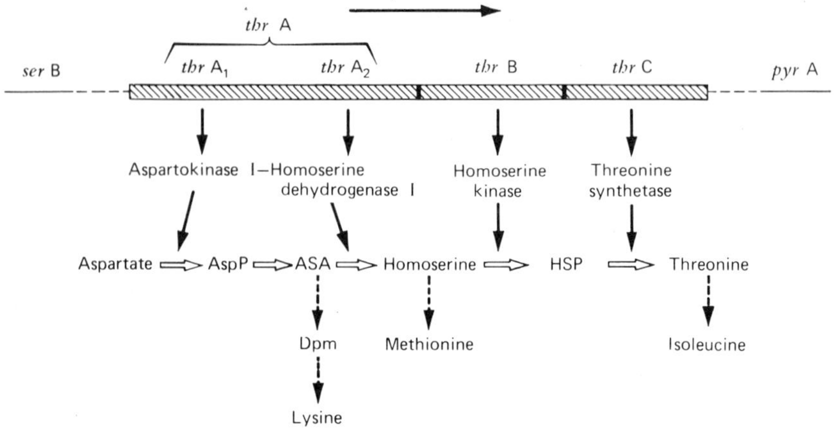

Fig. 3.7 The threonine operon. AspP, ASA, Dpm and HSP stand, respectively, for β-L-aspartylphosphate, L-aspartate semialdehyde, *meso*-diaminopimelate, and homoserine phosphate. Reproduced from ref. 27 with permission of the American Society for Microbiology.

Incubation of the enzyme with this chloroketone in the presence of $NADP^+$ and L-threonine inactivates the dehydrogenase, and leaves the aspartokinase activity intact with its characteristic inhibition by L-threonine, the tetrameric structure being conserved. Using radioactive AOC, we have determined that the inactivation of the dehydrogenase requires the covalent binding of one equivalent of the analog per subunit (54).

Structure of the *thrA* Gene The topological distribution of the two activities on the subunit has been confirmed by genetic analysis of the fine structure of the gene *thrA*, the structural gene for aspartokinase I-homoserine dehydrogenase I. It has been shown, by complementation studies using partial diploids, to be composed of two cistrons, with the operator proximal one ($thrA_1$) coding for the aspartokinase, whereas the operator distal one ($thrA_2$) is the structural region coding for the dehydrogenase (27) (Figure 3.7). The functional independence of the two domains shown by the isolation of active fragments is, moreover, illustrated by the properties of the mutants in the *thrA* gene. A large collection of point mutants have been obtained which have lost only one of the activities, being either of the AK^- HDH^+ or of the AK^+ HDH^- type, with a normal level of the remaining activity in all the cases tested. Moreover all the $thrA_1A_2$ mutants were found to be nonsense mutants. This suggests that a disorganization of one of the globules within one subunit does not necessarily alter the catalytic activity of the other one.

Intra- and Interprotomeric Molecular Interactions between the Different Regions of Aspartokinase I-Homoserine Dehydrogenase I

The Kinase Region is Buried in the Tetramer

DISSOCIATION IS NECESSARY FOR THE PROTEOLYTIC DESTRUCTION OF THE KINASE The proteolytic reaction of the tetrameric AK I-HDH I leads to a dimeric HDH I fragment (see Section 3.2). As shown in Figure 3.8, the rate of the proteolytic reaction is very strongly dependent on AK I-HDH I concentration. At high concentration, the enzyme is quite resistant to the action of subtilisin, the loss of aspartokinase activity is rather small, and a plateau is reached within one hour. Sodium dodecylsulfate gel electrophoresis confirms that only a small proportion of the original material is destroyed under these conditions. Upon dilution of the incubation mixture, proteolysis occurs until the kinase activity is completely destroyed. If only AK I-HDH I is diluted, the rate of proteolysis is much higher than in the experiment depicted. These results strongly suggest that the enzyme must first dissociate to be susceptible to the proteolytic attack. This hypothesis is

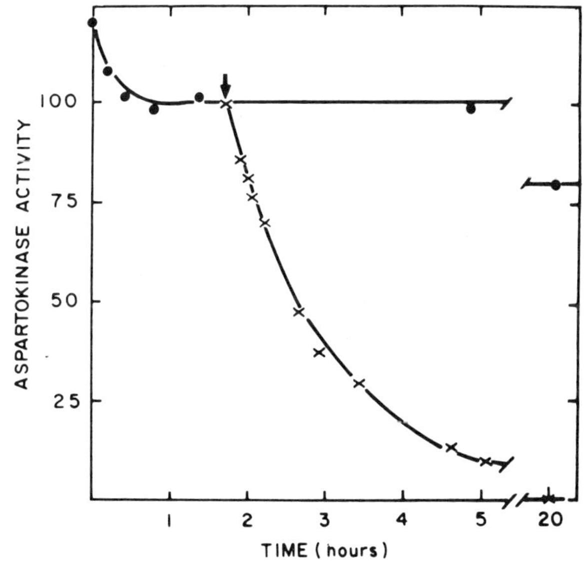

Fig. 3.8 Acceleration of the rate of proteolysis of AK I-HDH I by subtilisin upon dilution of the incubation mixture. At the time indicated by the arrow, the mixture (enzyme: 10 mg/ml; subtilisin: 0.4%, w/w) was diluted a hundred times. X: diluted sample; ●: undiluted sample. Reproduced from ref. 55 with permission of the European Journal Biochemistry.

supported by the following observations: in the presence of 0.15 *M* KCl, which stabilizes the tetrameric form, the proteolytic reaction is much slower; similarly in the presence of threonine where no dissociation occurs even at high dilutions, the protein is not digested (55).

Moreover, it is possible to show directly the existence of a dimeric intermediate retaining a functional aspartokinase region. When following the proteolytic reaction on polyacrylamide gels under nondenaturing conditions (Figure 3.9), a third band D′ is transiently present, as an intermediate between the native enzyme E and that of the 110,000 mol wt homoserine dehydrogenase fragment F. The D′ species was purified and found to have a molecular weight of approximately 170,000 and to yield on sodium dodecylsulfate gels a single band with the same migration as the native subunit. This shows that D′ is a dimer. It exhibits a lower level of aspartoki-

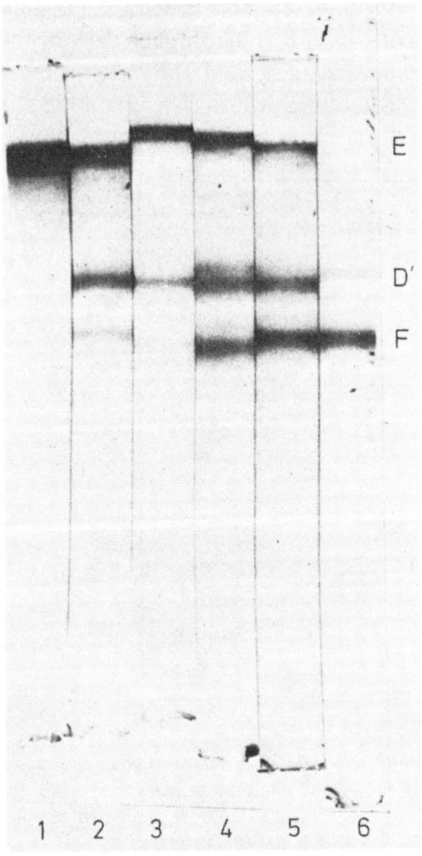

Fig. 3.9 Electrophoretic study of proteolysis: nondenaturing conditions. Activity stain. Starting from the left hand gel ($t = 0$), the gels correspond to aliquots taken at 30 min intervals. Reproduced from ref. 55 with permission of the European Journal of Biochemistry.

nase activity, has retained the homoserine dehydrogenase activity, albeit with a lessened threonine sensitivity (55).

Since the final fragment F has the same C-terminal as the polypeptide of the native protein, and since the chain of D′ has approximately the same molecular weight as that of the native protein, one is forced to the conclusion that the proteolytic reaction implies first the splitting of a short N-terminal fragment involved in the contact between the kinase regions in the tetramer, giving rise to an intermediate analogous to the D′ species.

Two models can be envisaged:

$$T \xrightarrow{\text{subtilisin}} T' \underset{}{\overset{\text{dissociation}}{\rightleftharpoons}} D' \xrightarrow{\text{subtilisin}} F$$

$$T \overset{\text{dissociation}}{\rightleftharpoons} D \xrightarrow{\text{subtilisin}} D' \xrightarrow{\text{subtilisin}} F$$

where T is the native tetramer, D the dimer resulting from the dissociation of T, F the homoserine dehydrogenase fragment, and T′ a slightly proteolyzed (on the N-terminal side) tetrameric form that dissociates to the characterized D′ intermediate.

The two models have a common feature: the substrate of the protease in the step leading to the destruction of the aspartokinase region is the D′ species; therefore the aspartokinase region is not accessible to the protease in the tetramer (55).

SULFHYDRYL GROUP ACCESSIBILITY AND DISSOCIATION Modification of a single sulfhydryl group by SHTP results in the loss of aspartokinase activity and the desensitization of homoserine dehydrogenase activity, in which cases the enzyme remains tetrameric (see Section 3.2). Modification by less specific reagents such as *p*-mercuribenzoate, *N*-ethylmaleimide, dithiobisnitrobenzoic acid (DTNB), iodoacetamidosalicylic acid, or iodoacetic acid leads to the same result. However, with these reagents, more sulfhydryl groups are titrated and the tetramer is transformed into a mixture of an active homoserine dehydrogenase aggregate and an active homoserine dehydrogenase dimeric species, suggesting that sulfhydryl groups are somehow involved in the intersubunit association areas (56).

Figure 3.10 is an example of the course of titration by DTNB and of the concomitant events. Out of the 12 —SH groups present per subunit, seven are titratable by DTNB, with the titration being accompanied by the destruction of the kinase and the desensitization of the dehydrogenase. It is noteworthy that during the titration by DTNB and all other sulfhydryl reagents tested, the remaining aspartokinase activity is fully sensitive to

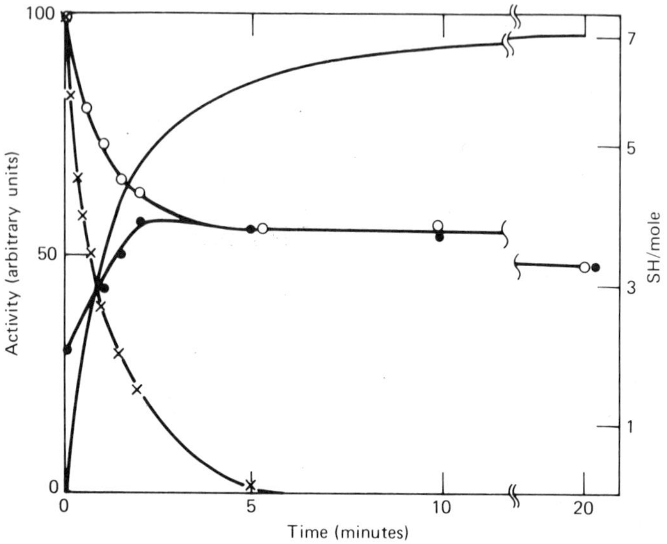

Fig. 3.10 Titration of available -SH groups of AK I-HDH I: concomitant inactivation of aspartokinase (X) and desensitization of homoserine dehydrogenase. The activity of the latter is measured without (O) and with (●) threonine. The solid line without symbols refers to the number of sulfhydryl groups titrated per mole of enzyme (71).

threonine. In the presence of 2 m*M* threonine, essentially no sulfhydryls can be titrated (6).

The above experiment was performed in the presence of 150 m*M* KCl. If the concentration of K^+ ions is reduced to 5 m*M*, very different results are obtained: although the final number of titratable sulfhydryls is the same, titration is much more rapid, and threonine protects only four sulfhydryl groups; moreover, titration of the three titratable —SH groups in the presence of threonine does not result in the loss of kinase or desensitization of dehydrogenase. The same qualitative and quantitative results are obtained if threonine is replaced by homoserine *plus* ATP (57).

Very early it was recognized by sucrose gradient centrifugation techniques that potassium ions and threonine stabilize the tetrameric form of the protein. In the absence of either of these ligands, the sedimentation coefficient diminishes notably; this is accompanied by a partial loss of homoserine dehydrogenase activity and by its desensitization (2). This question has been reexamined by Mackall and Neet (58) who studied the pH dependance of dissociation and suggested that, at alkaline pH's, the dissociation might involve the titration of sulfhydryl groups. This interpretation is in keeping with the observation reported earlier in the same section that AK I-HDH I dissociates when treated with sulfhydryl reagents.

Furthermore, the loss of aspartokinase activity and the desensitization of homoserine dehydrogenase activity undergone by the protein at pH 9 (59) are definitely accompanied by the formation of two interchain disulfide bonds per monomer. All the effects are prevented by 5 m*M* dithiothreitol (60). The same agents, namely L-threonine or aspartate *plus* K^+, protect against the effects of alkaline pH (59) and against dissociation (58). This is taken as a strong indication that sulfhydryl groups are involved in the association areas between dimers in the native tetramer.

The rates of inactivation of aspartokinase and of desensitization of the dehydrogenase upon —SH-titration by iodoacetamidosalicylic acid (Figure 3.11) (55) or by alkaline treatment (Figure 3.12 *a* and *b*) (60) are related to the dissociation of the protein, as shown by the protein concentration-dependent decays.

An extensive study has recently been made following the action of iodoacetic acid (IAA) on the two enzymatic activities and identifying the specific sulfhydryl groups involved in the association areas (61, 62). The reaction of IAA with AK I-HDH I leads to the inactivation of the aspartokinase activity, resulting from a pseudo-first order process with a τ ½ = 75 min. In contrast, the inactivation of the dehydrogenase results from two pseudo-first order events having a different rate constant (τ ½ respectively equal to 15 min and 16 h). While the aspartokinase remains as usual

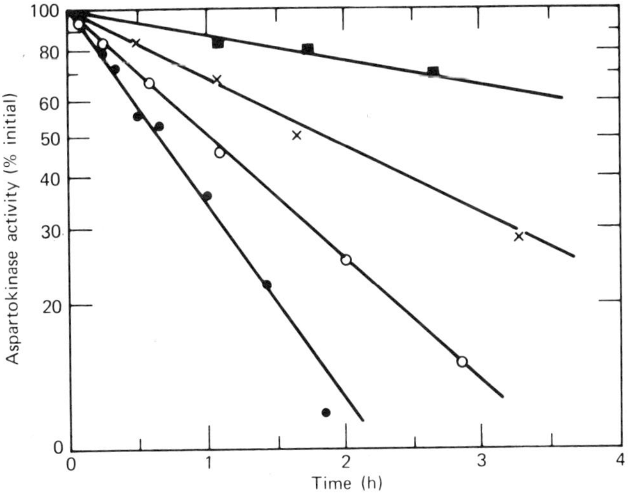

Fig. 3.11 Inactiviation of aspartokinase by 4-iodoacetamidosalicylate as a function of protein concentration. Iodoacetamidosalicylate (2.5 m*M*) was incubated with enzyme at the following concentrations: (■) 7.5 mg/ml (X) 2.5 mg/ml; (O) 0.5 mg/ml; (●) 0.1 mg/ml. Reproduced from ref. 55 with permission of the European Journal of Biochemistry.

sensitive during its inactivation process, the dehydrogenase becomes gradually insensitive to the feedback inhibitor with a kinetics similar to the one observed for its inactivation (Figure 3.13).

The kinetics of ^{14}C incorporation from radioactive IAA is pseudo-first order with τ ½ = 105 min. From Figure 3.14 an unexpected result appears: the alkylation of *a single —SH per tetramer* suffices for the complete desensitization of homoserine dehydrogenase and for almost all the effect observed on the homoserine dehydrogenase activity.

Tryptic maps of the carboxymethylated protein, carried out at different stages of the alkylation, show that this single —SH belongs to the peptide TNb, which is located in the aspartokinase moiety of the bifunctional enzyme (see illustration to section on structural studies, Section 3.2) (63). It is noteworthy that at this point, only 25% of the aspartokinase activity has been lost. At a later stage a second equivalent of the peptide TNB is labeled (Figure 3.15). Then, other peptides are alkylated, namely T85 (belonging

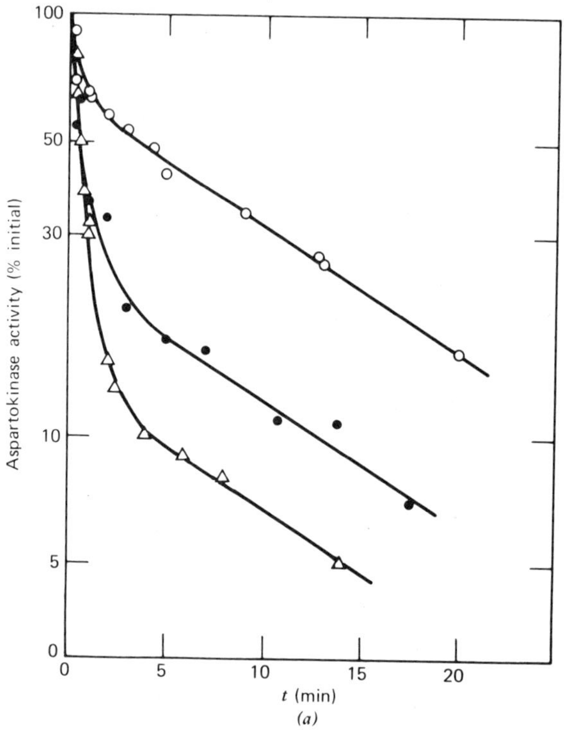

Fig. 3.12 (*a*) Inactivation of the aspartokinase at pH 9. ○: 15 μg/ml; ●: 5.3 μg/ml; Δ: 2 2.2 μg/ml. Reproduced from ref. 60 with permission of the European Journal of Biochemistry.

also to the aspartokinase domain) and TNa and T30 (belonging to the dehydrogenase domain) (63). The amount of radioactivity incorporated in each sulfhydryl group of these peptides at the end of carboxymethylation is similar, showing that when the reaction is complete, only half of the subunits of the native enzyme have been alkylated: actually, in keeping with earlier observations (56), gel filtration reveals that half of the protein migrates as an aggregate (Mr > 10^6) and half as a dimer (Mr ~ 180,000). The distribution of the radioactivity is indeed asymmetric: only the aggregate is radioactive, the dimer being barely labeled (Figure 3.16). The alkylation of the enzyme then is associated with a dissociation and an aggregation of the alkylated dimers, while the nonalkylated dimers are stable under this form. A study of the dissociation of the protein shows that it occurs only when 2 TNb per tetramer are titrated. It is difficult to decide from these studies whether the dissociation is caused by the alkylation of a

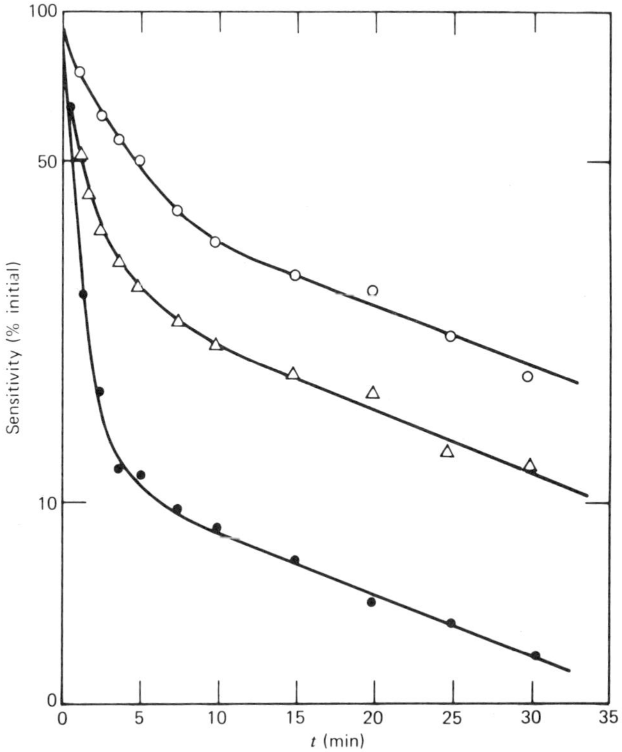

(b)

μg/ml. (b) Densensitization of the dehydrogenase at pH 9. O: 17.5 μg/ml; Δ: 5.5 μg/ml; ●:

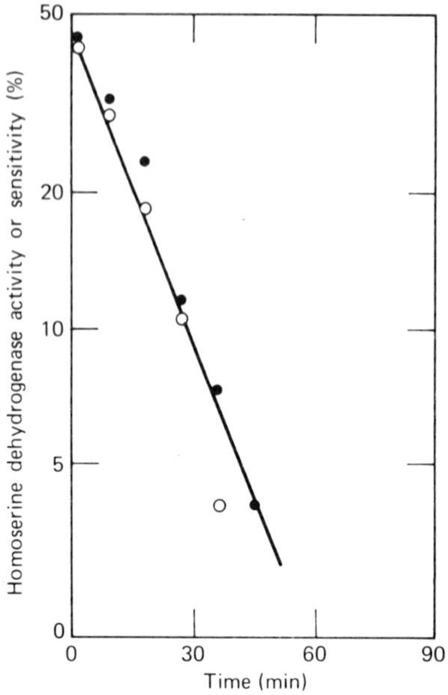

Fig. 3.13 Kinetics of homoserine dehydrogenase inactiviation and desensitization toward L-threonine by iodoacetic acid. O: Desensitization; ●: inactivation (61).

third cysteinyl residue on the alkylated tetramer, or if the weakening of the interaction forces between aspartokinase domains leads to dissociation and to the exposition of new cysteinyl residues.

IMMUNOCHEMICAL EVIDENCE Antibodies have been elicited in rabbits against native AK I-HDH I and its two fragments AK I and HDH I. The whole protein, AK I-HDH I, and its proteolytic fragment HDH I react as well with the antibodies raised against AK I-HDH I and HDH I, as seen by complement fixation and passive hemagglutination (64). This indicates that most of the antigenic determinants present on AK I-HDH I are also present on its proteolytic fragment HDH I. A protein, used as immunogen, elicits antibodies directed against the surface exposed to the environment (65). Thus the surface of the whole protein AK I-HDH I and of its fragment HDH I are immunochemically indistinguishable. In consequence in the protein the homoserine dehydrogenase domain, as represented by the HDH I fragment, is mainly located at the outside of the molecule (64).

The aspartokinase fragment, AK I, is recognized by antibodies raised against the whole protein AK I-HDH I (passive hemagglutination and complement fixation experiments) (64). But the reaction of AK I with antibodies raised against AK I-HDH I can be inhibited almost as well by the HDH I proteolytic fragment as by AK I-HDH I itself (inhibition of the passive hemagglutination reaction) (64). The reaction between AK I and antibodies raised against AK I-HDH I is then essentially due to antigenic determinants present on the "overlap" region (see Section 3.2 and illustration to section on structural studies of Section 3.2). Thus in the whole protein the "overlap" region is exposed, but the rest of the aspartokinase domain is essentially inaccessible to antibodies (64).

The experiments described so far have been carried in the absence of the allosteric effector, threonine. Both the reaction of AK I-HDH I with antibodies raised against the AK I fragment (64, 66) and the reaction of AK I with antibodies raised against AK I-HDH I (64) are considerably lowered in the presence of threonine. This indicates that in the presence of threonine all the antigenic determinants present on the aspartokinase region, including those on the "overlap," are "buried" in the AK I-HDH I molecule.

PROTEOLYSIS OF THE FULLY CROSS-LINKED TETRAMER Cross-linking of enzyme subunits with bifunctional reagents yields information about the

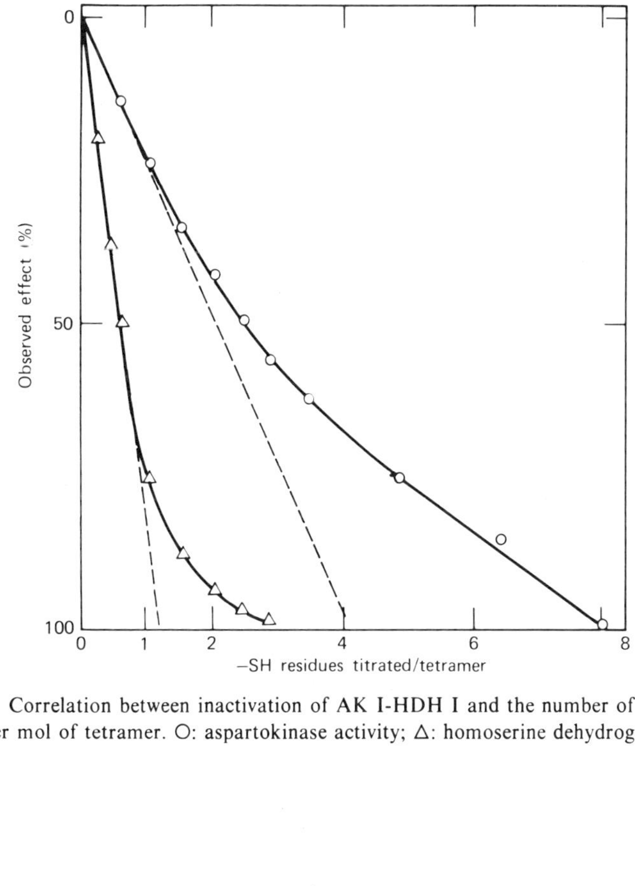

Fig. 3.14 Correlation between inactivation of AK I-HDH I and the number of -SH residues titrated per mol of tetramer. O: aspartokinase activity; Δ: homoserine dehydrogenase activity (62).

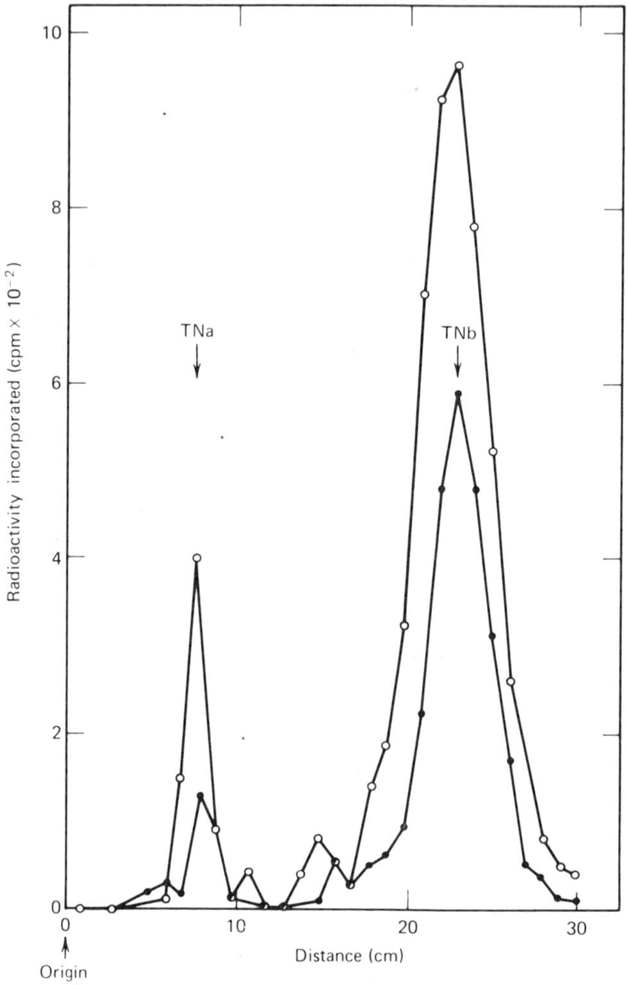

Fig. 3.15 Chromatography (after electrophoresis) of the tryptic digest of AK I-HDH- I carboxymethylated with radioactive IAA. ●: when the protein has lost 25% of its aspartokinase activity; O: when it has lost 50% of its aspartokinase activity (62).

number of subunits in an oligomeric protein and has actually been used to confirm the tetrameric nature of AK I-HDH I (55). In favorable cases the technique may provide information on the nature of the subunit contacts allowing one to differentiate between heterologous and isologous interactions (67).

A typical cross-linking experiment using AK I-HDH I and dimethylsuberimidate is illustrated in Figure 3.17 in which the kinase and the dehydro-

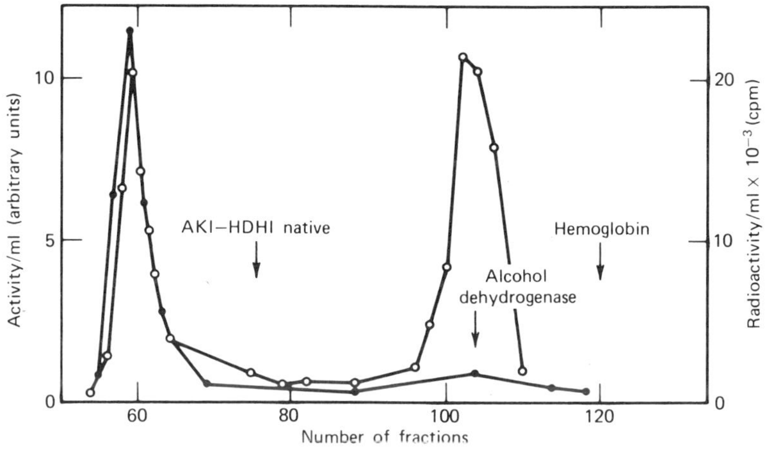

Fig. 3.16 Molecular weight determination at the end of carboxymethylation. Filtration on an agarose column. O: homoserine dehydrogenase activity; ●: radioactivity incorporated (62).

genase activities are followed in function of the reaction time. Both activities are partially destroyed during the reaction and are partially desensitized to L-threonine. Figure 3.18 illustrates the extent of cross-linking as determined after dodecylsulfate gel electrophoresis. The appearance of the tetramer (band D) was parallel to the loss and desensitization of the activities. Both the dimer and the trimer seem to be equally important as intermediates in the progression from monomer to tetramer. Thus, on this

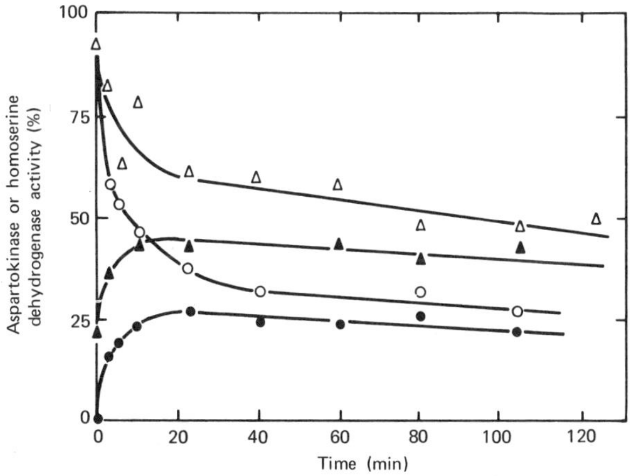

Fig. 3.17 Cross-linking of AK I-HDH I subunits with dimethylsuberimidate: effect on the catalytic activities. The activities are tested, in the absence (O, Δ) or presence (●, ▲) of 10 mM L-threonine. Circles: aspartokinase; triangles: dehydrogenase (55).

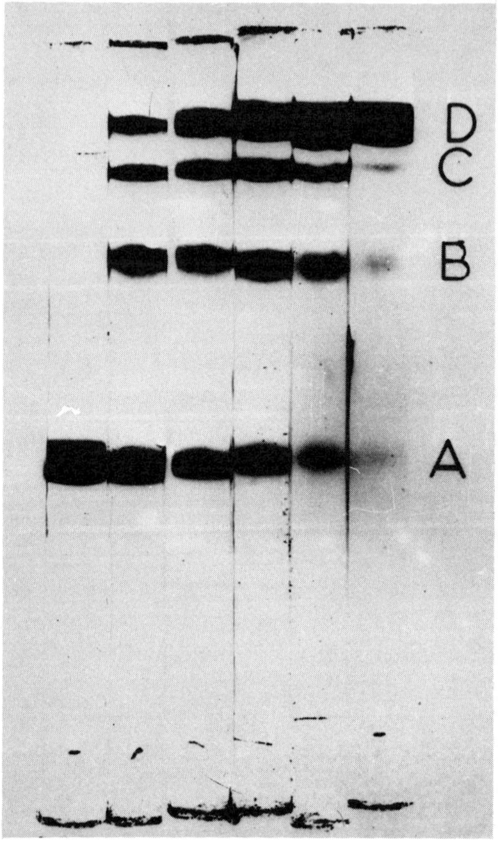

Fig. 3.18 Kinetics of cross-linking of the subunits: dodecylsulfate gels. From left to right, the reaction times were 0, 2, 5, 10, 20, and 60 min. Upon calibration, bands A, B, C, D correspond respectively to the monomer, dimer, trimer, and tetramer (55).

basis alone, one cannot distinguish between isologous and heterologous interactions (67). However, proteolysis of the cross-linked enzyme gives interesting indications on the nature of the association areas in the tetramer. Cross-linked AK I-HDH I was obtained in an experiment analogous to that described in Figure 3.18. Figure 3.19 shows that the protein is mostly cross-linked as a tetramer (band D), although some higher aggregates are also present. The effect of the proteolytic attack on the cross-linked enzyme is qualitatively analogous to that reported below for the native enzyme, namely a total loss of the residual aspartokinase activity and little change in homoserine dehydrogenase activity. The analysis of the products of proteolysis on sodium dodecylsulfate gels gave more interesting results however: as shown in Figure 3.19, the tetrameric cross-linked enzyme is first transformed into a species having the same migration as band B, probably corresponding to a cross-linked dimer. The final proteolysis product (band 1) has a molecular weight of 110,000, exactly that expected for the dimeric homoserine dehydrogenase fragment (2 × 55,000).

Since the ultimate product after limited proteolysis of the cross-linked tetramer is a cross-linked dimeric homoserine dehydrogenase fragment, we can conclude that the only way to cross-link dimers into a cross-linked tetramer is by linking their aspartokinase regions together; otherwise, a cross-linked tetrameric homoserine dehydrogenase of 220,000 mol wt (4 × 55,000) should have been obtained after proteolysis (see Figure 3.20). However, the existence of an intermediate dimer is somewhat surprising since cross-linking makes impossible any spontaneous dissociation of the molecule. The explanation is likely to be that the proteolytic reaction implies a preliminary splitting of a short fragment involved in the joining of

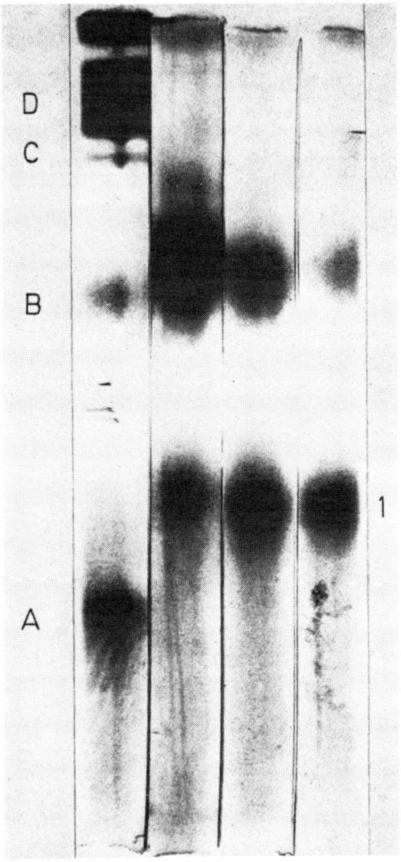

Fig. 3.19 Proteolysis of the fully cross-linked tetramer. The gel on the extreme left corresponds to the mixture subjected to proteolysis ($t = 0$); the three other gels correspond respectively to periods of proteolysis of 30 min, 2 h and 4 h. Bands A, B, C, D correspond to the same species as in Figure 3.18. Band 1 corresponds to a species of mol wt = 110,000 (55).

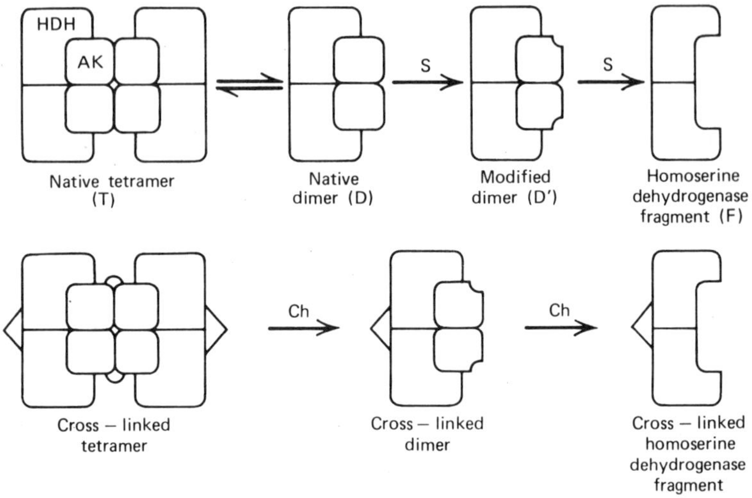

Fig. 3.20 Model for the organization of the subunits in AK I-HDH I. In the drawing representing the native tetramer (upper left) the two catalytic regions of one subunit have been noted. HDH = homoserine dehydrogenase region; AK = aspartokinase region. The kinase region represented as "buried" in the tetramer is much more exposed when the latter dissociates. The preliminary splitting of an amino-acid terminal peptide is shown as a notch in the kinase region. The proteolytic enzymes used are noted S for subtilisin and Ch for α-chymotrypsin. The enzyme has been shown as a planar structure for simplicity. Reproduced from 55 with permission of the European Journal of Biochemistry.

the kinase regions, to give rise to an intermediate analogous to the D′ species. In conclusion, the most likely structures for aspartokinase I-homoserine dehydrogenase I are the isologous planar and the isologous tetrahedral tetramers. The existence of isologous associations was indeed previously suggested by the dissociability of the protein to the dimer (2).

A Model of the Protein The two catalytic regions of aspartokinase I-homoserine dehydrogenase I are profoundly different: the aspartokinase is quite susceptible to disruption by any number of agents including heat, —SH reagents, and proteolysis, where the dehydrogenase region is little affected by any of these agents. The existence of discrete "globules" corresponding to separate domains, although supported by indirect and circumstancial evidence (in the absence of x-ray diffraction spectra), provides a satisfactory interpretation for the independent fragments that have been obtained and for the inter- and intrasubunit interactions in the native molecule.

The model shown in Figure 3.20 is consistent with both the dimeric nature of the homoserine dehydrogenase fragment and the tetrameric nature of the aspartokinase fragment (55). It is also compatible with the data presented above; namely the effects of dissociation on sulfhydryl groups accessibility and on the rate of proteolysis and the immunochemical evidence. The nearly normal Hill coefficient (n = 3.6 instead of 4 for the native enzyme) of the AK I fragment argues that it retains the allosteric properties essentially intact and hence the oligomeric contacts leading to the tetrameric structure. Indeed, the major conformational changes characteristic of the native enzyme (see below) may be largely confined to this region.

Using an approach quite different from ours, Mackall and Neet (58) have proposed a model of the enzyme essentially similar to the one in Figure 3.20.

The Binding of Ligands to Aspartokinase I-Homoserine Dehydrogenase I

The bifunctional protein has numerous ligands. In addition to its substrates, L-aspartate, ATP, homoserine, aspartate semialdehyde, NADP, and NADPH the allosteric effector must bind to the protein.

NADP and NADPH Binding Three methods have been used: equilibrium dialysis, fluorescence (68), and circular dichroism (69).

We studied the binding of the oxidized coenzyme by means of gel filtration of the protein, using (carboxyl-^{14}C) NADP on Sephadex G-50 equilibrated with the radioactive ligand. The linear Scatchard plot indicates a simple Michaelian type of variation with ligand concentration. Extrapolation leads to a value of 4 ± 0.25 mole of NADP bound per mole of enzyme, and the dissociation constant is close to 12 μM. The variation of homoserine dehydrogenase activity with pyridine nucleotide concentration is also Michaelian, the apparent dissociation constant being 10 times higher (K_m NADP = 110 μM) (2).

The absorption at 340 nm was used to detect the reduced coenzyme upon elution from Sephadex columns equilibrated with the ligand. At saturating concentration of NADPH 2.7 to 4.1 molar equivalents were bound. These values do not vary significantly in the presence of 1 mM threonine or with varying the KCl concentration from 150 mM to 650 mM, or in the presence of L-homoserine.

The change in molecular ellipticity of an extrinsic dichroic band at 348 nm caused by NADPH binding was used to titrate the protein with NADPH to a 15-fold higher ligand concentration than was possible with fluorimetric methods. The number of sites found, 4 ± 0.4 per tetramer (69), was in agreement with the fluorimetric determination, based on the considerable increase of the blue fluorescence of the coenzyme upon addition of aspartokinase I-homoserine dehydrogenase I.

ATP Binding We have seen that SHTP, a substrate of the aspartokinase activity, can be covalently bound to the protein with a stoichiometry of one equivalent per polypeptide chain.

The direct determination of the stoichiometry is difficult because of the low affinity of ATP for the protein, and the errors are fairly large. Ehrlich and Takahashi have attempted these measures, using AT*P*-γ-^{32}P in the absence of metal (70). The best computer fit of the data gives 4.6 ± 1.6 sites per tetramer and a dissociation constant of 2.8 ± 1.0 m*M*. However, the maximum number of molecules of ATP actually found is 2.5 per tetramer.

Aspartate Semialdehyde and Homoserine Binding No direct measurement has been attempted because of the instability of the aldehyde and because the affinity of the protein for homoserine is so low that the stoichiometry of its binding cannot be determined by equilibrium dialysis. We have seen that the stoichiometry can be deduced indirectly by the use of the affinity label analog, 2-amino-4-oxo-5-chloropentanoic acid: one equivalent per monomer inactivates the dehydrogenase. The fact that covalent labeling requires the presence of pyridine nucleotide indicates a compulsory order of binding in keeping with the fact that coenzyme binding does not need homoserine and is not increased in its presence. Kinetic analysis demonstrates that the analog forms a reversible complex with the protein, prior to irreversible binding (54).

Threonine Binding by the Native Enzyme and by the Isolated Active Fragments (Domains) The experiments using L-^{14}C-threonine were performed by equilibrium dialysis and gel filtration (68). Figure 3.21 shows that, in the presence of 0.15 *M* KCl, saturation of the binding sites is reached at free ligand concentration in the 0.3 to 0.5 m*M* range, where 8.1 ± 0.4 moles of threonine are bound to 1 mole of enzyme. Under these conditions, only 10 to 15% of the ligand is bound to the protein, but the value at saturation was found to be reproductible within 5%. Treatment with —SH reagents that desensitize the homoserine dehydrogenase activity and destroy the aspartokinase activity yields a modified protein that no longer binds threonine. Table 3.4 shows that at saturation the number of molar equivalents of bound threonine is essentially eight in the presence of 0.15 *M* or 0.5 *M* KCl. However, at nonsaturating concentration of free threonine, the extent of binding depends on a number of factors, especially the concentration of KCl. High concentrations of KCl, which diminish the binding of threonine (Figure

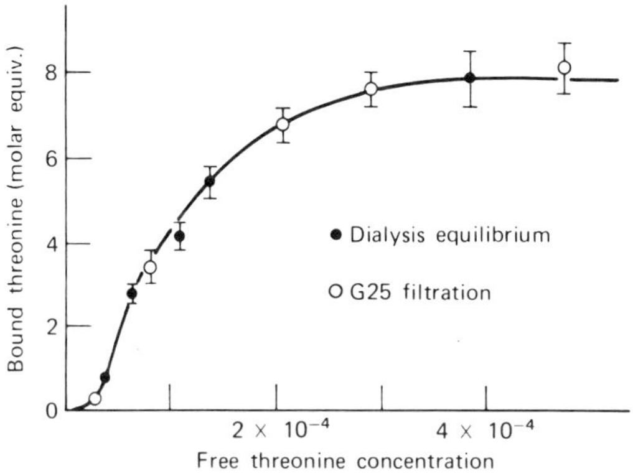

Fig. 3.21 Binding of L-threonine to AK I-HDH- I From ref. 71.

3.22*a*), have been found to displace in a very similar manner the inhibition curves of homoserine dehydrogenase (12) and aspartokinase (26) activities. In the presence of aspartate, the binding of threonine is also found to decrease (68) (Figure 3.22*b*), a result that could be expected, since the inhibition of the aspartokinase is competitive towards aspartate (1). As seen in Figure 3.21, threonine binding is cooperative at high potassium ion concentration. The sigmoidicity of the binding curves disappears when K^+ concentration is lowered to 50 m*M* (55), as shown in the corresponding Scatchard plot (Figure 3.23). This plot was consistent with eight equal affinity sites for threonine, a result that is not easily interpretable (55).

Table 3.4 Binding of Threonine to Aspartokinase I-Homoserine Dehydrogenase I (71)

Additions	Concentration of Free L-Threonine (m*M*)	Molar Equivalents of Bound Threonine
0.15 *M* KCl	0.16	6.9
	0.38	8
	0.86	7.6
0.5 *M* KCl	0.20	3.8
	0.50	6.9
0.15 *M* KCl + 25 μ*M* NADPH	0.37	7.5
0.15 *M* KCl + *N*-ethylmaleimide treated protein	0.30	0

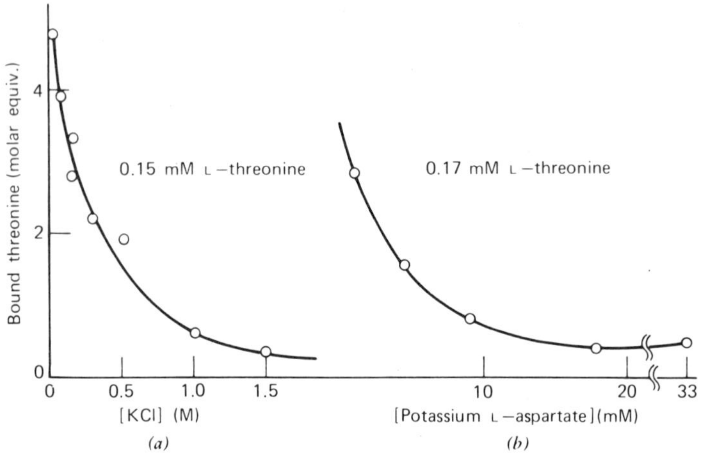

Fig. 3.22 Influence of KCl (*a*) and aspartate (*b*) on the binding of L-threonine (68).

It is important to mention at this point that the homoserine dehydrogenase fragment does not bind threonine (71) and that only four binding sites were found on the tetrameric aspartokinase fragment (72).

The Catalytic Sites of Aspartokinase I-Homoserine Dehydrogenase I: Identification of Four of the Eight Threonine Binding Sites

Ryzewski and Takahashi (73) have observed that the aspartokinase activity of AK I-HDH I is selectively inactivated by Co(III) incorporation. As noted before, aspartokinase activity requires Mg^{2+} as a metal cofactor. Nuclear magnetic resonance studies (70) using Mn^{2+} instead of Mg^{2+} indicate that the metal serves as a bridge between the enzyme and the ATP. Since Co(III) will substitute for Mg^{2+} for aspartokinase activity, Ryzewski and Takahashi incubated the enzyme in the presence of Co(II) and oxidized *in situ* to a Co(III) complex by H_2O_2. The Co(III) is not removable by dialysis and is released by incubation with dithiothreitol. The Co(III) enzyme has no aspartokinase activity, but retains 60% of the original dehydrogenase activity. The kinetics of aspartokinase inactivation follows closely that of Co(III) incorporation, which saturates at one metal atom per subunit, at which point the residual homoserine dehydrogenase is still inhibited by threonine. This suggests that the integrity of the aspartokinase site is not required for the threonine binding linked to the homoserine dehydrogenase inhibition.

Limited proteolysis (48) of this derivative leads to an homoserine dehydrogenase fragment devoid of Co(III), showing that cobalt is fixed onto the aspartokinase moiety of the protein (75).

Using the same *in situ* oxidation method Wright, Feldman, and Takahashi (74) have then prepared affinity labeled substrate and inhibitor adducts. Enzyme-Co(III) ATP, enzyme-Co(III)-aspartate, and enzyme-Co(III)-threonine have thus been produced. The formation of the latter adduct leads the authors to conclude that threonine inhibits the kinase by binding in the first coordination sphere of the catalytic metal ion cofactor in keeping with nuclear magnetic resonance studies carried out with Mn(II) and [1-^{13}C-threonine] (75). The latter studies also indicated that the metal site and one threonine site were adjacent, the distance between the metal and the carboxylcarbon of threonine being 0.44 nm. There were also quaternary adducts produced when ATP, aspartate, and threonine were all present during the oxidation by H_2O_2: enzyme-Co(III)-ATP-aspartate and enzyme-Co(III)-ATP-threonine. This is at first sight a paradoxical result, since the presence of threonine was expected to prevent the access of aspartate to the active site. However, the sum of the number of aspartate and threonine molecules incorporated per active site is one (74). This shows that aspartate and threonine cannot occupy the kinase active site

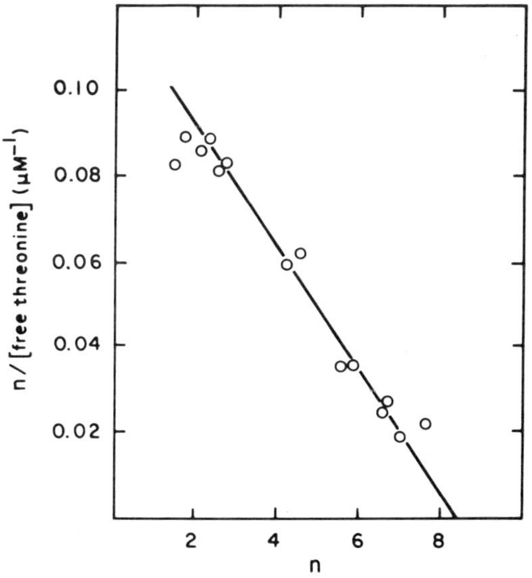

Fig. 3.23 Scatchard plot of L-threonine binding at 50 m*M* KCl. Reproduced from ref. 55 with permission of the The European Journal of Biochemistry.

simultaneously and is consistent with the direct competitive inhibition of aspartate by threonine observed in steady-state kinetic studies (see below) and in the direct binding studies.

The ternary enzyme-Co(III)-threonine adduct had four binding sites per tetramer (76), half that of the native tetramer (71, 55). This derivatized enzyme was shown by equilibrium dialysis to still bind one equivalent of ^{14}C-L-threonine per subunit: the binding and the inhibition of the residual homoserine dehydrogenase activity were still cooperative in the presence of high K^+, whereas the Scatchard plot was linear with low K^+ (Figures 3.24 and 3.25). This behavior is qualitatively found with the native enzyme. Threonine still protected the —SH groups of the enzyme-Co(III)-threonine adduct, produced changes in the circular dichroism comparable to those observed with the native enzyme, and also protected against proteolytic digestion. Thus, at least some of the major conformational changes are ascribable to the binding of threonine to the four nonderivatized sites. The inhibition of dehydrogenase by kinase substrates (5) was shown not to be due to the binding of aspartate and ATP to the kinase active sites but to their binding within the dehydrogenase domains of the enzyme (76).

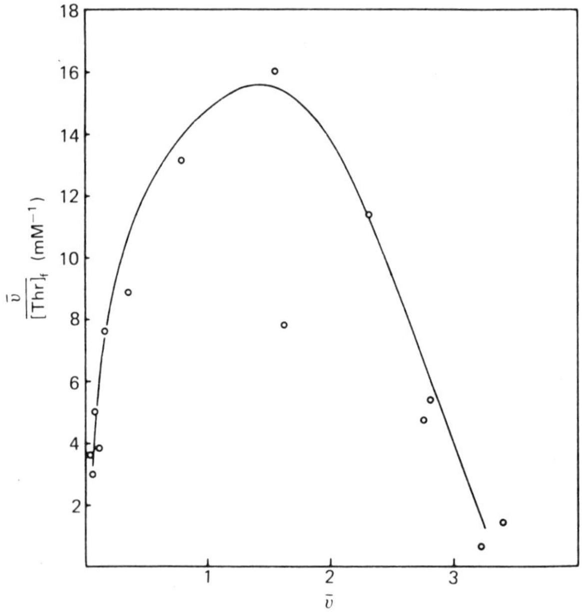

Fig. 3.24 Cooperative binding of L-threonine to AK I-HDH- I-Co (III)-threonine complex at 150 m*M* KCl. Reprinted with permission from *Biochemistry*, **16**, 1541 (1977). Copyright 1977 by the American Chemical Society.

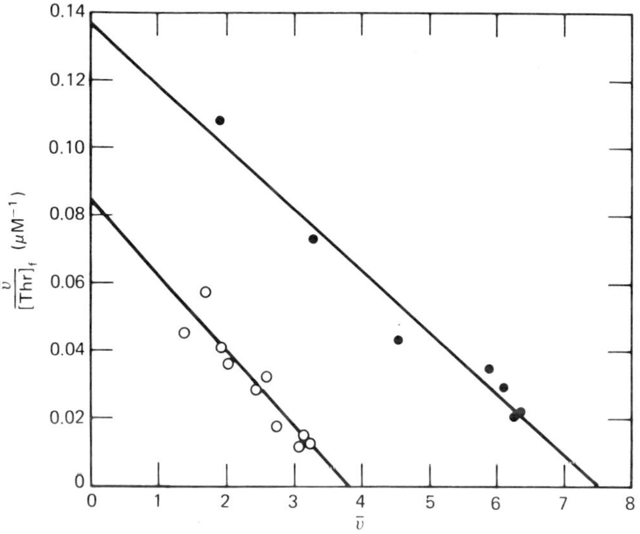

Fig. 3.25 The same binding as in Figure 3.24, but with 10 m*M* KCl only. ●: native enzyme; O: affinity labeled enzyme. Reprinted with permission from *Biochemistry*, **16**, 1541 (1977). Copyright 1977 by the American Chemical Society.

Recently, the same group of workers has subjected the enzyme-Co(III)-ATP complex, labeled with ^{57}Co to proteolytic digestion. As with the native enzyme, the kinase domain was selectively destroyed; the Co(III) affinity label was absent from the HDH I fragment. Gel filtration of the trypsinized complex, followed by ion exchange chromatography, allowed the isolation of a C-terminal arginine tryptic peptide containing 0.8 ATP and 0.9 Co per arginine, in 17% yield (77). Its composition is:

$$Asx_2, Thr_{1\text{-}2}, Ser_2, Glx_3, Gly_4, Ala_2, Val, Leu, Arg$$

Unfortunately the progress made in the elucidation of the total sequence of AK I-HDH I in our laboratory has not yet allowed us to identify this peptide.

When the fluorescent probe 1,N (6) etheno adenosine-5′-triphosphate (ϵ-ATP), an ATP analogue, was bound at the kinase active site by Co(III) affinity labeling, and the dehydrogenase active site was saturated with the NADPH analog, thionicotinamide adenine dinucleotide phosphate (TNADPH), it was observed that the latter quenched the fluorescence of ϵ-ATP. It was checked that the two analogues were functioning as substrates. The distance between ϵ-ATP and the quencher was determined to be 2.9 $\pm$ 0.4 nm. In the presence of threonine, the distance increased to 3.6 $\pm$ 0.5 nm.

It was concluded that threonine binding either increased the intersite distance by 0.7 nm or caused a reorientation of the probe at the dehydrogenase site (78).

Kinetics at Equilibrium

Table 3.5 summarizes some of the kinetic parameters of the two reactions catalyzed by aspartokinase I-homoserine dehydrogenase I under standard conditions at 27°. For comparison the data are given for the two derived fragments AK I and HDH I.

Both reactions require K^+ ions for activity (2, 12, 26). In addition K^+ is important for the integrity of the quaternary structure. K^+ is known to be necessary for the activity of many kinases, but its requirement for a dehydrogenase is somewhat atypical. Broussard et al. (79) have found that this requirement can be obviated by low concentrations of ATP-Mg. Even with ATP-Mg, no dehydrogenase activity is measurable in the total absence of K^+. Experiments performed in the reverse direction (HS → ASA) show that

Table 3.5 Some Kinetic Parameters of the Reactions Catalyzed by Aspartokinase I-Homoserine Dehydrogenase I and by the Two Derived Fragments (Domains) (71)

	Aspartokinase I-Homoserine Dehydrogenase I	Aspartokinase I Fragment	Homoserine Dehydrogenase I Fragment
Km L-aspartate	1500 μM	6000	
Km ATP	180 μM	3000	
Km L-aspartate semialdehyde	120 μM		200 μM
Km L-homoserine	6500 μM		Not measured
Km NADP	110 μM		Not measured
Km NADPH	40 μM		60 μM
Turnover number, aspartokinase (mole aspartyl phosphate synthesized × sec^{-1} × mole $enzyme^{-1}$)	57	28	
Turnover number, homoserine dehydrogenase forward (mole NADPH oxidized × sec^{-1} × mole $enzyme^{-1}$)	400		400
Reverse (mole NADP reduced × sec^{-1} × mole $enzyme^{-1}$)	33		

with 5 m*M* K^+, only 10% of the maximum activity is measurable; maximum activity can be reached with the addition of either 100 m*M* KCl or 50 μ*M* ATP-Mg (57).

The homoserine dehydrogenase I activity has a typical Michaelian behavior, that is exhibits no homotropic cooperative effects of its substrates. At first no cooperative effects were detected for aspartokinase I, using aspartate as the substrate (1). However, this result was obtained at a high K^+ concentration which, as we shall see below, favors the same active form (R form) of the enzyme as aspartate. Therefore, Wampler and Westhead (26) have tested the aspartokinase at low K^+ concentrations. Cooperative effects of aspartate were then observed. The same result was obtained by Patte (80) when the activity was measured at a high threonine concentration where the inactive (T form) is favored.

Inhibition by L-Threonine The inhibition of aspartokinase I (1) and homoserine dehydrogenase I (2) activities by L-threonine, the end product of the pathway, was recognized early. The inhibition of the kinase activity is competitive versus aspartate (1) and ATP-Mg (1). The inhibition of the dehydrogenase is noncompetitive versus homoserine (2).

The inhibition by threonine of both activities displays homotropic cooperative effects (2, 81) (Figures 3.2 and 3.3). In conjunction with the presence or absence of substrate cooperative effects (see above), aspartokinase I and homoserine dehydrogenase I are allosteric enzymes respectively of the K and the V type (82).

The concentrations of L-threonine necessary to inhibit 50% of the maximal activity are, in our standard assay procedures, of 0.3 m*M* for the kinase and 0.5 m*M* for the dehydrogenase. At low K^+ concentrations threonine exerts its inhibition on homoserine dehydrogenase at lower concentrations, and the inhibition is of the noncooperative type (57).

Regulation by L-Threonine is Associated with Conformational Changes of the Protein

Difference U.V. Spectroscopy The spectrophotometric comparison of two identical solutions of protein, one in a phosphate buffer containing 150 m*M* KCl and the other in the same buffer containing 0.15 *M* KCl *plus* 0.5 m*M* L-threonine shows a difference spectrum in the 250 to 310 nm range (Figure 3.26). The complexity of the spectra indicates that several aromatic residues are involved. From these spectra, two conformations of the protein have been postulated: a T form, characterized by the fact that the aromatic residues are buried, and an R form, with exposed aromatic residues (83).

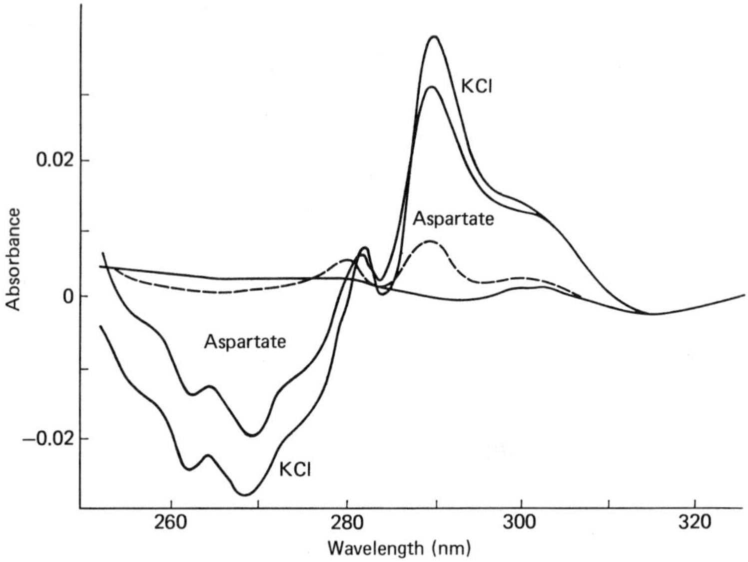

Fig. 3.26 Difference spectra in the presence of threonine. Baseline: reference and sample, no addition. Dashed line: reference, no addition; sample, 0.5 m*M* L-threonine. Aspartate spectrum: reference, 15 m*M* L-aspartate; sample, 15 m*M* L-aspartate and 0.5 m*M* L-threonine. K spectrum: reference, 0.15 *M* KCl; sample, 0.15 *M* KCl and 0.5 m*M* L-threonine. Reproduced from ref. 83 with permission of the European Journal of Biochemistry.

The same spectrophotometrically defined T form can be obtained also in the absence of KCl. The equilibrium can be shifted back to the R form by addition of aspartic acid.

Fluorimetric Studies The change in the absorption spectrum of the protein upon addition of threonine in the presence of 0.15 *M* KCl is associated with a quenching and a slight shift of the protein fluorescence (68). When the latter is excited at wavelengths below 280 nm, threonine quenches the fluorescence only if aspartate or potassium ions are present (Figure 3.27); the advantage of these findings is described later in the study of the allosteric equilibrium.

Fluorescence Energy Transfer When the complex formed between AK I-HDH I and NADPH is excited at 290 nm, the excitation is transferred from an aromatic residue of the protein to NADPH, and an emission band is observed at 450 nm. The addition of threonine strongly quenches the fluorescence transfer (Figure 3.28) indicating a probable increase of distance between the excited aromatic residue and the NADPH, since threonine affects neither the affinity of NADPH nor the total number of equivalents of the coenzyme bond (68, 83).

Circular Dichroism Circular dichroic studies have shown that the spectrum of the protein is characterized by a complex overlap of negative bands. These bands arise from aromatic transitions. The binding of threonine to the enzyme results in a decrease of the magnitude of the ellipticity of the bands (Figure 3.29). The change of ellipticity at 280 nm was used to study the conformational change of the protein upon addition of threonine (69). The curve is sigmoidal and can be superimposed on that obtained using other physicochemical methods (68, 83). The circular dichroism change at 280 nm is thus essentially measuring the state function.

Hydrogen Exchange with the Solvent Measurement of the hydrogen exchangeable with tritiated water in the presence or the absence of the allosteric effector results, at threonine saturation, in a 40% increase in the

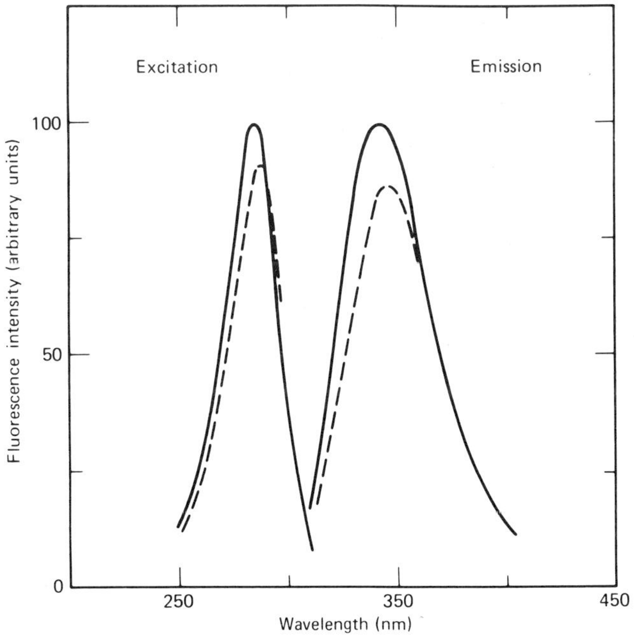

Fig. 3.27 Fluorescence of the protein. The excitation emission observed at 345 nm and emission (excitation at 282 nm) spectra were recorded (continuous line). Then 1 m*M* L-threonine was added and the same measurements repeated (broken line). Reproduced from ref. 68 with permission of the European Journal of Biochemistry.

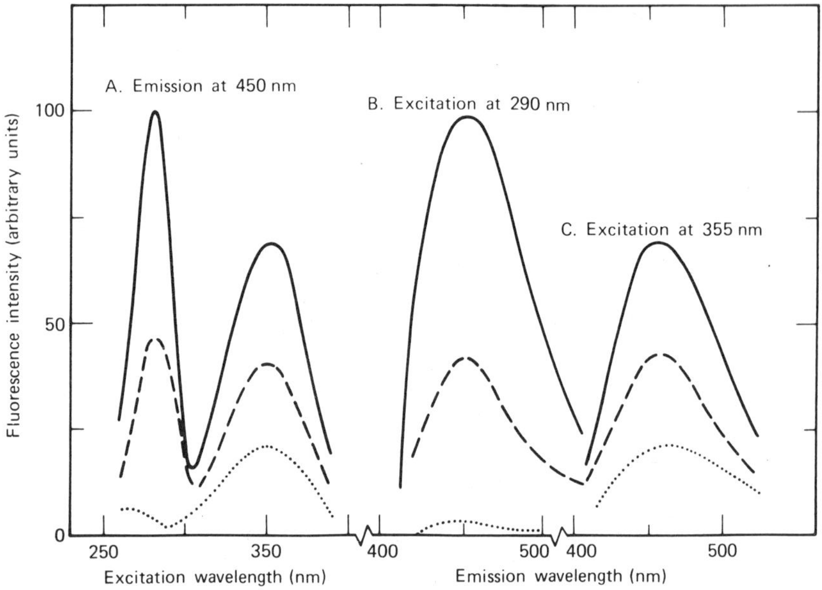

Fig. 3.28 Fluorescence of the protein-NADPH complex. the fluorescence spectra of a solution of 1.5 μM protein and 3 μM NADPH were recorded before (continuous line) and after (broken line) addition of 1 mM L-threonine. Corrections were made for absorption. Emission was observed at 450 nm for the excitation spectrum (A); the excitation wavelength was 290 nm for the transfer emission spectrum (B), and 355 nm for the emission spectrum of NADPH (C). The dotted line represents the fluorescence of 3 μM NADPH under the same conditions. Reproduced from ref. 68 with permission of the European Journal of Biochemistry.

number of tightly sequestered hydrogens (84). The results show an increase in core size upon the binding of threonine indicating a tighter structure with the hydrophobic interior of the enzyme becoming less accessible to the solvent, which agrees with the ultraviolet spectral changes reported above.

The Allosteric Equilibrium

The difference spectra (68, 83) and the effects on the fluorescence of the protein (68, 83) can be used to define two conformations of the protein. Let T be the form with buried chromophores (low absorption at 209 nm and high absorption at 289 nm), and R the form with exposed chromophores; a straightforward interpretation of the data is that T is stable in the absence of added ligand or in the presence of threonine, and R is stable in the presence of potassium ions or of aspartate. As a consequence of this model difference spectra similar to these should be obtained in the absence of added threonine when samples of protein in buffer containing KCl or aspartate are compared with similar solutions with no added ligands. This is found to be the case (83). Aspartate and K^+, at least, are structurally unrelated, and it is *a priori* likely that the three ligands under study have different binding sites. Short range perturbations of the aromatic residues of the protein would then be expected to have very different effects on the ultraviolet absorption when aspartate or K^+ are bound. The similarity of the actual spectra implies that the ligands act by shifting an equilibrium between conformations of the protein rather than merely modifying the environment of their respective binding sites.

Advantage has been taken of the easiness and sensitivity of the fluorimetric measurements for studying the effects of intermediate concentrations of the ligands and attempting to describe more quantitatively the changes of the protein conformations. Figure 3.30 shows that upon adding threonine to enzyme solutions containing 150 m*M* KCl, the variations of absorbance and of fluorescence are coordinated. It can be assumed that both techniques detect the same phenomenon. Figures 3.31 and 3.32 describe the effects of the three ligands on the equilibrium. The ordinates $R/R + T$ or $T/R + T$ represent the fraction of the protein in the R or T

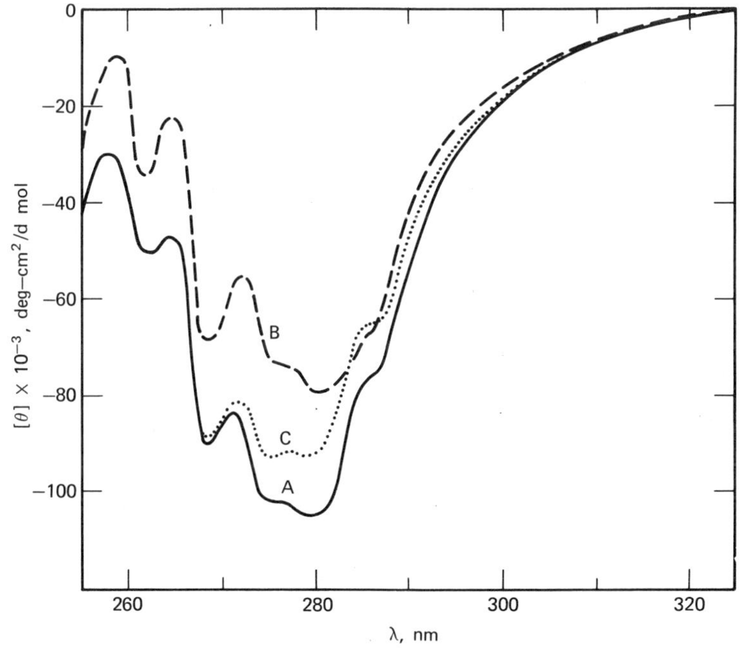

Fig. 3.29 Near UV circular dichroic spectra of AK I-HDH I. A: free enzyme; B: enzyme + 2 m*M* L-threonine; C: enzyme + 20 m*M* L-aspartate. Reprinted with permission from *Biochemistry*, **9**, 2776 (1970). Copyright 1970 by the American Chemical Society.

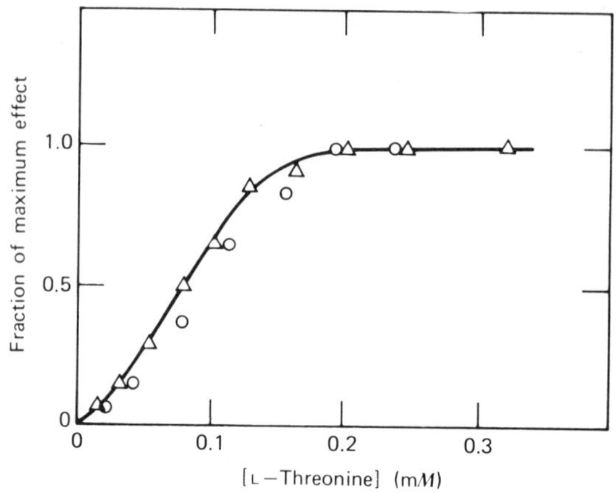

Fig. 3.30 Correlation between the spectroscopic effects of threonine. Successive amounts of L-threonine are added to the protein in the presence of 0.15 *M*KCl at 25°C. The absorbance at 291 nm (O) and the intensity of the fluorescence excited at 280 nm and observed at 335 nm (Δ) are followed. protien concentrations: 11.5 μ*M* and 0.5 μ*M* respectively. Reproduced from ref. 83 with permission of the European Journal of Biochemistry.

states. It can be seen in Figure 3.31 that aspartate and K^+ give positive homotropic effects (the curves are sigmoidal) and positive heterotropic effects (they help each other), but they give negative heterotropic effects with threonine. When threonine is added to protein solutions containing various amounts of the other ligands (Figure 3.32), similar features are observed.

As a first approximation, the above experimental facts may be visualized by a two state model

$$R \underset{\text{Aspartate, } K^+}{\overset{\text{Threonine}}{\rightleftharpoons}} T$$

A strong correlation may be established between the measurements of absorption, of fluorescence, and of threonine binding, all of which are compatible with the existence of two states. Table 3.6 shows that a similar correlation exists with the catalytic and regulatory properties of the protein. The competitive inhibition of the aspartokinase activity by threonine (K type) can be accounted for by the preferential binding of aspartate to the active *R* form. In contrast, the inhibition of the dehydrogenase activity and the analogous activation by K^+ do not seem to be due to an effect of the *R*-*T* conformation change on the binding of the substrates, for a noncompetitive and incomplete inhibition are observed (4). They may be due to an actual requirement for K^+, or more likely, to an improper configuration of

Table 3.6 General Properties of the R and T Conformations[a]

Property	*T* Form	*R* Form	Remarks
Binding of ligands:			
(Dissociation constants)			Shifts the equilibrium towards:
Threonine	$K_T = 30\ \mu M$	$K_R \geq 1\ \mathrm{m}M$[b]	*T* form
Aspartate	$K_T \geq 20\ \mathrm{m}M$	$K_R = 3\ \mathrm{m}M$	*R* form
K^+ ions	Low affinity	High affinity	*R* form
NADPH	$K_T = 0.3\ \mu M$	$K_R = 0.3\ \mu M$	no detectable effect
Spectrophotometric properties:			
Relative fluorescence intensity	100	116	Excitation at 280 nm, emission at 335 nm
Tryptophan and tyrosine residues	Buried	Exposed	Difference spectrum
Enzymatic activities:			
Aspartokinase	None	Active	*T* does not bind the substrate
Homoserine dehydrogenase	10 to 15%	100%	Different configurations of the active sites

[a] Reprinted from ref. 83 with permission of the European Journal of Biochemistry.

[b] Binding of threonine at saturation of KCl or aspartate.

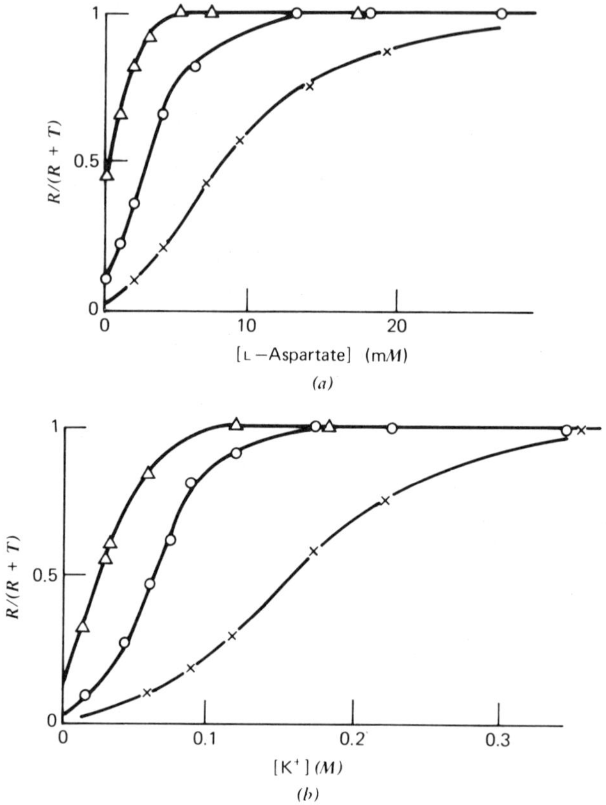

Fig. 3.31 Effects of aspartate and of K^+ on the allosteric equilibrium. The values of $R/R + T)$ in the presence of various concentrations of ligands are derived from measurements of the fluorescence intensity of the protein. (*a*): effects of aspartate. The following supplements are added: O: none; Δ, 30 m*M* KCl; X, 40 μ*M* L-threonine. (*b*): effects of K^+. The following supplements are added: O, none; Δ, 2 m*M* L-aspartate; X, 40 μ*M* L-threonine. Reproduced from ref. 83 with permission of the European Journal of Biochemistry.

the dehydrogenase active sites. Although the binding constants for NADPH for the *R* and the *T* forms are similar, the fluorescence of the complex with the *T* form is much smaller (83) and indicates that the environment of the bound nucleotide is different.

Presteady State Kinetics. Relaxations of the Allosteric Equilibrium

The first fast kinetic experiments were performed by Barber and Bright (85) studying the early kinetic events during the homoserine dehydrogenase reaction: when threonine was rapidly mixed with AK I-HDH I an initial lag in the onset of threonine inhibition was observed. The rate constants for the lag times were dependent on threonine concentration but independent of protein concentration and therefore interpreted as reflecting an isomerization of active to inactive enzyme. These results were confirmed and extended by Janin and Iwatsubo (86) in the authors' laboratory and by Bearer and Neet (87). The latter authors noted, however, that in contrast to the relaxation times, which are strongly temperature-dependent (86), the kinetic Hill coefficients obtained from standard spectrophotometric measurements of threonine inhibition of homoserine dehydrogenase activity were temperature-independent. They thus rule out a kinetic contribution of

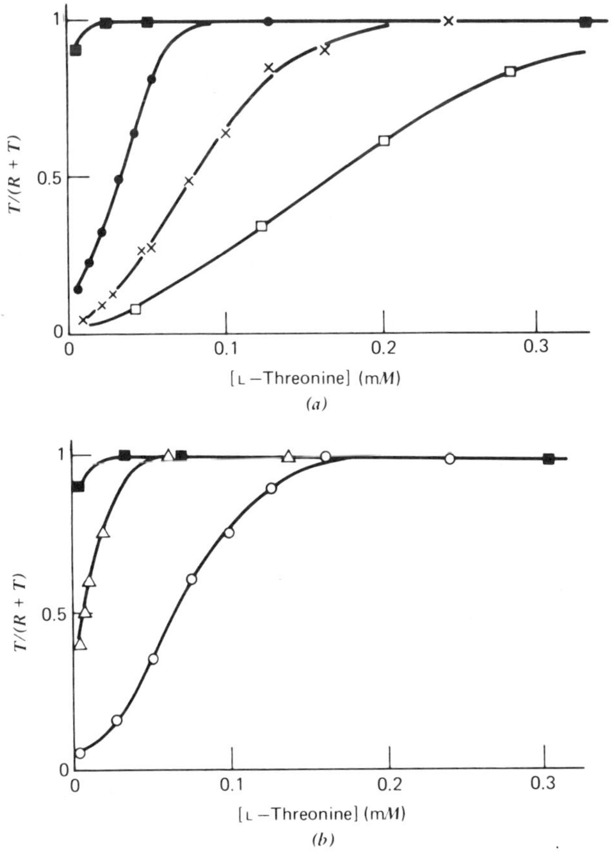

Fig. 3.32 Effects of threonine on the allosteric equilibrium. L-threonine is added to a solution of protein containing the following supplements: (*a*) ■, none; ●, 60 m*M* KCl; X, 0.15 *M* KCl; □, 0.45 *M* KCl. (*b*) ■, none; △, 5 m*M* L-aspartate; ○, 20 m*M* L-aspartate. Reproduced from ref. 83 with permission of the European Journal of Biochemistry.

the inhibition lag to the observed cooperativity. In addition, they find no correlation between the threonine concentration causing 50% of inhibition of maximal homoserine dehydrogenase activity and the concentration at half the maximal rate of transition. Furthermore, the Hill coefficient measured from direct threonine binding is identical to that for threonine inhibition of homoserine dehydrogenase steady state activity. They conclude that the cooperative inhibition of homoserine dehydrogenase by threonine is best described by rapid equilibrium binding of threonine, with subsequent site-site interactions, independent of the rate of isomerization.

Janin and Iwatsubo (86) have compared the kinetics of prestationary inhibition by threonine with those of the variations of spectroscopic parameters linked to the protein itself (Table 3.7). Although the first type of measurement involves the presence of the substrates of the dehydrogenase, NADPH, and aspartate semialdehyde, the kinetics are similar in all cases studied. NADPH addition has actually no effect on the $R \rightleftharpoons T$ isomerization upon addition of threonine in the presence of 0.15 *M* KCl; moreover this reaction could be followed by using the fluorescence of bound NADPH, which changes considerably with the conformation of the protein. This was taken as a straightforward indication that the spectroscopic effects observed

Table 3.7 Rate Constant of the R to T Reaction in the Presence of 0.15 M KCl[a,b]

	Rate Constants with Threonine			
Effect Observed	0.3 m*M* (sec^{-1})	1 m*M* (sec^{-1})	3 m*M* (sec^{-1})	30 m*M* (sec^{-1})
Quenching of the fluorescence of the protein	1.5	3.2	7.0	30
Absorption at 291 nm (increase)	1.4	—	7.5	33
Absorption at 266 nm (decrease)	—	—	7.2	30
Quenching of the fluorescence of the protein-NADPH complex	1.6	—	6.5	25
Prestationary inhibition of the dehydrogenase activity	2.2	3.7	8.0	—

[a] Reprinted from ref. 86 with permission of the European Journal of Biochemistry.
[b] All rate constants are measured at 25°C. The fluorescence of the protein-NADPH complex is excited at 340 nm and observed at 450 nm.

in the study of the protein itself correlate strongly with the effects on the catalytic activities and are related to the regulation of aspartokinase I-homoserine dehydrogenase I. Although the above stopped flow experiments demonstrate a single rate determining step in the *R*-*T* conformation change, a careful study of the effects of the ligands suggests strongly that other processes take place, which must be of small amplitude or yield relaxations faster than the 5 msec limit of the stopped-flow technique. Since the equilibrium is strongly temperature-dependent (83), Janin and Iwatsubo applied the temperature-jump technique in the absence of ligands, and by raising the temperature from 20° to 28°, they were able to detect two relaxations (86). The first one, R_1, was large ($\tau_1 \sim 70$ msec), and its characteristics were reminiscent of the relaxation detected in the stopped-flow apparatus; the decrease of the fluorescence intensity and of the light absorption at 265 nm, as well as the increase of the absorption at 297 nm, expresses an increase of the relative concentration of the T form, expected from its higher enthalpy (85). The second, R_2, has a smaller relative amplitude and a sign opposite to that of R_1 in the fluorescence and in the absorption measurements at both wavelengths. R_2 is much faster ($\tau_2 \sim 1$ msec). R_2, as well as R_1, is independent of protein concentration. Since both relaxations are observed in the absence of threonine or aspartate, binding of these ligands is irrelevant. These results and the large energies of activation (20 kcal/M for R_1 and 25 kcal/M for R_2) of both relaxations indicate that the relaxation of the *R*-*T* equilibrium goes through *two* isomerization steps and suppose the existence of at least one conformation S of the protein that equilibrates fast with one of the major conformations and slowly with the other. Although the scheme:

$$R \underset{k_r}{\overset{k_1}{\rightleftharpoons}} S \underset{k_u}{\overset{k_3}{\rightleftharpoons}} T$$

explains the relaxation data satisfactorily, these are insufficient to demonstrate its validity since the rate constants cannot be determined in the absence of a direct measurement of the *R*/*S* and *T*/*S* ratios, and their contributions to τ_1 and τ_2 cannot be assigned without too many assumptions.

The kinetic studies of Bearer and Neet (87) revealed the presence of a lag of threonine inhibition of the aspartokinase activity. This lag is different from the homoserine dehydrogenase transient lag, its half-time being of the order of minutes as compared to seconds for the dehydrogenase. Furthermore, the initial velocity of aspartokinase activity (at about 10 sec) is inhibited, whereas the initial velocity of the dehydrogenase (at about 100 msec) is not. Since the K_I for inhibition of the aspartokinase initial velocity (after 10 sec) is 0.6–1 m*M*, similar to that of homoserine dehydrogenase steady state inhibition, the authors suggest that the same threonine set of sites is responsible for inhibition of the initial aspartokinase velocity and the homoserine dehydrogenase steady state activity. The data show that the inhibition of the initial aspartokinase activity by threonine reaches a maximum of 85–95% inhibition, whereas the further inhibition in the steady state after the slow transient continues to completion at 3–5 m*M*.

Structural Studies in Progress

Sequence Studies The amino acid sequence of AK I-HDH I is in progress in our laboratory. The polypeptide chain has been cleaved by cyanogen bromide, and the cyanogen bromide peptides have been purified and sequenced (63, 49, 88, 45, 46). Also the tryptic (and chymotryptic) peptides containing methionine, tryptophan, and cysteine residues have been sequenced and have allowed the alignment of a larger number of cyanogen bromide peptides (45). Of great help in the determination of the primary structure is the existence of the aspartokinase fragment and the homoserine dehydrogenase fragment, which has allowed the division of the polypeptide chain in three regions (Figure 3.33) (63). Thus a peptide common to the two

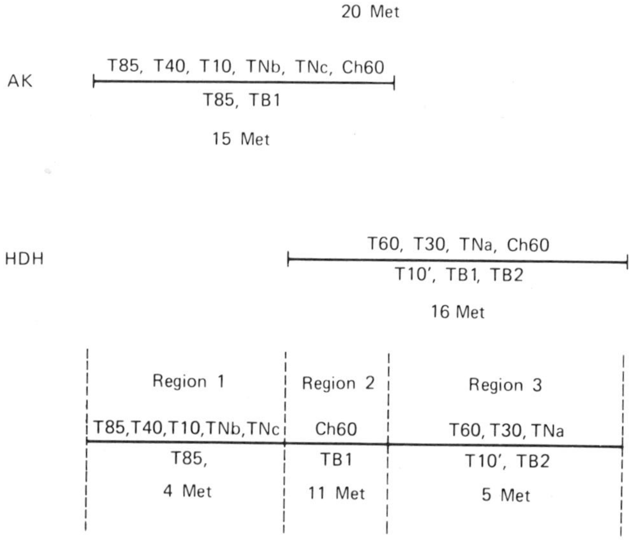

Fig. 3.33 Distribution of methionine residues and of cysteine and tryptophan-containing peptides in the three regions of AK I-HDH I. Peptides above the line are the carboxymethyl-cysteine containing peptides; the ones below the line are the tryptophan containing peptides (63).

fragments is in region 2 while a peptide present only in one of the fragments can be assigned to either region 1 or 3. It is striking that region 2 is extremely rich in methionine residues (55% of the methionines are clustered in 20% of the sequence). This is also the case in another bifunctional protein, the lactose repressor, and may be of some significance.

About 75% of the amino-acid sequence of AK I-HDH I is now known.

Recently the DNA sequence of the *thrA* gene has been undertaken in our laboratory and should help to obtain the total sequence of AK I-HDH I. As shown in Section 3.4, this sequence is of great interest for the comparison of AK I-HDH I and AK II-HDH II.

Crystal Structure AK I-HDH I has been crystalized in the presence of threonine and NADPH (89). The tetragonal crystals have an asymmetric unit of 1×10^6 Å^3 and belong to the space group I422 or $I4_1 22$. Assuming that the solvent occupies 53% of the total volume, this asymetric unit will contain a tetrameric molecule (89). The symmetry of this crystal form does not, therefore, express the possible symmetrical arrangement of the four subunits. So far attempts to obtain crystals better suited for x-ray crystallography have been unsuccessful.

3.3 ASPARTOKINASE II-HOMOSERINE DEHYDROGENASE II

Aspartokinase II-Homoserine Dehydrogenase II is also a Bifunctional Enzyme

Localization of the Two Functions on One Polypeptide Chain The first indications that the two activities are carried by the same protein were obtained by Patte et al. (3). First, they showed that the two characters were cotransducible. Second, they obtained a mutant that had lost simultaneously aspartokinase II and homoserine dehydrogenase II; spontaneous revertants of this mutant that had recovered homoserine dehydrogenase II had also recovered aspartokinase II. Third, after about a 100-fold purification they could not separate the two activities.

AK II-HDH II was first obtained in a pure state by Falcoz-Kelly et al. (7), who showed that the aspartokinase and homoserine dehydrogenase activities remain associated through a 400-fold purification and that the protein sediments as a single, sharp, and symmetrical peak in the ultracentrifuge. Its molecular weight was determined by equilibrium sedimentation (7, 90) to be 185,000 ± 10,000, and its sedimentation coefficient is $s^{\circ}_{20,w} = 7.6 (\pm 0.2) \times 10^{-13}$ sec (7).

The next important step was to determine whether or not this protein, which carries two enzymatic activities, is made of identical subunits. The experimental evidence leading to the conclusion that AK II-HDH II is a bifunctional enzyme is the following (90): the subunit molecular weight, as determined by polyacrylamide gel electrophoresis in the presence of sodium dodecyl sulfate, is 88,000, and a single band is obtained. Electrophoresis on polyacrylamide gels in the presence of 8 *M* urea of AK II-HDH II also shows a single band. AK II-HDH II is thus a dimer composed of subunits of the same molecular weight and charge (90).

The amino terminal sequence of AK II-HDH II is unique and is

$$\overset{1}{\text{Ser}}\text{-Val-Ile-Ala-}\overset{5}{\text{Gln}}\text{-Ala-Gly-Ala-}\overset{9}{\text{Lys}}$$

Its C-terminus is —Leu—Leu—COOH.

The cysteine containing peptides obtained after complete tryptic digestion of AK II-HDH II have been studied (90): their amino acid composition and N-terminal residue were determined. The native protein molecule possesses 10 to 12 cysteine residues (Table 3.8). If its subunits are identical one would expect a maximum value of 10 cysteine containing peptides, and if they are different a maximum value of 20. Eight to 10 different cysteine containing peptides have been obtained in agreement with the hypothesis of identity of subunits.

These results prove that AK II-HDH II is a dimer and strongly support the conclusion that the two subunits are identical. This conclusion is confirmed by the fragmentation pattern obtained upon proteolysis of AK II-HDH II (see next paragraph). This implies that the two catalytic activities are carried by the same polypeptide chain and that AK II-HDH II, as AK I-HDH I, is a bifunctional enzyme.

Some Properties of Aspartokinase II-Homoserine Dehydrogenase II

GENERAL Neither aspartokinase II nor homoserine dehydrogenase II is inhibited by L-methionine (3), S-adenosylmethionine (7), L-threonine (3), or combinations, two by two, of methionine, threonine, isoleucine, and lysine (80). There is no cooperative binding of the substrates for either of the activities (80). Table 3.9 shows some kinetic parameters of the reactions catalyzed by aspartokinase II and homoserine dehydrogenase II.

In contrast to AK I-HDH I, the substrates of one of the activities of AK II-HDH II do not affect the other. ATP and aspartate do not inhibit homoserine dehydrogenase II, homoserine and NADPH do not inhibit aspartokinase II (3). NADPH, a good protector against thermal inactiva-

Table 3.8 Amino Acid Composition of Aspartokinase II-Homoserine Dehydrogenase II, Mr Polypeptide Chain = 88 000[a]

Amino Acids	Aspartokinase II-Homoserine Dehydrogenase II (residues/chain)
Tryptophan	16
Lysine	23
Histidine	19
Arginine	58–59
Carboxymethylcysteine	9–10
Aspartate	73
Threonine	26
Serine	60
Glutamate	93
Proline	19–20
Glycine	58
Alanine	82
Valine	65
Methionine	8–9
Isoleucine	37
Leucine	100
Tyrosine	19–20
Phenylalanine	27

[a] Reprinted from ref. 90 with permission of the European Journal of Biochemistry.

Table 3.9 Some Kinetic Parameters of the Reactions Catalyzed by Aspartokinase II-Homoserine Dehydrogenase II (7)

Km L-aspartate		2.1 mM
Km ATP		1.9 mM
Km NADPH		0.15 mM
Km NADH (also can be used in the dehydrogenase reaction)		0.32 mM
Km aspartate semialdehyde		0.19 mM
Turnover number, homoserine dehydrogenase (mole NADPH oxidized $\times$ sec^{-1} $\times$ mole enzyme^{-1})		67
Turnover number, aspartokinase (mole aspartyl phosphate synthetized $\times$ sec^{-1} $\times$ mole enzyme^{-1})	155	14

tion of homoserine dehydrogenase II, does not protect aspartokinase II against such a treatment (3).

SULFHYDRYL GROUP ACCESSIBILITY Figure 3.34 shows the course of —SH titration of AK II-HDH II by dithiobisnitrobenzoic acid (DTNB) (91). In the presence of DTNB, aspartokinase II is almost totally inactivated while only 50% of the homoserine dehydrogenase activity is lost. In the presence of 20 mM ATP the kinetic of titration of —SH groups and of aspartokinase inactivation are slowed down in a parallel manner (91). Ten to 12 —SH groups (out of 20, see Table 3.8) are titrated per enzyme molecule (91): about eight cysteine residues are thus inacessible.

Fractionation into Domains

As we have seen in Section 3.2, in the case of AK I-HDH I, one way to study the organization of a protein is to subject it to limited proteolysis and try to obtain (a) fragment(s) still endowed with some of the properties of that protein. If successful, this technique permits an experimental approach to the domains of a large protein.

Upon proteolysis by subtilisin (or chymotrypsin) of AK II-HDH II, apartokinase activity gradually disappears and is finally completely destroyed, whereas homoserine dehydrogenase activity is essentially con-

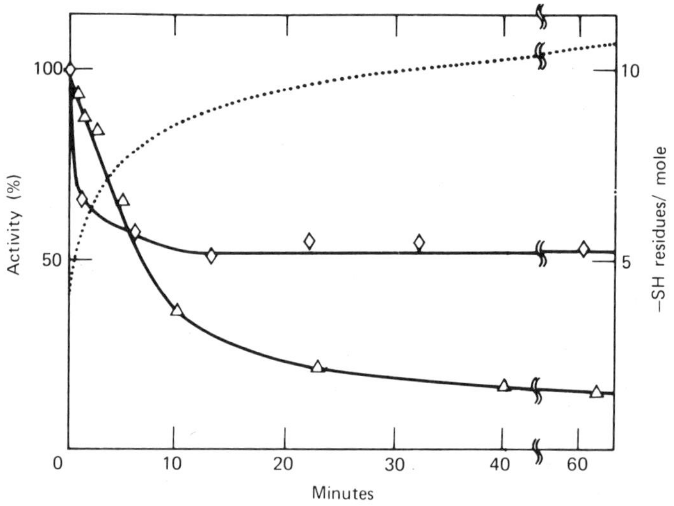

Fig. 3.34 Titration of the -SH groups of AK II-HDH II (91). Titration is followed by adsorption at 410 nm (---), aspartokinase activity (Δ) and homoserine dehydrogenase activity (◆). The protein concentration is 1.7 μM, and the DTNB concentration is 50 μM.

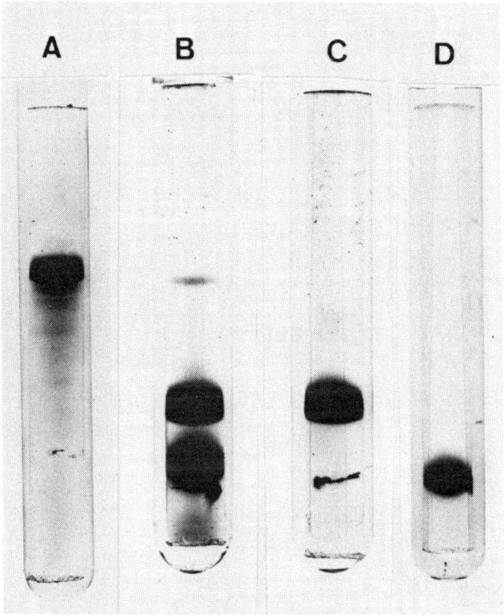

Fig. 3.35 Gel electrophoresis on 5% polyacrylamide gels in the presence of sodium dodecyl sulfate. (A) AK II-HDH II, (B) AK II-HDH II after subtilisin proteolysis, (C) purified homoserine dehydrogenase fragment, (D) purified small proteolytic fragment. Reproduced from ref. 92 with permission of the American Society of Biological Chemists.

served (92). The ligands aspartate, homoserine, and potassium chloride considerably slow down the rate of proteolysis, but the overall result is the same, that is in all cases when aspartokinase activity has reached 1% to 5% of its original value, at least 80% of the dehydrogenase activity is conserved (92).

During proteolysis of AK II-HDH II, as followed by gel electrophoresis in the presence of sodium dodecyl sulfate, the band corresponding to AK II-HDH II (Mr = 88,000) gradually disappears, while two bands corresponding to two fragments of 37,000 and 24,000 mol wt respectively appear simultaneously (Figures 3.35, A and B). When proteolysis is completed there is one molar equivalent of the Mr = 37,000 fragment released from one molar equivalent of AK II-HDH II. This shows that the small fragment does not derive from the large one, and thus that the two fragments originate from nonoverlapping regions of the polypeptide chain.

The two proteolytic fragments of AK II-HDH II have been separated by gel filtration under nondenaturing conditions. The elution profile shows two peaks of 280 nm absorbing material; homoserine dehydrogenase activity is associated only with the first peak. The first peak corresponding to material of 70,000 mol wt represents the larger fragment (Figure 3.35 C), and the second peak corresponding to material of 50,000 mol wt represents the smaller fragment (Figure 3.35 D).

Thus a fragment carrying the homoserine dehydrogenase activity can be obtained by limited proteolysis. It is a dimer (of 2 × 37,000) and appears to be homogeneous using different criteria: equilibrium sedimentation, gel electrophoresis in the presence of either sodium dodecyl sulfate or 8 M urea, N-terminal residue analysis. The smaller fragment, although inactive, has retained enough structure to be a dimer; it arises from a different part of the polypeptide chain. The small fragment can thus be thought of as a first approximation for an experimental study of (part of) the aspartokinase domain.

The present results provide a model of the quaternary organization of AK II-HDH II (Figure 3.36) by correlating the fragmentation pattern to the original structure. The native enzyme is viewed as composed of at least two compact domains, each formed by the tight association of two identical regions.

The homoserine dehydrogenase fragment (HDH II fragment) can be considered as a good experimental model of the homoserine dehydrogenase domain of AK II-HDH II when isolated, that is when it does not interact with the rest of the molecule. The comparison of some of the properties of AK II-HDH II and its HDH II fragment may yield some information about the role and importance of the interactions between domains of the protein and is presented below.

Refolding Studies

Multifunctional proteins are thought to have originated from the fusion of genes coding for different proteins. The question then is: how does a mul-

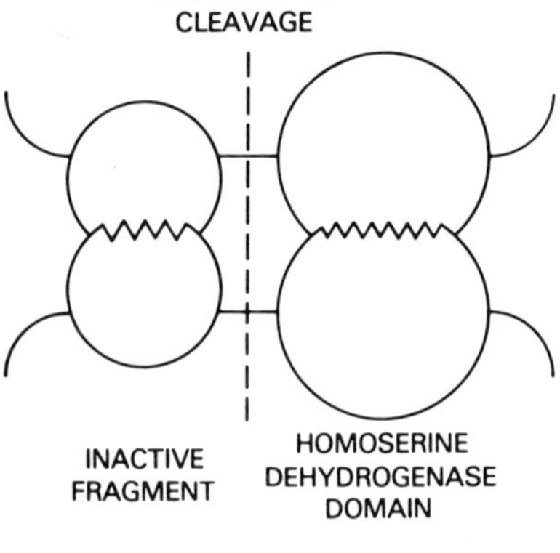

Fig. 3.36 Model of the quaternary structure of the aspartokinase II-homoserine dehydrogenase II molecule showing features known about this protein: identical subunits, existence of two nonoverlapping proteolytic fragments, dimeric nature of both fragments, weak or no association between the dimeric fragments, and active homoserine dehydrogenase fragment. The relative position of the two fragments with respect to the NH_2 and COOH termini of the chain is not known. Reproduced from ref. 92 with permission of the American Society of Biological Chemists.

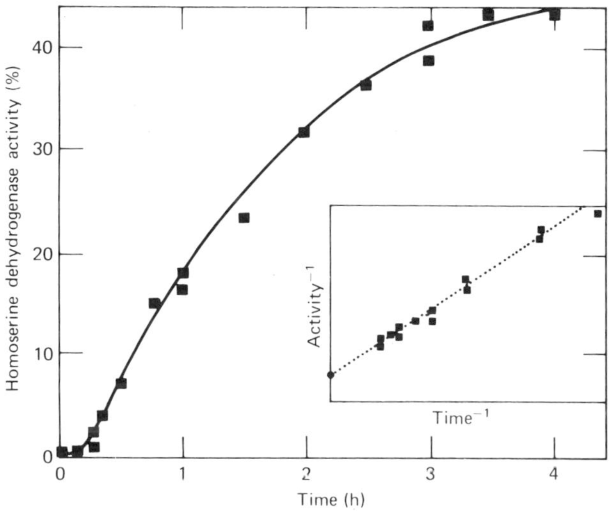

Fig. 3.37 Renaturation of AK II-HDH II measured by the reappearance of its dehydrogenase activity (93). AK II-HDH II concentration is 2.7×10^{-8} *M*. *Inset*: second order representation of these data. (●) gives the activity measured for the native enzyme at the same concentration. Note that this activity is the same as that which would be obtained at an infinite time, as estimated from the double reciprocal plot. Thus the maximal yield of renaturation of AK II-HDH II is not far from 100%; the native enzyme is not stable enough at very low concentration, and it would require at least 20 h to approach such a yield.

tifunctional protein fold or refold after unfolding? Does each of the domains refold independently of the others or not? AK II-HDH II appears to be well suited for refolding studies. It is a large protein folded up in (at least) two domains. Since it is bifunctional, refolding can be monitored by its two different activities. The kinetics of refolding are unambiguous since it is a dimer (the simplest quaternary structure). The existence of the homoserine dehydrogenase fragment, also dimeric, permits a comparison of the folding of the homoserine dehydrogenase domain when isolated and when interacting with the other domains of AK II-HDH II.

Refolding of AK II-HDH II AK II-HDH II has been renatured after denaturation in 6 *M* guanidine hydrochloride (93). Figure 3.37 shows the time course of reappearance of dehydrogenase activity upon renaturation. The activity versus time curve shows a lag phase at early times. At longer times (after about 30 min) the reappearance of activity is governed by a bimolecular reaction (Figure 3.37 inset). This indicates that the reassociation of two chains to form a dimer is the rate limiting step. The second order association rate constant is about 10^3 M^{-1} sec^{-1}.

The presence of a lag phase at early times of renaturation (Figure 3.37), which does not depend significantly on protein concentration, indicates that another reaction takes place prior to the formation of active dimeric species. This reaction presumably corresponds to some kind of (partial) folding of the monomer into species able to associate and yield active dimers.

Renaturation of AK II-HDH II can also be monitored following the aspartokinase activity (93). The kinetics of reappearance of aspartokinase activity are the same as those of homoserine dehydrogenase activity (lag phase at early times, followed by a bimolecular reaction) (Figure 3.38).

The refolding of AK II-HDH II can thus be described by the following mechanisms:

$$2\,M_U \rightleftharpoons 2\,M_F \rightleftharpoons D$$

where M_U is the unfolded state of the monomer, M_F a (partially) folded monomer, and D the native dimer. The lag phase could correspond to the time necessary to form enough M_F so that the first reaction $M_U \rightarrow M_F$ is no longer rate limiting. It is interesting to note that the same steps are apparently involved in the formation of native structure in the regions of AK II-HDH II responsible for the two activities. The fact that the reappearance of both activities is linked to the formation of the dimer shows that monomeric species are, at most, weakly active. Thus the complete fold-

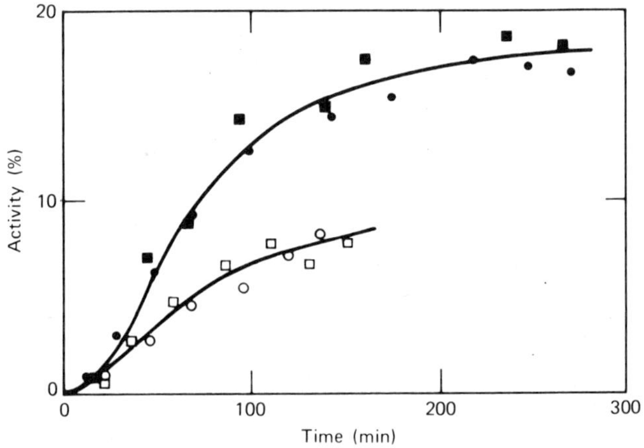

Fig. 3.38 Renaturation of AK II-HDH II measured by the reappearance of both kinase (■, □) and dehydrogenase (●, O) activities. Ak II-HDH II concentration is 1.8×10^{-8} M. Upper curve, and 9×10^{-9} M, lower curve (93).

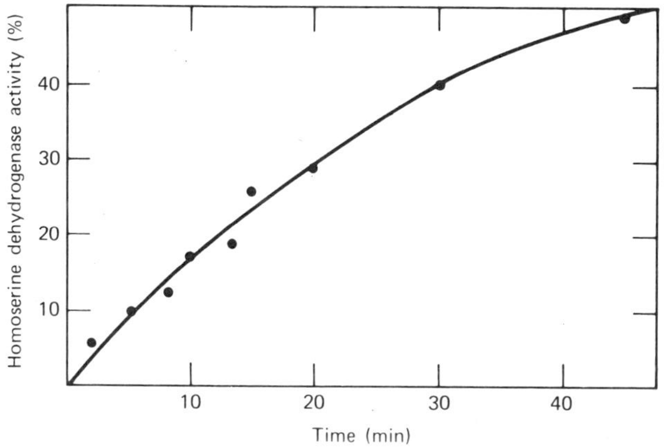

Fig. 3.39 Renaturation of the proteolytic HDH II fragment measured by the reappearance of its dehydrogenase activity. HDH II fragment concentration is 4×10^{-9} *M* (93).

ing of each polypeptide chain is achieved only upon formation of the quaternary structure.

Refolding of the Homoserine Dehydrogenase Proteolytic Fragment Figure 3.39 shows that the HDH II proteolytic fragment is also capable of refolding after complete unfolding in 6 *M* guanidine hydrochloride (93). No lag phase can be observed (it has to be shorter than 2 min if it exists) and the reappearance of homoserine dehydrogenase activity is again governed by a bimolecular reaction with an association rate constant of about 2×10^4 M^{-1} sec^{-1}. It is worth noting that this association rate constant is about 20 times higher than that of the entire enzyme.

The proteolytic HDH II fragment is not only endowed with homoserine dehydrogenase activity but also it can refold completely after denaturation in 6 *M* guanidine hydrochloride. Thus the compact region corresponding to this fragment in the whole protein, the homoserine dehydrogenase domain, is an independent folding unit. As for AK II-HDH II, monomeric species of the HDH II fragment are, at most, weakly active, since the reappearance of activity depends on the formation of dimeric species.

The monomers of the HDH II fragment reassociate into dimers about 20 times faster than the monomers of AK II-HDH II. This suggests that the process that limits the reassociation into dimers in AK II-HDH II and in the HDH II fragment is not the same. In AK II-HDH II the rate of formation of native structure in both the kinase and dehydrogenase regions is limited by common processes: this (together with the fact that the HDH II fragment is renatured faster that AK II-HDH II) reveals the existence of some interaction between the kinase and dehydrogenase region of AK II-HDH II important for the refolding of the monomer and for the reassociation into dimers.

3.4 COMPARISON OF THE TWO ASPARTOKINASES-HOMOSERINE DEHYDROGENASES

Knowing that there are in the same organism two different bifunctional enzymes carrying the same two activities, one wonders immediately how similar these two proteins are and what their evolutionary origin is. As soon as aspartokinase II and homoserine dehydrogenase II were discovered in 1967 (3) this question was asked, and several attempts have been made since to answer it. We present here the similarities and differences of the two aspartokinases-homoserine dehydrogenases.

Biochemical and Chemical Comparison

Enzymatic Activities Both enzymes possess sites for the same substrates, aspartate, homoserine, aspartate semialdehyde, ATP, NADP, and NADPH. In addition, AK I-HDH I can bind threonine, its allosteric inhibitor Section 3.2.

The apparent affinity constants of the two enzymes for their substrates are quite similar, but the turnover numbers are different (Tables 3.5 and 3.9).

In the cases of AK I-HDH I, the substrates of one of the two activities affect the other one, either as inhibitors or as protectors against inactivation by different treatments (Section 3.1). Such interactions are not demonstrable in the case of AK II-HDH II (Section 3.3.).

Aspartokinase I and homoserine dehydrogenase I are inhibited allosterically by threonine, but there is no such inhibitor for aspartokinase II and homoserine dehydrogenase II (this is confirmed by the fact that the growth of mutants, devoid of AK I-HDH I, is not totally stopped in the presence of methionine, in agreement with the fact that repression is the only demonstrable regulation mechanism).

Molecular Weight The subunit molecular weight of AK I-HDH I is 86,000 ± 3000 (Section 3.2), that of AK II-HDH II is 88,000 ± 3000 (Section 3.3). The subunits of the two enzymes have slightly different migrations upon electrophoresis on polyacrylamide gels in the presence of sodium dodecyl sulfate. This was evidenced by running the two proteins on the same gel (90). This indicates that the subunit molecular weight of the two enzymes may be slightly different (this migration difference might also be due to slight differences in sodium dodecyl sulfate binding).

The subunit molecular weight of AK I-HDH I and AK II-HDH II are thus very similar, but AK I-HDH I is a tetramer in its native state, and AK II-HDH II is a dimer.

Sulfhydryl Group Accessibility Modification of cysteinyl group of AK I-HDH I by DTNB leads to the loss of aspartokinase activity; about 50% of the homoserine dehydrogenase I is lost, and the remaining activity is desensitized (Section 3.2). The effect on AK II-HDH II of the modification of cysteinyl groups by DTNB is very similar, leading to the complete loss of aspartokinase II activity and to the loss of only 50% of homoserine dehydrogenase II (Section 3.3).

Amino Acid Composition and Tryptic Maps The tryptic maps of the two enzymes show no striking similarity (although tryptic maps of such large polypeptide chains are difficult to compare because of the large number of peptides). Thus, if the two proteins have some sequence homology, this rules out the possibility that they have long stretches of identical sequences.

Comparison of their amino acid composition (Tables 3.3 and 3.8) according to the statistical analysis of Harris and Teller (94) indicates 65 to 90% sequence homology, assuming that their analysis is valid even for large polypeptide chains where the difference in amino acid composition could be expected to average out.

Peptides Containing Cysteinyl Residues The amino acid sequence of the cysteine containing peptides of AK I-HDH I is known (Section 3.2). The amino acid composition and N-terminal residue of several of the cysteine containing peptides of AK II-HDH II have been determined (90). No similarity between cysteine containing peptides of the two enzymes can be detected.

The Domains of the Aspartokinases-Homoserine Dehydrogenases Both proteins can be proteolyzed and lead to fragments carrying the homoserine dehydrogenase activity, although the HDH II fragment is shorter than the HDH I fragment (Mr = 37,000 and 55,000 respectively) (Sections 3.2 and 3.3). This shows that in both proteins a large part, and in particular (a large part of) the kinase domain, is not necessary for the expression of homoserine dehydrogenase activity (in the case of AK I-HDH I, the presence of the kinase domain may be necessary for the homoserine dehydrogenase activity to be inhibited).

Thus for both proteins the dehydrogenase domain is quite independent of the kinase domain for the expression of activity. In addition, in the case of AK II-HDH II, the homoserine dehydrogenase domain can refold independently of the rest of the protein (Section 3.3).

Also in AK I-HDH I (a large part of) the homoserine dehydrogenase domain is not necessary for the expression and inhibition of aspartokinase activity (Section 3.2). Whether this is also true in the case of AK II-HDH II remains an open question.

Immunochemical Comparison

With Antigens in Their Native State

COMPARISON OF AK I-HDH I AND AK II-HDH II The first immunological study of AK I-HDH I and AK II-HDH II was carried by Kaminski et al. (95) who, after eliciting antibodies in rabbits against AK I-HDH I and AK II-HDH II, used precipitation by double diffusion to detect antigen-antibody reactions. Figure 3.40 shows that in homologous reaction (i.e., AK I-HDH I reacting with antiserum raised against AK I-HDH I and AK II-HDH II reacting with antiserum raised against AK II-HDH II) each of the antigen antibody systems forms an intense line of precipitation. No heterologous reaction can be detected, and, as seen on Figure 3.40, the precipitation lines formed in homologous antigen-antibody reactions grow and extend independently from each other and do not deviate at the proximity of the wells containing heterologous reactants. Thus there is no

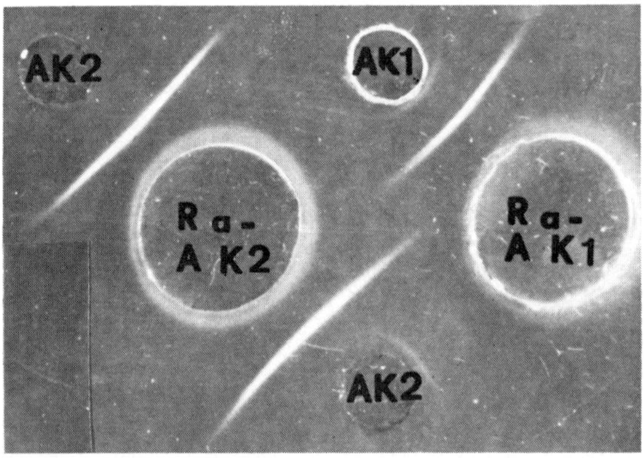

Fig. 3.40 Reaction of AK I-HDH I and AK II-HDH II (AK I and AK II respectively on the picture) with rabbit antiserum against AK I-HDH I (Ra-AK I) and against AK II-HDH II (Ra-AK II) as observed by precipitation by double diffusion. Reproduced from ref. 95 with permission of the European Journal of Biochemistry.

reactions between one of the bifunctional proteins and antiserum directed against the other one.

The same conclusion was reached by Zakin et al. (96) who could not detect any immunological similarity between native AK I-HDH I and AK II-HDH II by complement fixation, passive hemagglutination, and binding to a specific immunoadsorbent gel.*

COMPARISON OF THE TWO PROTEOLYTIC FRAGMENTS HDH I AND HDH II The results obtained with the proteolytic fragments are strictly comparable with those obtained with the entire enzymes. The HDH I fragment is not recognized by antibodies raised against either native AK II-HDH II or native HDH II fragment, and the HDH II fragment is not recognized by antibodies raised against either native AK I-HDH I or native HDH I fragment as seen by passive hemagglutination and complement fixation (96).

With Antigens in Their Denatured State

COMPARISON OF THE DENATURED AK I-HDH I AND AK II-HDH II Since, surprisingly, no immunological cross-reactivity could be detected between native AK I-HDH I and AK II-HDH II, Zakin et al. (96) undertook an immunological comparison of the two proteins when denatured, hoping to reveal a cross-reactivity between the two proteins.

Antibodies were elicited in rabbits against the denatured proteins, that is denatured in 6 *M* guanidine hydrochloride and alkylated with *N*-ethyl maleimide (NEM) (96). For immunological assays the proteins were denatured in 6 *M* guanidine, alkylated with iodoacetic acid, and the carboxymethylcysteine derivatives were citraconylated in order to become soluble.

There is an immunological similarity between denatured AK I-HDH I and AK II-HDH II (the antigen-antibody reaction was measured by complement fixation and passive hemagglutination) (Table 3.10). Indeed, the antisera directed against denatured AK I-HDH I and AK II-HDH II react efficiently with the heterologous antigen, denatured AK II-HDH II, and AK I-HDH I respectively. This reaction is specific of the two aspartokinases-homoserine dehydrogenases since none of the antisera reacts with any of the other denatured proteins tested (Table 3.10). It is also specific of the denatured state since the antiserum directed against one of the two

* A cross-reaction between the native aspartokinases homoserine dehydrogenases based on immunoadsorbant experiments has been reported (P. Truffa-Bachi, N. Guiso, G. N. Cohen, J. Thèze, and B. Burr (1974), *Proc. Natl. Acad. U.S.A.* **72,** 1268. This technique may lead to artefactual results, and these experiments could not be reproduced.

Table 3.10 Reaction of Denatured AK I-HDH I and Denatured AK II-HDH II with Antisera Directed Against Denatured AK I-HDH I and Denatured AK II-HDH II (96)[a]

	Titer of the Antisera	
Antigen	Antidenatured AK I-HDH I	Antidenatured AK II-HDH II
Denatured AK I-HDH I	1/6561	1/729
Denatured AK II-HDH II	1/729–1/2187	1/6561
Denatured homoserine kinase	1/27	1/81
Denatured β_2 subunit of tryptophan synthetase	1/81	1/27
Denatured ribonuclease A	1/3	1/3

[a] The denatured proteins used as controls are homoserine kinase, β_2 subunit of tryptophan synthetase of *E. coli*, and bovine pancreatic ribonuclease A.

aspartokinases-homoserine dehydrogenases in its native state does not react with the other one when denatured (96).

COMPARISON OF THE DENATURED PROTEOLYTIC FRAGMENTS HDH I AND HDH II The denatured fragments were prepared as the denatured proteins but they do not have to be citraconylated to be soluble. This is an improvement since citraconylation probably masks some of the antigenic determinants.

The results obtained with the denatured fragments are very similar to those obtained with the denatured proteins. Each denatured fragment is recognized by both the antisera raised against denatured AK I-HDH I and AK II-HDH II, that is against the homologous and heterologous entire protein as measured by passive hemagglutination and complement fixation (Figure 3.41).

Discussion As we have seen above, the cross-reaction between AK I-HDH I and AK II-HDH II (and between HDH I and HDH II) strictly depends on the denatured state of these proteins both as immunogens and as antigens. The common antigenic determinants involved in the cross-reaction are thus completely masked in the native proteins.

Since denatured proteins are assumed not to have any defined and stable conformation, an immunochemical resemblance between two denatured proteins reflects some similarity in their amino acid sequences. The following evidence strongly supports this hypothesis in the case of the aspartokinases-homoserine dehydrogenases: the immunochemical reaction of the denatured HDH II proteolytic fragment with antiserum raised against denatured AK I-HDH I is the same as that of the pool of its cyanogen bromide peptides (97). This shows that only local structural elements are involved in the immunochemical behavior of denatured proteins. The experiments described here show that there is some sequence homology between AK I-HDH I and AK II-HDH II (and between the two homoserine dehydrogenases domains as represented by HDH I and HDH II). Although the extent of this sequence homology is not known, it provides a strong argument in favor of a common origin for the two proteins.

The absence of cross-reactivity between native species does not imply that

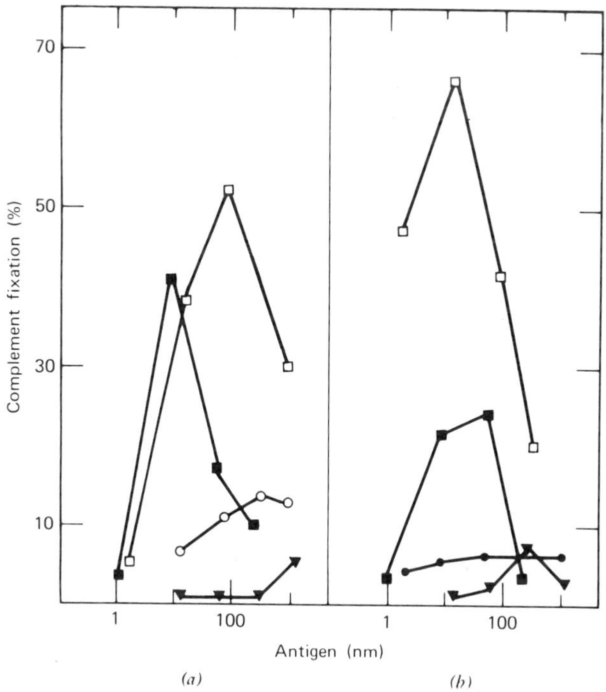

Fig. 3.41 Complement fixation assays of the reaction of the antisera against denatured (A) AK I-HDH I and (B) AK II-HDH II with denatured (■) fragment HDH I and (□) fragment HDH II. Also shown are the results given by (▼) denatured β_2 subunit of tryptophan synthetase. Denatured bovine pancreatic ribonuclease A gives no reaction with either antiserum (results not shown). The dilutions of the antisera are 1/200 for the reaction with the fragment HDH I in (A), 1/100 for that with the fragment HDH II in (B), and 1/50 for all other reactions. Serum from nonimmunized rabbits (dilution 1/50 for all other reactions. Serum from nonimmunized rabbits (dilution 1/50) does not react with denatured (○) fragment HDH II, and (●) fragment HDH I. From (96).

this sequence homology is low. Indeed in a few cases, proteins having homologous amino acid sequences do not show any immunochemical cross-reactivity in their native state. In two such cases, of high sequence homology, the proteins were shown to immunologically cross-react when denatured (98, 99). Several possibilities may account for the lack of cross-reactivity between the native aspartokinases-homoserine dehydrogenases. Portions of the amino acid sequences that are highly homologous may reside in the interior of the native molecules (and for large proteins the ratio of the number of internal residues to the number of external residues is high). Also there may be highly homologous sequences at the surface that are in a slightly different conformation on the two proteins and thus are not recognized by heterologous antibodies. Finally, there may be regions on the surface whose sequences are homologous but which are not immunogenic.

Concluding Remarks

The existence of two bifunctional enzymes carrying the same two activities in the same organism raises several questions. What is the selective advantage of bifunctionality for the aspartokinases-homoserine dehydrogenases? Do they derive from a common ancestor? And, if so, was the common ancestor bifunctional, or did the gene fusion occur later?

The two aspartokinases-homoserine dehydrogenases have some similarities and some differences, but they show some sequence homology, which supports the hypothesis that they derive from a common ancestor. At this stage of knowledge AK II-HDH II appears to be simpler; it is not inhibited by its end product, and it is a dimer and not a tetramer. Further comparative studies of these two enzymes may lead to the explanation of some of their functional differences, especially as regards allosteric regulation, in terms of structural differences.

In the case of AK I-HDH I one can argue that the end product, threonine, when in excess in the cell, inhibits the two activities simultaneously and that this is a selective advantage that maintains a bifunctional enzyme. In the case of AK II-HDH II no such argument holds. Since the reaction catalyzed by the two aspartokinases homoserine dehydrogenases are not sequential, one could also think that one or the two bifunctional enzymes form a multienzyme complex with aspartate semi-aldehyde dehydrogenase and maybe with other enzymes of the threonine and methionine biosynthetic pathways. Finally, in many bacterial species the aspartokinase and homoserine dehydrogenase activities are carried by different enzymes which are regulated in a variety of ways.

Thus, although AK I-HDH I was one of the first well demonstrated multifunctional enzymes, it is still not clear why it is bifunctional.

REFERENCES

1 E. R. Stadtman, G. N. Cohen, G. Le Bras and H. de Robichon-Szulmajster (1961), *J. Biol Chem.* **236,** 2033.

2 J. C. Patte, G. Le Bras, T. Loviny, and G. N. Cohen (1963), *Biochim. Biophys. Acta* **67,** 16.

3 J. C. Patte, G. Le Bras, and G. N. Cohen (1967), *Biochim. Biophys. Acta* **136,** 245.

4 G. N. Cohen, J. C. Patte, and P. Truffa-Bachi (1965) *Biochem. Biophys. Res. Comm.* **13,** 546.

5 J. C. Patte, P. Truffa-Bachi, and G. N. Cohen (1966), *Biochim. Biophys. Acta* **128,** 426.

6 P. Truffa-Bachi, R. van Rapenbusch, J. Janin, C. Gros, and G. N. Cohen (1968), *Eur. J. Biochem.* **5,** 73.

7 F. Falcoz-Kelly, R. van Rapenbusch, and G. N. Cohen (1969), *Eur. J. Biochem.* **8,** 146.

8 M. Freundlich (1963), *Biochem. Biophys. Res. Comm.* **10,** 277.

9 G. N. Cohen and J. C. Patte (1963), *Cold Spring Harbor Symp. Quant. Biol.* **28,** 513.

10 J. C. Patte and G. N. Cohen (1965), *Biochim. Biophys. Acta* **99,** 557.

11 J. Thèze, D. Margarita, G. N. Cohen, F. Borne, and J. C. Patte (1974), *J. Bacteriol.* **117,** 133.

12 G. N. Cohen, J. C. Patte, P. Truffa-Bachi, C. Sawas, and M. Doudoroff (1965), in *Mécanismes de régulation des activités cellulaires chez les microorganismes*, C.N.R.S., Paris, pp. 243–253.

13 G. N. Cohen and M. L. Hirsch (1954), *J. Bacteriol.* **67,** 182.

14 M. L. Hirsch and G. N. Cohen (1954), *Biochim. Biophys. Acta* **15,** 560.

15 A. Szentirmai, M. Szentirmai, and H. E. Umbarger, (1968), *J. Bacteriol.* **95,** 1672.

16 J. E. Brenchley and L. S. Williams (1975) *Ann. Rev. Microbiol.* **29,** 251.

17 H. J. Teas (1950), *J. Bacteriol.* **59,** 93.

18 H. J. Teas, N. H. Horowitz, and M. Fling (1948), *J. Biol. Chem.* **172,** 651.

19 H. de Robichon-Szulmajster, Y. Surdin, and R. K. Mortimer (1966), *Genetics* **53,** 609.

20 R. L. Cafferata and M. Freundlich (1969), *J. Bacteriol.* **97,** 193.

21 G. N. Cohen, R. Y. Stanier, and G. Le Bras (1969), *J. Bacteriol.* **99,** 791.

22 S. Black and N. G. Wright (1955), *J. Biol. Chem.* **213,** 27.

23 S. Black and N. G. Wright (1955), *J. Biol. Chem.* **213,** 51.

24 P. Truffa-Bachi and G. N. Cohen (1970), *Methods in Enzymology*, Vol. 17A, Academic Press, New York, pp. 694–699.

25 F. Falcoz-Kelly and G. N. Cohen, (1970), *Methods in Enzymology*, Vol. 17A, Academic Press, New York, pp. 699–702.

26 D. E. Wampler and E. G. Westhead (1968), *Biochemistry* **7,** 1661.

27 J. Thèze and I. Saint-Girons (1974), *J. Bacteriol.* **118,** 990.

28 I. Saint-Girons and D. Margarita (1975), *J. Bacteriol.* **124,** 1137.

29 J. F. Gardner and D. H. Smith (1975), *J. Bacteriol.* **124,** 161.

30 E. J. Johnson, G. N. Cohen, and I. Saint-Girons (1977), *J. Bacteriol.* **129,** 66.

31 H. Hennecke, A. Böck, J. Thomale, and G. Nass (1977), *J. Bacteriol.* **131,** 943.

32 M. Casadaban (1976), *J. Mol. Biol.* **104,** 541.

33 I. Saint-Girons (1978), *Mol. Gen. Genetics* **162,** 95.

34 M. Cohn, G. N. Cohen, and J. Monod (1953), *CR* **236,** 746.

35 G. N. Cohen and F. Jacob (1959), *CR* **248,** 3490.

36 R. J. Rowbury and D. D. Woods (1966), *J. Gen. Microbiol.* **42,** 155.

37 B. J. Bachmann, K. B. Low, and A. L. Taylor (1976), *Bacteriol. Rev.* **40,** 116.

38 A. Ahmed, (1973), *Molec. Gen. Genet.* **123,** 299.

39 A. Dautry-Varsat and D. Margarita, unpublished results.

40 W. L. Starnes, P. Munk, S. B. Maul, G. N. Cunningham, D. J. Cox, and W. Shive (1972), *Biochemistry* **11,** 677.

41 F. Falcoz-Kelly, J. Janin, J. C. Saari, M. Véron, P. Truffa-Bachi, and G. N. Cohen (1972), *Eur. J. Biochem.* **28,** 507.

42 P. Truffa-Bachi. R. van Rapenbusch, J. Janin, C. Gros, and G. N. Cohen (1969), *Eur. J. Biochem.* **7,** 401.

43 R. B. Clark and J. W. Ogilvie (1972), *Biochemistry* **11,** 1278.

44 D. E. Wampler (1972), *Biochemistry* **11,** 4428.

45 L. Sibilli (1977), Thèse de Doctorat ès-Sciences, Paris.

46 P. Cossart (1977), Thèse de Doctorat ès-Sciences, Paris.

47 J. Janin, P. Truffa-Bachi, and G. N. Cohen (1967), *Biochem. Biophys. Res. Comm.* **26,** 429.

48 M. Véron, F. Falcoz-Kelly, and G. N. Cohen (1972), *Eur. J. Biochem.* **28,** 520.

49 P. A. Briley, L. Sibilli, M. A. Chalvignac, P. Cossart, G. Le Bras, A. De Wolf, and G. N. Cohen, (1978), *J. Biol. Chem.*, **253,** 8867.

50 M. E. Goldberg (1969), *J. Mol. Biol.* **46,** 441.

51 D. M. Bonner, J. A. De Moss, and S. E. Mills (1965), in V. Bryson and H. J. Vogel, eds. *Evolving Genes and Proteins*, Academic Press, New York, p. 305.

52 P. Truffa-Bachi and H. D'A. Heck (1971), *Biochemistry* **10,** 2700.

53 E. Khedouri, P. M. Anderson, and A. Meister (1966), *Biochemistry* **6,** 3533.

54 C. G. Hirth, M. Véron, C. Villar-Palasi, N. Hurion, and G. N. Cohen (1975), *Eur. J. Biochem.* **50,** 425.

55 M. Véron, J. C. Saari, C. Villar-Palasi, and G. N. Cohen (1973), *Eur. J. Biochem.* **38,** 325.

56 G. N. Cohen, J. C. Patte, P. Truffa-Bachi, and J. Janin (1967), in *Regulation of nucleic acid and protein biosynthesis*, Elsevier Publishing Company, Amsterdam, pp. 357–365.

57 P. Truffa-Bachi, J. M. Costrejean, M. C. Py, and G. N. Cohen (1974), *Biochimie* **56,** 215.

58 J. C. Mackall and K. E. Neet (1973), *Biochemistry* **12,** 3483.

59 P. Truffa-Bachi, G. Le Bras, and G. N. Cohen (1966), *Biochim. Biophys. Acta* **128,** 450.

60 Y. Jacques and P. Truffa-Bachi (1976), *Eur. J. Biochem.* **62,** 485.

61 E. Fontan and P. Truffa-Bachi (1978), *J. Biol. Chem.* **253,** 2754.

62 E. Fontan and P. Truffa-Bachi (1978), *J. Biol. Chem.* **253,** 2758.

63 L. Sibilli, P. Cossart, M. A. Chalvignac, P. A. Briley, J. M. Costrejean, G. Le Bras, and G. N. Cohen (1977), *Biochimie* **59,** 943.

64 M. Zakin and A. Dautry-Varsat, manuscript in preparation.

65 J. J. Crumpton (1974), in M. Sela, ed., *The Antigens*, Vol. 2, Academic Press, New York, pp. 1–72.

66 J. M. Costrejean, N. Guiso, D. B. Cowie, G. N. Cohen, and P. Truffa-Bachi (1975), *Eur. J. Biochem.* **50,** 431.

67 F. H. Carpenter and K. T. Harrington (1972), *J. Biol. Chem.* **247,** 5580.

68 J. Janin, R. van Rapenbusch, P. Truffa-Bachi, and G. N. Cohen, (1969), *Eur. J. Biochem.* **8,** 128.

69 H. d'A. Heck and P. Truffa-Bachi (1970), *Biochemistry* **9,** 2776.

70 R. S. Ehrlich and M. Takahashi, (1973), *Biochemistry* **12,** 4309.

71 P. Truffa-Bachi, M. Véron, and G. N. Cohen (1974), in G. B. Fasman, ed., *CRC Critical Reviews in Biochemistry*, Vol. 2, Chemical Rubber Company Press, Inc., Cleveland, pp. 379–415.

72 J. M. Costrejean and P. Truffa-Bachi, manuscript in preparation.

73 C. Ryzewski and M. Takahashi, (1975), *Biochemistry* **14,** 4482.

74 J. K. Wright, J. Feldman, and M. Takahashi, (1976), *Biochemistry* **15,** 3704.

75 A. Tilak, K. Wright, S. Damle, and M. Takahashi (1976), *Eur. J. Biochem.* **69,** 249.

76 J. K. Wright and M. Takahashi (1977), *Biochemistry* **16,** 1541.

77 J. K. Wright, J. Feldman, and M. Takahashi (1976), *Biochem. Biophys. Res. Comm.* **72,** 1456.

78 K. Wright and M. Takahashi (1977), *Biochemistry* **16,** 1548.

79 L. Broussard, G. N. Cunningham, W. L. Starnes, and M. Shive (1972), *Biochem. Biophys. Res. Comm.* **46,** 1181.

80 J. C. Patte (1967), Thèse de Doctorat ès-Sciences, Paris.

81 J. C. Patte and G. N. Cohen (1964), *CR* **259,** 1255.

82 J. Monod, J. Wyman, and J. P. Changeux, (1965), *J. Mol. Biol.* **12,** 88.

83 J. Janin and G. N. Cohen (1969), *Eur. J. Biochem.* **11,** 520.

84 M. Takahashi and E. W. Westhead (1971), *Biochemistry* **10,** 1700.

85 E. D. Barber and H. J. Bright (1968), *Proc. Natl. Acad. Sci. US* **60,** 1363.

86 J. Janin and M. Iwatsubo (1969), *Eur. J. Biochem.* **11,** 530.

87 C. F. Bearer and K. E. Neet, manuscript in preparation.

88 P. Cossart-Gheerbrant, L. Sibilli-Weill, P. A. Briley, M. A. Chalvignac. G. Le Bras, and G. N. Cohen (1978), *Biochim. Biophys. Acta* **535,** 206–215.

89 J. Janin (1974), *FEBS-Letters* **45,** 318.

90 A. Dautry-Varsat, L. Sibilli-Weill, and G. N. Cohen (1977), *Eur. J. Biochem.* **76,** 1.

91 F. Falcoz-Kelly (1971), Thèse de Doctorat ès-Sciences. Paris.

92 A. Dautry-Varsat and G. N. Cohen (1977), *J. Biol. Chem.* **252,** 7685.

93 A. Dautry-Varsat and J. R. Garel, (1978), *Proc. Natl. Acad. Sci. U.S.A.* **75,** 96.

94 C. E. Harris and D. C. Teller (1973), *J. Theor. Biol.* **38,** 347.

95 M. Kaminski, F. Falcoz-Kelly, P. Truffa-Bachi, J.-C. Patte, and G. N. Cohen (1969), *Eur. J. Biochem.* **11,** 278.

96 M. Zakin, J. R. Garel, A. Dautry-Varsat, G. N. Cohen, and G. Boulot (1978), *Biochemistry*, **17,** 4318.

97 L. Sibilli and M. Zakin, unpublished results.

98 N. Arnheim. J. Sobel, and R. Canfield (1971), *J. Mol. Biol.* **61,** 237.

99 R. Arnon and E. Maron (1971), *J. Mol. Biol.* **61,** 225.

Chapter 4

ANTHRANILATE SYNTHASE: RELATIONSHIPS BETWEEN BIFUNCTIONAL AND MONOFUNCTIONAL ENZYMES

HOWARD ZALKIN

Department of Biochemistry
Purdue University
West Lafayette, Indiana

4.1 INTRODUCTION

Anthranilate synthase catalyzes the first specific reaction in the biosynthetic pathway to tryptophan in microorganisms and plants. The pathway is outlined in Figure 4.1. In all (1) but rare cases (2) the enzyme is subject to end product inhibition by tryptophan. Anthranilate synthase assumes a variety of quaternary structures and aggregational states in different organisms. In all prokaryotic and several eukaryotic microorganisms the enzyme contains nonidentical subunits (1), whereas in one known case, *Euglena gracilis*, the enzyme is monomeric (3). Anthranilate synthase is monofunctional in some organisms and multifunctional in others. A number of multifunctional anthranilate synthases have been characterized. In such enzymes the anthranilate synthase activity is associated with the second, third, or third and fourth enzyme activities of the tryptophan pathway (1). This article focuses on key properties of the bifunctional anthranilate synthase: anthranilate 5-phosphoribosylpyrophosphate phosphoribosyltransferase (*P*-ribosyl-

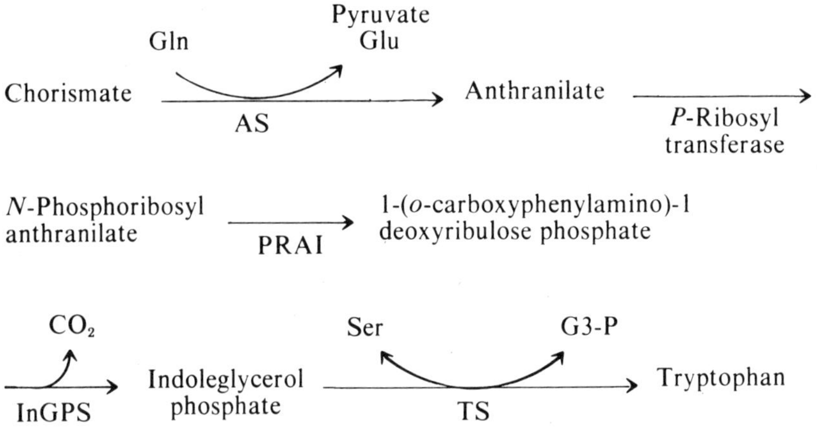

Fig. 4.1 Tryptophan biosynthetic pathway. The abbreviations are: Gln, glutamine; Glu, glutamate; Ser, serine; G3-P, glyceraldehyde 3-P; AS, anthranilate synthase; P-ribosyltransferase, anthranilate 5-phosphoribosylpyrophosphate phosphoribosyltransferase; PRAI, phosphoribosylanthranilate isomerase; InGPS, indoleglycerol phosphate synthase; TS, tryptophan synthase.

transferase) from *Salmonella typhimurium* and *Escherichia coli* and the relationship of this bifunctional enzyme to certain of the monofunctional anthranilate synthases.

4.2 NOMENCLATURE FOR PROTEIN SUBUNITS

Similar to other glutamine amidotransferases, anthranilate synthase can utilize NH_3 in addition to glutamine. All well characterized bacterial anthranilate synthases contain nonidentical protein chains and exhibit a well defined structure-function relationship. A protein chain of approximately 60,000 to 80,000 mol wt catalyzes the NH_3-dependent synthesis of anthranilate but lacks the site for glutamine. This reaction is shown in equation 1. In keeping with

$$\text{chorismate} + NH_3 \rightarrow \text{anthranilate} + \text{pyruvate} \quad (1)$$

the original nomenclature of Ito and Yanofsky (4) this chain is designated anthranilate synthase component I (AS I). AS I is the product of the *E. coli* *trpE* gene. In monofunctional anthranilate synthase a second protein chain of approximately 15,000–24,000 mol wt binds glutamine and functions in amide transfer to AS I. This subunit is designated anthranilate synthase component II (AS II). AS I and AS II interact to form an oligomer. The glutamine-dependent reaction of the AS I·AS II oligomer is shown in equation 2.

$$\text{Chorismate} + \text{glutamine} \rightarrow \text{anthranilate} + \text{pyruvate} + \text{glutamate} \quad (2)$$

In multifunctional anthranilate synthases AS II is covalently joined to other proteins. In *E. coli* and *S. typhimurium* AS II is joined to P-ribosyltransferase. In *E. coli* the bifunctional AS II-P-ribosyltransferase of approximately 65,000 mol wt is the product of the *trpD* gene. The only reported exception of the AS I·AS II subunit model is in *Euglena gracilis* (3).

4.3 SUBUNIT COMPOSITION

Monofunctional Anthranilate Synthase

As shown in Table 4.1 many monofunctional anthranilate synthases have been described. Most of the molecular weight estimations were by gel filtra-

Table 4.1 Monofunctional Anthranilate Synthases

Organism	Enzyme Purity[a]	Molecular Weight: Anthranilate Synthase	Molecular Weight: AS I	Molecular Weight: AS II	Molecular Weight: *P*-ribosyl-transferase	Subunit Composition[c]	Subunit Association[d]	Reference
Aeromonas formicans	E	~220,000	[b]—	[b]—	~67,000	[b]—	N	5
Proteus morganii	E	~140,000	—	~20,000	~67,000	$(I)_2(II)_2$	D	5
Proteus vulgaris	E	~140,000	—	~20,000	~67,000	$(I)_2(II)_2$	D	5
Pseudomonas acidovorans	E	~155,000	~71,000	~18,000	[b]—	$(I)_2(II)_2$	D	6
Pseudomonas testosteroni	E	~155,000	~71,000	~18,000	[b]—	$(I)_2(II)_2$	D	6
Erwinia carotavora	E	~140,000	—	—	~67,000	$(I)_2(II)_2$	N	5
Erwinia hafniae	E	~140,000	—	—	~67,000	$(I)_2(II)_2$	N	5
Serratia marcescens	HP	141,000	60,000	21,000	45,000	$(I)_2(II)_2$	N	7, 8, 9
Serratia marinorubra	E	~140,000	—	—	~45,000	$(I)_2(II)_2$	N	5
Enterobacter liquefaciens	E	~140,000	—	—	~45,000	$(I)_2(II)_2$	N	5
Acinetobacter calcoaceticus	E	~86,000	~70,000	~14,000	~46,000	(I)(II)	D	10, 11
Pseudomonas putida	HP	75,000	63,000	21,700	—	(I)(II)	D	12, 13
Pseudomonas aeruginosa	E	[b]—	~64,000	~18,000	—	(I)(II)	D	6

Bacillus subtilis	PP, HP	~84,000	~67,000	~20,000	[e]	(I)(II)	D	14, 15
Bacillus licheniformis	PP	—	>100,000	~24,000	—	—	D	15
Bacillus coagulans	PP	—	~80,000	~18,000	—	—	D	15
Bacillus pumilus	PP	~86,000	~80,000	~23,000	~45,000	(I)(II)	D	15, 16
Bacillus mascerans	PP	—	~62,000	~24,000			D	15
Bacillus alvei	PP	~90,000	~58,000	~15,000	~54,000	(I)(II)	D	15, 16, 17
Chromobacterium violaceum	PP	~86,000	—	—	~52,000	—	N	18
Staphylococcus aureus	PP	~65,000	—	—	~30,000	—	N	19
Clostridium butyricum	E, PP	~84,000	~127,000[f]	~15,000	—	—	D	20, 21
Euglena gracilis	HP	80,000[g]	—	—	~325,000[h]	I-II[g]	N	3, 22

[a] Purity of preparations used for molecular weight estimation: HP, highly purified or homogeneous; PP, partially purified; E, crude extract.

[b] Not reported.

[c] In most cases these designations are estimations from molecular weight data or by analogy to closely related enzymes: I, AS I; II, AS II.

[d] Dissociation (D) or nondissociation (N) of subunits upon gel filtration.

[e] Molecular weight is comparable to that from *B. pumilus* and *B. alvei* (16).

[f] Probably an aggregate of $(63{,}000)_2$.

[g] Monomeric AS I and AS II appear to be fused into a single polypeptide chain.

[h] Multifunctional $(155{,}000)_2$.

tion or sucrose gradient centrifugation using crude enzyme preparations. The main exception is the 21,700 mol wt for homogeneous *P. putida* AS II determined by sodium dodecyl sulfate (SDS) gel electrophoresis and confirmed by amino acid sequence (13). Molecular weight values are useful as a criterion of enzyme similarity and also to suggest the probable subunit structure. Except for the enzymes from *Aeromonas formicans* and *Euglena gracilis* the monofunctional anthranilate synthases fall into two major groups: tetrameric enzymes having approximately 140,000 to 155,000 mol wt and dimeric enzymes approximately 65,000 to 90,000 mol wt. For both the dimeric and tetrameric enzymes the AS I subunit is, in most cases, approximately 60,000 to 80,000 mol wt and the AS II subunit approximately 15,000 to 24,000 mol wt. Molecular weight estimations of *P*-ribosyltransferase are shown for comparison with the bifunctional anthranilate synthase: *P*-ribosyltransferase enzymes. There appear to be two main classes of *P*-ribosyltransferase enzymes: those with approximately 67,000 mol wt and those of approximately 45,000 mol. wt. For several organisms the sum of the molecular weights of AS II and *P*-ribosyltransferase is approximately 65,000, which is similar to that for bifunctional AS II: *P*-ribosyltransferase.

In most cases the AS I and AS II subunits of the dimeric enzymes undergo facile association-dissociation. Dissociation of subunits occurs readily during gel filtration and introduces uncertainty into the molecular weight estimation for the oligomeric state. Dissociation is facilitated by high ionic strength and by low protein concentration (15, 24) while glutamine promotes association of subunits (25). Tetrameric anthranilate synthases from species of *Pseudomonas* and *Proteus* also appear to exhibit subunit dissociation during gel filtration, but dissociation is less complete as compared to the dimeric enzymes. Since free AS I was not detected in extracts of *Proteus* species dissociation is tentative and other explanations for free AS II are possible. In other tetrameric anthranilate synthases and two of the dimeric enzymes subunit dissociation does not occur upon gel filtration. A possible physiological basis for this phenomenon is discussed in Section 4.5.

The best characterized tetrameric anthranilate synthase is from *Serratia marcescens*. Subunit association in this enzyme is extraordinarily avid. The only preparative separation presently available is gel filtration in 60% formic acid (23). A stoichiometry of $(\text{AS I})_2(\text{AS II})_2$ was assigned based on staining of the separated protein chains on SDS gels, intramolecular cross linking experiments, number of binding sites for the substrate chorismate, glutamine affinity analog 6-diazo-5-oxonorleucine, and feedback inhibitor tryptophan (7, 26). It appears likely that all tetrameric anthranilate synthases of approximately 140,000 mol wt are similar in structure and

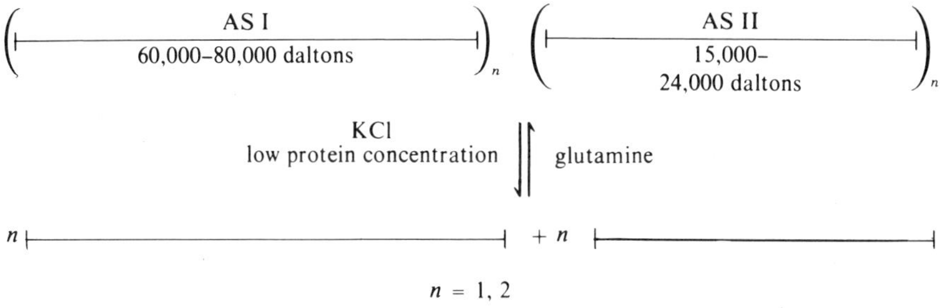

Fig. 4.2 Schematic representation of subunit composition and aggregation state of monofunctional anthranilate synthases. Most dimeric ($n = 1$) and some tetrameric ($n = 2$) anthranilate synthases exhibit association-dissociation. Other monofunctional anthranilate synthases, as noted in Table 4.1, do not dissociate.

function to the enzyme from *Serratia marcescens*. Most of the data for anthranilate synthase in Table 4.1 can be accommodated by the simple scheme shown in Figure 4.2.

Multifunctional Anthranilate Synthase

A number of multifunctional anthranilate synthases have been described and are listed in Table 4.2. The anthranilate synthase is associated with *P*-ribosyltransferase in bacteria, with indoleglycerol-*P* synthase (InGPS) in yeast and with phosphoribosylanthranilate isomerase:indoleglycerol-*P* synthase (PRAI-InGPS) in fungi. The best characterized of the bacterial enzymes is from *S. typhimurium*. Homogeneous *S. typhimurium* anthranilate synthase:*P*-ribosyltransferase has a molecular weight of 280,000 and contains two 64,000 dalton chains of AS I, and two chains of bifunctional AS II:*P*-ribosyltransferase of 65,000 mol wt. The molecular weights of AS I and the native anthranilate synthase:*P*-ribosyltransferase were determined by sedimentation equilibrium centrifugation while that for AS II:*P*-ribosyltransferase was estimated by SDS gel electrophoresis. A substantial fraction of enzyme is particulate and is usually not purified (36, 37). Particulate anthranilate synthase:*P*-ribosyltransferase has recently been shown to be attached to the inner membrane fraction of *E. coli* (J. J. Reiners and H. Zalkin, unpublished). This particulate enzyme is released by repeated washing with buffers of low ionic strength. The enzyme released from membrane fragments is identical to that initially in the soluble fraction (37). *P*-ribosyltransferase activity is latent in particulate anthranilate synthase:*P*-ribosyltransferase and is restored in the solubilized enzyme. The significance of the membrane-bound form of anthranilate synthase:*P*-

Table 4.2 Multifunctional Anthranilate Synthases

Organism	Enzyme Activities[a]	Purity	Molecular Weight			Subunit	Reference
			Native	AS I	AS II-X		
Salmonella typhimurium	AS-PRT	HP	280,000[c]	64,000	65,000	$(I)_2(II\text{-}PRT)_2$	27, 28, 40
Enterobacter aerogenes	AS-PRT	HP	~170,000	—[b]	—[b]	—[b]	29
Escherichia coli	AS-PRT	HP	~260,000[c]	~60,000	~60,000	$(I)_2(II\text{-}PRT)_2$	30
Citrobacter freundii	AS-PRT	E	~250,000[c]	—	—	$(I)_2(II\text{-}PRT)_2$	5
Citrobacter ballerupensis	AS-PRT	E	~250,000[c]	—	—	$(I)_2(II\text{-}PRT)_2$	5
Enterobacter cloacae	AS-PRT	E	~250,000[c]	—	—	$(I)_2(II\text{-}PRT)_2$	5
Erwinia dissolvens	AS-PRT	E	~250,000[c]	—	—	$(I)_2(II\text{-}PRT)_2$	5
Aspergillus nidulans	AS-PRAI-InGPS	E	—[b]	—	—	—	31
Neurospora crassa	AS-PRAI-InGPS	HP	~330,000	70,000	94,000	$(I)_2(II\text{-}PRAI\text{-}InGPS)_2$	32, 33
Claviceps spec. SD58	AS-PRAI-InGPS	PP	200,000–400,000	—	—	—	34
Saccharomyces cerevisiae	AS-InGPS	E	—	—	—	—	31, 35

[a] The abbreviations are given in Figure 4.1 except for PRT, anthranilate 5-phosphoribosylpyrophosphate phosphoribosyltransferase.

[b] Not reported.

[c] A membrane-bound form is also present.

[d] The composition of bifunctional anthranilate synthase: *P*-ribosyltransferases of molecular weight approximately 250,000 is assumed to be similar to the *S. typhimurium* enzyme.

ribosyltransferase is unknown, but, as noted in Table 4.2, it has been detected in all bacterial species except for *Enterobacter aerogenes*. Although the data are less complete for other bacteria the enzyme from *E. coli* appears virtually identical to the *S. typhimurium* enzyme in subunit size and oligomeric composition. Molecular weight differences may not be significant since empirical techniques (gel filtration, sucrose gradient centrifugation) were used for all the microbial sources except *S. typhimurium*. It appears likely that all bacterial anthranilate synthase:*P*-ribosyltransferase enzymes have the composition (AS I)$_2$(AS II:*P*-ribosyltransferase)$_2$.

A simplified schematic representation of bifunctional anthranilate synthase:*P*-ribosyltransferase is given in Figure 4.3. With respect to the bifunctional AS II:*P*-ribosyltransferase chain, an AS II function of molecular weight approximately 23,000 is contained within the NH_2-terminal third of the protein. The *P*-ribosyltransferase function is contained within the CO_2H-terminal two thirds of the chain. Evidence for the assignment of function to structure is given in Section 4.4. The molecular weight of 65,000 for the bifunctional AS II:*P*-ribosyltransferase is similar to the sum of molecular weights of monofunctional AS II plus *P*-ribosyltransferase in several organisms (Table 4.1). The anthranilate synthase:*P*-ribosyltransferases from *S. typhimurium* and *E. coli* are similar to the monofunctional *S. marcescens* enzyme by requiring strongly denaturing conditions for separation of subunits.

The best characterized of the eukaryotic multifunctional anthranilate synthases is from *Neurospora crassa* (32, 33). The characterization has been made difficult by the persistence of trace amounts of very active protease in highly purified preparations. The *Neurospora* enzyme contains nonidentical chains of approximately 70,000 and 94,000 mol wt. One subunit exhibits NH_3-dependent activity (AS I), and the other is an AS II-PRA-InGPS trifunctional chain. Although it has not been reported which function should be ascribed to each of the subunits it appears likely that the 70,000 dalton subunit is the NH_3-dependent AS I and the 94,000 dalton subunit is the

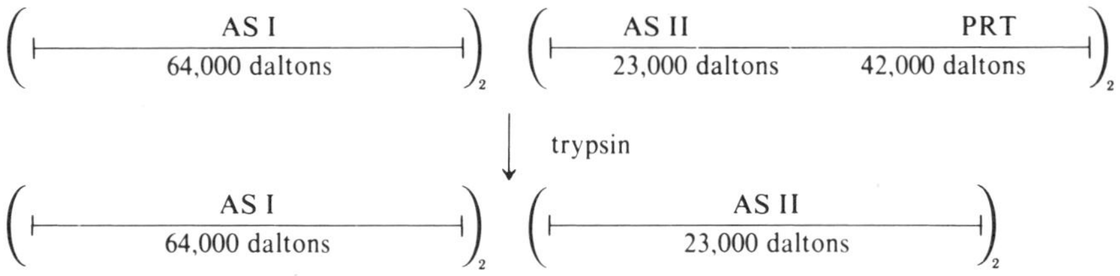

Fig. 4.3 Schematic representation of subunit composition of *S. typhimurium* bifunctional anthranilate synthase: P-ribosyltransferase. The scheme shows the results of limited trypsin digestion. The abbreviations are defined in Fig. 4.1 and Table 4.1.

trifunctional AS II-PRAI-InGPS (32, 33). AS-PRAI-InGPS from a number of other fungi (31, 34) may be similar to the enzyme from *Neurospora*. Little information is available on the bifunctional AS-InGPS from *Saccharomyces cerevisiae* (31, 35).

4.4 CATALYTIC AND REGULATORY PROPERTIES OF BIFUNCTIONAL ANTHRANILATE SYNTHASE:P-RIBOSYLTRANSFERASE

Assignment of Functions to Protein Chains

Sites for chorismate, NH_3, and tryptophan can be assigned to the AS I subunit of the bifunctional (AS I)$_2$(AS II:*P*-ribosyltransferase)$_2$ based on the NH_3-dependent synthesis of anthranilate by homogeneous *S. typhimurium* (27) and *E. coli* (38) AS I. Binding of one equivalent of tryptophan promotes a conformation change (27) that inhibits enzyme activity. The native enzyme contains two AS I chains and binds two equivalents of chorismate and tryptophan (28).

Biochemical (23, 37, 39, 40) and genetic (41, 42) evidence indicates that an approximately 23,000 dalton NH_2-terminal segment of AS II:*P*-ribosyltransferase contains the AS II function, and a nonoverlapping approximately 42,000 dalton CO_2H-terminal segment contains the *P*-ribosyltransferase activity. Mild proteolysis of anthranilate synthase:*P*-ribosyltransferase with trypsin specifically digests a large CO_2H-terminal segment of AS II:*P*-ribosyltransferase leaving an unmodified glutamine- and NH_3-dependent anthranilate synthase that lacks *P*-ribosyltransferase activity. This result is shown schematically in Figure 4.3. This scheme emphasizes structural similarity between trypsin-digested bifunctional anthranilate synthase and the monofunctional enzymes. All catalytic and regulatory properties of glutamine- and NH_3-dependent anthranilate synthase are virtually identical in the trypsin-digested and native enzymes (37, 43). Amino acid sequencing studies (23) indicate no digestion of AS I and AS II at the NH_2-termini. Identical results are obtained by trypsin digestion of monomeric AS II:*P*-ribosyltransferase chains (40). The sole effect of trypsin digestion of intact anthranilate synthase:*P*-ribosyltransferase is to convert the 65,000 dalton AS II:*P*-ribosyltransferase to an approximately 23,000 dalton AS II. Genetic analyses fully support the model shown in Figure 4.3. Mutations in an operator proximal segment of *E. coli trpD* coding for the NH_2-terminal portion of AS II:*P*-ribosyltransferase result in loss of glutamine-dependent anthranilate synthase activity (42). Strains with deletions that remove the operator distal two thirds of *trpD* coding for the CO_2-terminal two thirds of AS II:*P*-ribosyltransferase retain glutamine-dependent anthranilate synthase and lack *P*-ribosyltransferase. Moreover, deletion strains lacking the first third of *trpD* are missing AS II function but retain *P*-ribosyltransferase activity (41). These results thus support the view that the AS II and *P*-ribosyltransferase activities of the bifunctional protein are associated with distinct, nonoverlapping regions of the protein chain as shown in Figure 4.3.

The foregoing results, summarized in Figure 4.3, indicate that trypsin digestion effectively converts the bifunctional *S. typhimurium* and *E. coli* anthranilate synthase:*P*-ribosyltransferase to a Serratia-like monofunctional anthranilate synthase. This conclusion is supported by results of amino acid sequencing (23). There is 60% sequence identity in the first 25 residues of *S. typhimurium* and *S. marcescens* AS I and 73% sequence identity in the first 51 AS II residues from these organisms. Affinity labeling experiments with the glutamine analog 6-diazo-5-oxonorleucine also support the view that the only site for reaction of glutamine is in the AS II segment of AS II:*P*-ribosyltransferase (37, 44).

More recent experiments by Miozzari and Yanofsky (45) unequivocally confirm the conclusion of distinct, nonoverlapping regions for AS II and *P*-ribosyltransferase that are joined by a hexapeptide "connector or hinge" in *E. coli*. The CO_2H-terminal residue of *S. marcescens* AS II is a Lys that is conserved in *E. coli* AS II and immediately precedes a hexapeptide connector that serves to fasten the *P*-ribosyltransferase. It is thus likely that one effect of trypsin on *E. coli* AS II:*P*-ribosyltransferase is to cleave after this Lys and clip off *P*-ribosyltransferase.

Reaction Mechanism

The anthranilate synthase reaction mechanism appears to be similar for the monofunctional and bifunctional enzymes. The mechanism of Nagano et al. (44) has been previously reviewed (1). An additional step is now included in view of more recent information. In this mechanism (Figure 4.4) chorismate binds to AS I and glutamine to AS II. Direct binding measurements establish that each substrate binds independently of the other (24, 28). A key feature is the formation of a covalent glutaminyl-AS II intermediate followed by amide transfer to yield a chorismate$\cdot NH_3 \cdot$AS I complex and a γ-glutamyl-AS II thiolester intermediate. Reaction of the γ-glutamyl-enzyme intermediate with hydroxylamine yields γ-glutamylhydroxamate (44, 46). Binding of chorismate stimulates formation of the covalent intermediate(s) greater than 25-fold and provides evidence that chorismate binding precedes formation of the covalent intermediates. Pabst et al. (47) and Kane et al. (25) first provided evidence for the formation of a noncovalent

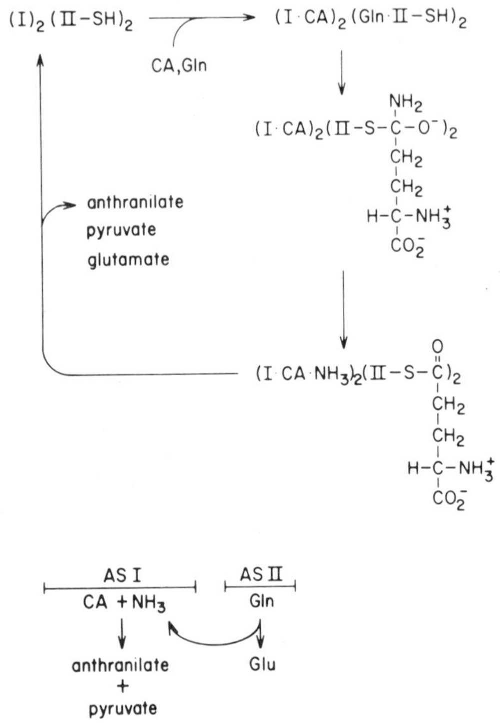

Fig. 4.4 Anthranilate synthase mechanism. AS I and AS II are designated I and II-SH, respectively. A tetrameric monofunctional nondissociating enzyme is shown. The mechanism applies equally well to bifunctional anthranilate synthase: *P*-ribosyltransferase. In the first step chroismate (CA) and glutamine (GLN) bind in random order to AS I and AS II, respectively. In the second step a hypothetical glutaminyl-enzyme covalent intermediate is formed. The amide is then transferred to the NH_3 site on AS I leaving a γ-glutamyl thioester-AS II intermediate. The last step shows anthranilate formation on AS I and release of all products. The amide transfer from glutaminyl-AS II to AS I is shown schematically below. In the absence of glutamine and independent of AS II, AS I catalyzes the NH_3-dependent synthesis of anthranilate.

anthranilate synthase:*P*-ribosyltransferase·glutamine complex in the absence of chorismate. Direct binding measurements confirm binding of 1 equivalent of glutamine to *P. putida* AS II or the AS I:AS II complex in the absence of chorismate (24). Since the K_D for glutamine binding in the absence of chorismate is similar to the K_m for glutamine in the synthesis of anthranilate it is unlikely that chorismate exerts any effect on formation of the enzyme glutamine Michaelis complex. However, chorismate is clearly required to form the covalent intermediate. There is no information on the order of product release. As stated above no differences in the anthranilate synthase mechanism have been detected in the monofunctional and bifunctional enzymes so for this reason the *P*-ribosyltransferase function has been omitted from the scheme shown in Figure 4.4.

A limited study of initial velocity and product inhibition patterns for two forms of *S. typhimurium* *P*-ribosyltransferase, anthranilate synthase:*P*-ribosyltransferase or AS II:*P*-ribosyltransferase, suggested a sequential mechanism (48).

Mechanism for Feedback Inhibition

Feedback inhibition of anthranilate synthase by tryptophan is apparently physiologically significant since it has been retained in a wide variety of organisms. The mechanism for inhibition is particularly interesting in the bifunctional anthranilate synthase:*P*-ribosyltransferase; both activities are subject to inhibition. Inhibition of anthranilate synthase results from binding of 1.0 equivalent of tryptophan per AS I with a dissociation constant of 40 μM (27). Inhibition of AS I by tryptophan is competitive with chorismate and noncompetitive with NH_3 (38, 49). As expected there is no cooperativity for tryptophan inhibition of the monomeric AS I and no cooperativity for tryptophan binding. Binding of tryptophan results in a conformation change that decreases the rate of reaction of sulfhydryl groups with DTNB (27). The close correspondence between change in conformation and inhibition suggests that inhibition may result from the conformation change that accompanies binding of tryptophan to a regulatory site. The major effects of aggregation of AS I into the native (AS I)$_2$(AS II:*P*-ribosyltransferase)$_2$ multifunctional enzyme are as follows (28):

1 The K_D for tryptophan binding is reduced from 40 μM to 0.9 μM.
2 There is a corresponding increase in affinity for chorismate binding.
3 The binding of tryptophan acquires cooperativity. The Hill coefficient for tryptophan binding changes from 1.0 to 1.4 indicative of cooperativity in the aggregate. The Hill coefficient for inhibition exhibits a corresponding change from 1.0 to 1.6–1.8.
4 The Hill coefficient for chorismate binding remains 1.0. Equilibrium dialysis experiments with anthranilate synthase:*P*-ribosyltransferase show that binding of chorismate and tryptophan are mutually exclusive (28).

Thus binding of tryptophan to a regulatory site on AS I provokes a conformation change that prevents chorismate from binding to AS I and secondarily prevents formation of the covalent glutaminyl-AS II:*P*-ribosyltransferase intermediate.

The best evidence that the tryptophan regulatory site is distinct from the catalytic site comes from analyses of several *E. coli* anthranilate synthase:*P*-ribosyltransferase feedback-defective mutant strains (47). Strain MT^R with a point mutation in *trpE* (coding for AS I) makes anthranilate synthase:*P*-ribosyltransferase having the normal K_m of 5 μM for chorismate but having a 150-fold reduced affinity for tryptophan. Strain EM54 had a tenfold increased K_m for chorismate, but the enzyme was completely insensitive to tryptophan inhibition, meaning that the K_i for tryptophan was more than 1000-fold higher than in the wild type. In these two cases the regulatory site has clearly been mutationally altered with lesser or no effect on the catalytic site. Pabst et al. reported additional evidence for a distinct regulatory site. Anthranilate synthase activity in strain MT^S revertant AP4 was inhibited by tryptophan strictly noncompetitively with respect to chorismate in contrast to a "pseudocompetitive" type of inhibition in the wild type in which double reciprocal plots were nonlinear due to cooperativity. In this mutant, cooperativity between tryptophan sites was abolished, and simple noncompetitive inhibition indicates that substrate and inhibitor can bind simultaneously to distinct sites. Earlier evidence for desensitization of anthranilate synthase to tryptophan inhibition by Mg^{2+} has been reviewed (1). Desensitization of regulatory proteins has been used to argue for distinct catalytic and regulatory sites.

Several studies (30, 39, 47, 48, 50) have explored the mechanism for inhibition of *P*-ribosyltransferase by tryptophan in bifunctional anthranilate synthase:*P*-ribosyltransferase. Most of the evidence summarized below favors the view that binding of tryptophan to the AS I regulatory site provokes a conformation change that inhibits not only anthranilate synthase but also *P*-ribosyltransferase of the bifunctional enzyme. The kinetics for inhibition are complex and highly unusual; inhibition is most easily interpreted as resulting from changes in conformation. For example, maximal inhibition by tryptophan requires saturating concentrations of both *P*-ribosyltransferase substrates, anthranilate, and *PP*-ribose-*P* (48). Inhibition is partial (30, 39, 47, 48, 50). Maximal inhibition by tryptophan when saturated with substrates is 60–70%. Half maximal inhibition requires approximately 1 μM tryptophan (47, 48), a concentration equivalent to its K_D for binding to aggregated AS I. (AS I)$_2$(AS II:*P*-ribosyltransferase)$_2$ binds only two equivalents of tryptophan, one per AS I subunit (28); binding to *P*-ribosyltransferase sites in the aggregate has not been detected. Chorismate competitively reverses tryptophan inhibition of *P*-ribosyltransferase (47, 48). The kinetic constant for this effect of chorismate is identical to the K_m for chorismate in the anthranilate synthase reaction (47). Thus all the foregoing effects of tryptophan and chorismate on *P*-ribosyltransferase activity are in accord with the mutually exclusive binding of these ligands to the AS I subunit of the aggregate resulting in conformation changes that either inhibit the two activities or reverse inhibition. Analyses of several of the anthranilate synthase feedback defective mutant strains (47), mentioned above, further support this model for inhibition of *P*-ribosyltransferase. Two strains, each with a point mutation in *trpE* (AS I), exhibit concomitantly altered inhibition of anthranilate synthase and *P*-ribosyltransferase. The K_i's for tryptophan inhibition of anthranilate synthase and *P*-ribosyltransferase are increased 150-fold and 90-fold, respectively in the feedback resistant strain MTR. Conversely the K_i's for inhibition of the two reactions are decreased 40- to 50-fold in strain MT^S which is hypersensitive to feedback inhibition. *P*-Ribosyltransferase in strain EMS4 is completely resistant to tryptophan inhibition as is anthranilate synthase.

If binding of tryptophan to the AS I subunit of the aggregate provokes two conformation changes that cause inhibition of catalysis on the AS I and AS II-*P*-ribosyltransferase subunits some treatments might uncouple only one of these tryptophan dependent inhibitory changes. The effect of Mg^{2+} on tryptophan inhibition can be explained according to this view. Mg^{2+} antagonizes tryptophan inhibition of anthranilate synthase (36). At 1 m*M* Mg^{2+}, tryptophan inhibition of anthranilate synthase is reduced by 75% while tryptophan inhibition of *P*-ribosyltransferase is maximal (48). A similar result is obtained by varying pH. At pH 9, 100 μM tryptophan inhibits anthranilate synthase 100% and *P*-ribosyltransferase not at all. At pH 6.5 the inhibition of anthranilate synthase is approximately 40% and *P*-ribosyltransferase approximately 55% (48).

The only possibly discordant observation relating to this mechanism for tryptophan inhibition is the weak inhibition of unaggregated AS II:*P*-ribosyltransferase ($<$ 20%) at very high tryptophan concentrations (40, 48). This inhibition may result from nonspecific binding and represent an *in vitro* artifact.

The *P*-ribosyltransferase of bifunctional AS II:*P*-ribosyltransferase can thus be viewed as a multimeric regulatory protein that contains a catalytic subunit (AS II:*P*-ribosyltransferase) and a regulatory subunit (AS I). The regulatory subunit happens to be a catalytic subunit for the other activity. This is not unlike the chorismate mutase catalytic site serving as a regulatory site for DAHP synthase in the *B. subtilis* multifunctional DAHP synthase:chorismate mutase (51).

Interaction between Sites

Kinetics and ligand binding experiments suggest a number of intra- and intersubunit site-site interactions in anthranilate synthase: *P*-ribosyltransferase. It is sufficient to emphasize interactions between sites on AS I and AS II:*P*-ribosyltransferase and between the two independent catalytic regions of AS II:*P*-ribosyltransferase. There are at least three clear examples for interactions between sites on AS I and sites on AS II:*P*-ribosyltransferase. The chorismate requirement for alkylation of the AS II glutamine site by 6-diazo-5-oxonorleucine or for glutaminase activity (44) indicates that binding of chorismate to AS I promotes a conformation change that influences the reactivity of the AS II glutamine site. Likewise binding of tryptophan to a regulatory site on AS I inhibits *P*-ribosyltransferase activity (40, 47, 48) presumably by way of a conformation change. Finally, in *E. coli* mutant strain MT^{s}, *P*-ribosyltransferase activity is activated by chorismate (47). The concentration dependence for the activation of *P*-ribosyltransferase is similar to the K_m of chorismate for anthranilate synthase. However there is no chorismate-dependent 6-diazo-5-oxo-norleucine alkylation. Thus it appears that in this anthranilate synthase:*P*-ribosyltransferase having a genetically altered AS I, the normal chorismate dependent conformation change is not transmitted from AS I to the AS II glutamine site, but rather some sort of interaction is misdirected from AS I to the *P*-ribosyltransferase site on AS II:*P*-ribosyltransferase.

Pabst et al. (47) have reported indirect evidence for an interaction between the AS II glutamine site and the *P*-ribosyltransferase site. They noted that glutamine competitively reverses tryptophan inhibition of *P*-ribosyltransferase activity of *E. coli* anthranilate synthase:*P*-ribosyltransferase. Since noncovalent binding of glutamine to AS II does not influence tryptophan binding to AS I, bound glutamine may block transmission of the tryptophan-dependent conformation change to *P*-ribosyltransferase. It is thus apparent that even though the AS II and *P*-ribosyltransferase activities are functionally independent (37, 40, 41) they are capable of interaction with each other and with AS I.

Subunit Association

In the bifunctional anthranilate synthase:*P*-ribosyltransferase from *S. typhimurium*, *E. coli*, and *E. aerogenes* the AS I and AS II:*P*-ribosyltransferase subunits are tightly associated and cannot reversibly dissociate *in vivo* or *in vitro*. Unusually stringent conditions are required for separation *in vitro*. Thus far the only successful method for subunit separation on a preparative scale is gel filtration in 30% formic acid. This is in contrast to the weak affinity of AS I and AS II subunits of several of the monofunctional enzymes (Section 4.3).

It appears that the AS II segment of AS II:*P*-ribosyltransferase is instrumental in the assembly and maintenance of stable anthranilate synthase:*P*-

ribosyltransferase tetramers. Crude or purified AS I is a monomeric protein that exhibits no tendency to self-aggregate. The *P*-ribosyltransferase segment of the enzyme is obviously not involved in maintaining the tetrameric structure since a stable tetramer remains when the *P*-ribosyltransferase is removed by trypsin digestion. AS II:*P*-ribosyltransferase, on the other hand, exhibits a propensity to dimerize (40, 48). Grieshaber and Bauerle (40), however, have shown that the stable AS II:*P*-ribosyltransferase dimers that form *in vitro* cannot readily aggregate with AS I and therefore cannot be intermediates in the normal subunit assembly process. However, a brief heat treatment restored the capacity of a small fraction of AS II:*P*-ribosyltransferase dimers to aggregate with AS I. It appears likely that AS II:*P*-ribosyltransferase directs the assembly of anthranilate synthase:*P*-ribosyltransferase tetramers in one of two ways as shown in Figure 4.5. (*a*) AS I and AS II:*P*-ribosyltransferase may interact to form a transient dimer

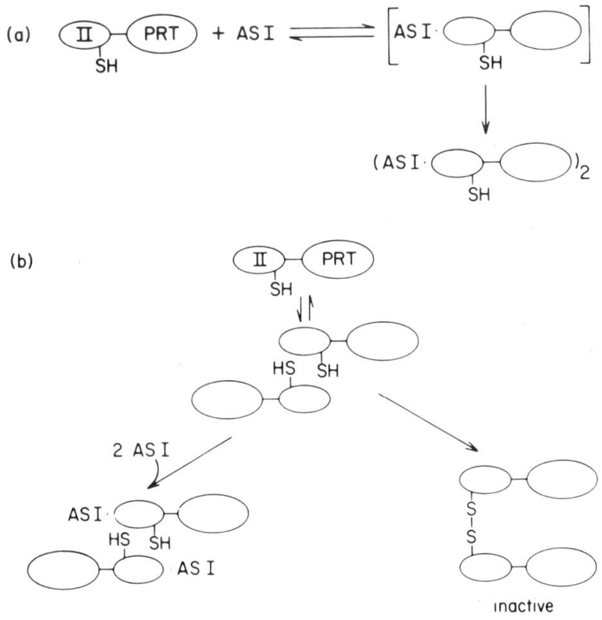

Fig. 4.5 Schemes for subunit assembly of anthranilate synthase: P-ribosyltransferase. The bifunctional AS II: P-ribosyltransferase is designated II-PRT. The AS II active site cysteine is shown as -SH. In scheme (*a*) AS I and AS II: P- ribosyltransferase first associate reversibly, followed by stable tetramer formation. In scheme (*b*) AS II: P-ribosyltransferase first self-associates reversibly followed by interaction with two subunits of AS I to form the tetramer. Disulfide formation yields a covalently linked dimeric form of AS II: P-ribosyltransferase that cannot assemble with AS I although it has P-ribosyltransferase activity. The covalently linked dimeric P-ribosyltransferase is refractory to trypsin digestion.

followed by rapid association of dimers to tetramer. The reversibly associating monofunctional anthranilate synthases (Table 4.1) provide a precedent for the first step of this route. (*b*) AS II:*P*-ribosyltransferase subunits may self-associate to form a noncovalent dimer that then aggregates with two AS I subunits. The formation *in vitro* of an AS II:*P*-ribosyltransferase disulfide linked by the reactive AS II active site cysteines prevents tetramer formation. We propose that the nonfunctional dimer characterized by Grieshaber and Bauerle is largely the disulfide. Holmes and Kane (52) have shown that air oxidation of the essential cysteine in *B. subtilis* AS II prevents interaction with AS I.

The properties of the very similar monofunctional *Serratia marcescens* anthranilate synthase are in accord with these views. The *Serratia* enzyme is tetrameric but is not associated with *P*-ribosyltransferase. The *Serratia P*-ribosyltransferase is monomeric and exhibits no tendency for self-association. Thus association appears to be a property of AS II. In this regard the properties of *P*-ribosyltransferase from *Erwinia carotovora* (53) are of interest. This *P*-ribosyltransferase protein chain has a molecular weight of 40,000 and is thus smaller than the 45,000 dalton *Serratia* enzyme. However, the *Erwinia P*-ribosyltransferase appears to have a tendency to self-associate since the native enzyme has a molecular weight of approximately 67,000. Amino acid sequence homology is obtained for these two proteins using an alignment that leaves a 12 residue NH_2-terminal "tail" for the *Erwinia* enzyme as shown in Figure 4.6. It is thus possible that the information for self-association of *Erwinia P*-ribosyltransferase resides in the 12 residues of the NH_2-terminus. It will be of interest to determine whether or not this "tail" sequence is in the CO_2H-terminus of *Serratia* AS II and is a determinant for assembly of the tetramer.

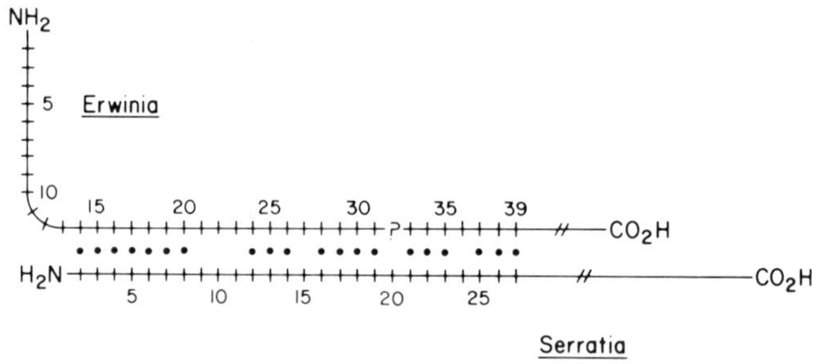

Fig. 4.6 Schematic representation of *Erwinia carotovora* and *Serratia marcescens* P-ribosyltransferases. The first 27 residues of the Serratia sequence are aligned with the *Erwinia* enzyme as shown, leaving a 12 residue "tail" for *Erwinia* P-ribosyltransferase. Dots indicate identical residues. The question mark is an unidentified residue. The *Erwinia* P-ribosyltransferase is approximately 5000 daltons smaller than the *Serratia* enzyme. Data are from Largen *et al.* (52).

4.5 KEY PROPERTIES OF MONOFUNCTIONAL ANTHRANILATE SYNTHASE

As noted in Section 4.3, AS I and AS II subunits of dimeric anthranilate synthases reversibly dissociate *in vitro*. In the case of *B. subtilis* (54) and *Acinetobacter calcoaceticus* (10) facile association-dissociation correlates with the capacity of AS II to function with two NH_3-dependent enzymes: AS I for synthesis of anthranilate and 4-aminobenzoate synthase for synthesis of the folate precursor 4-aminobenzoate. It is uncertain whether these cells contain sufficient AS II to saturate both AS I and 4-aminobenzoate synthase or whether AS II may shuttle between the two aminases. It is also uncertain whether AS II functions in other organisms in these two or in other glutamine-dependent reactions. There are no compelling reasons why the physiological role of reversible association-dissociation must be the catalytic function of AS II with multiple aminases. In fact Kane (55) has presented evidence for a role of association-dissociation in regulating expression of *trpX* the *B. subtilis* gene for AS II. According to this proposal free AS II signals an excess of AS II over AS I plus 4-aminobenzoate synthase and exerts negative control on expression of *trpX*. Alternatively free AS I and 4-aminobenzoate synthase signal a deficiency of AS II and exert positive control on expression of *trpX*.

The reaction mechanism and mechanism for tryptophan inhibition appear to be virtually identical in the monofunctional and bifunctional enzymes. The requirement for a chorismate·AS I·AS II complex for the glutaminase activity is similar to that of the bifunctional anthranilate synthase and indicates that chorismate·AS I is required to form the covalent glutamyl-AS II thioester intermediate (7, 24). Curiously the glutamine analog 2-amino-4-oxo-5-chloropentanoic acid but not 6-diazo-5-oxonorleucine can alkylate free *P. putida* AS II in the absence of chorismate·AS I (24). In this instance the reactivity of the chloroketone affinity analog does not faithfully mimic that of glutamine.

The amino acid sequence of *P. putida* AS II was recently completed (13). This AS II has 197 residues and a calculated molecular weight of 21,684. The key features of the sequence are: (*a*) three cysteine residues at positions 54, 79, and 140 and (*b*) a -Gly-Gly-Arg-Arg(CO_2H) CO_2H-terminus. Alkylation of cysteine-79 with 2-amino-4-oxo-5-chloropentanoic acid or iodoacetamide specifically inactivates the glutamine dependent reaction and implicates this residue in acyl enzyme formation. The sequence of the first 61 residues of *E. coli* AS II:*P*-ribosyltransferase has been reported (23) and is compared with *P. putida* AS II in Figure 4.7. A similarity of 23/61 residues is obtained using the alignment shown. The uniform distribution of similar residues suggests evolution from a common ancestral gene. It will be of interest to compare the sequence similarity around active site Cys-79.

	1 5 10 15
P. putida	Met-Leu- Leu -Met-Ile- Asp-Asn -Tyr- Asp-Ser-Phe-Thr-Tyr-Asn -Val-
E. coli	Ala-Asp- Ile- Leu -Leu-Leu- Asp-Asn -Ile- Asp-Ser-Phe-Thr-Tyr-Asn -Leu-
	5 10 15
	20 25 30
P. putida	Val-Gln-Tyr- Leu -Gly-Glu-Leu- Gly -Ala-Glu- Val -Lys-Val-Ile- Arg
E. coli	Ala-Asp-Gln- Leu -Arg-Ser-Asn- Gly -His-Asn- Val -Val-Ile-Tyr- Arg
	20 25 30
	35 40 45
P. putida	Asn -Asp-Glu-Met-Thr-Ile-Ala-Glu- Ile-Glu -Ala- Leu -Asp-Pro-Glu
E. coli	Asn -His- Ile- Pro- Ala-Gln-Thr-Leu- Ile-Glu -Arg- Leu -Ala-Thr-Met
	35 40 45
	50 55
P. putida	Arg - Ile-Val- Val - - - - - - - - - Ser-Pro-Gly-Pro -Cys-Thr- Pro -Ser
E. coli	Ser-Asn-Pro- Val -Leu-Met-Leu - Ser-Pro-Gly-Pro -Gly-Val- Pro -Ala
	50 55 60

Fig. 4.7 Amino acid sequence comparison of NH_2-terminal segments of *P. putida* AS II and *E. coli* AS II: P-ribosyltransferase. Dashed lines indicate a three residue deletion used to maximize similarity.

4.6 GENERAL CONSIDERATIONS CONCERNING POSSIBLE PHYSIOLOGICAL FUNCTIONS OF BIFUNCTIONAL ENZYMES

There is no evidence that bears directly on the question why some organisms utilize a bifunctional anthranilate synthase:*P*-ribosyltransferase and other organisms contain a monofunctional enzyme. There are, however, several observations and speculations that may be pertinent. First, a bifunctional enzyme may not exhibit a unique function. For example, mutations causing fusion of continguous genes may persist if not detrimental. Alternatively a bifunctional enzyme may represent an intermediate stage in an evolutionary pathway toward a system of unique function. On the other hand, there is evidence for special functions of multifunctional proteins. Gaertner and coworkers (56) have shown that enhanced catalytic efficiency can result from coordinate activation of all five activities of a multifunctional aromatic amino acid biosynthetic protein by the first substrate. Product formation by this multifunctional protein also exhibits a shorter lag than that expected for the sum of the individual reactions (57). In several cases multifunctional proteins ensure a small pool of an enzyme-bound metabolite and promote "channeling" (58–60). Other plausible roles of multifunctional proteins include: (*a*) a mechanism to coordinate the synthesis of multiple functions; (*b*) a shield for labile intermediates from water and a more rapid utilization. There is no evidence to suggest that any of these considerations apply to *S. typhimurium* or *E. coli* anthranilate synthase:*P*-ribosyltransferase. It will be of interest to compare the catalytic properties of *S. marcescens* *P*-ribosyltransferase with *E. coli* AS II:*P*-ribosyltransferase. The possibility remains open that all organisms having a monofunctional anthranilate synthase utilize AS II for synthesis of anthranilate and in other glutamine-dependent reactions.

4.7 RELATIONSHIP OF BIFUNCTIONAL ANTHRANILATE SYNTHASE: *P*-RIBOSYLTRANSFERASE TO MONOFUNCTIONAL ANTHRANILATE SYNTHASE

Divergent Evolution

Protein sequencing (23) and immunological cross-reactivity (61, 62) suggest structural similarity between bifunctional and monofunctional anthranilate synthases. The data in Table 4.3 summarize the number of identical residues in NH_2-terminal segments of AS I and AS II in monofunctional and bifunctional enzymes. These NH_2-terminal sequences do not include the active site cysteine. The data show that the NH_2-termini of *S. typhimurium* and *E. coli* AS I and AS II: *P*-ribosyltransferase are virtually identical. There is also 73–77% identity between NH_2-termini of *S. typhimurium* or *E. coli* AS

Table 4.3 Similarity in Amino Acid Sequences of Amino-Terminal Segments of Anthranilate Synthases from Some Bacteria[a]

	Number of Identical Residues	
Organisms Compared	AS I	AS II or AS II:*P*-ribosyltransferase
E. coli/S. typhimurium	21/25[b]	48/51
E. coli/S. marcescens	14/25	46/60
S. typhimurium/S. marcescens	14/25	37/51
E. coli/P. putida		23/61
S. marcescens/P. putida		23/60

[a] The table compares the number of identical residues in NH_2-terminal segments of AS I, AS II, and AS II:*P*-ribosyltransferase from various organisms. The data for *E. coli, S. typhimurium* and *S. marcescens* are from Li et al. (23). Data for *P. putida* are from Kawamura et al. (13).

[b] Designation gives identical residues, 21, to the total number compared, 25.

II: *P*-ribosyltransferase and *S. marcescens* AS II. These results thus support the divergent evolution of *S. marcescens* monofunctional AS II and AS II of bifunctional AS II: *P*-ribosyltransferase from a common ancestral gene. There is less sequence similarity of *P. putida* with *S. marscens* and *E. coli* AS II NH_2-termini. Yet the sequence similarities of monofunctional AS II from *P. putida* and *S. marcescens* (line 5) with the bifunctional AS II: *P*-ribosyltransferase from *E. coli* (lines 2, 4) are fairly comparable. It thus appears that the *P. putida* AS II has also evolved from the putative AS II ancestoral gene. A major conclusion is that there are no distinctive features in primary structure to distinguish monofunctional and bifunctional anthranilate synthases.

Mechanisms Relating Monofunctional and Bifunctional Anthranilate Synthases

In principle bifunctional proteins could have evolved from two monofunctional proteins or conversely monofunctional enzymes might arise from processing of a precursor bifunctional protein. The former possibility was favored by Bonner et al. (63) who proposed that 2 genes-2 polypeptides preceded 1 gene-1 enzyme in the evolution of tryptophan synthase and is supported by experimental demonstrations of gene fusion to yield multifunctional proteins (64–66). Fusion of contiguous *trpG* (AS II) and *trpD* (*P*-ribosyltransferase) genes is a plausible route for evolution of bifunctional AS II-*P*-ribosyltransferase in *E. coli*, *S. typhimurium*, and *E. aerogenes*.

We have considered several possibilities to account for the persistence of monofunctional anthranilate synthases. As noted in Section 4.5, function of the AS II subunit in multiple reactions as in *B. subtilis* (54) and *A. calcoaceticus* (10) requires a monofunctional AS II and thus a monofunctional anthranilate synthase. Two other possibilities have been considered. Recently completed DNA sequencing experiments indicate that *trpG* is contiguous with *trpD* and other *trp* operon genes in *S. marcescens* (45). Coordinate regulation suggests the same organization for species of Pseudomonas and Acinetobacter (67). In these organisms translation of polycistronic mRNA would yield monofunctional *trpG* and *trpD* proteins, AS II and *P*-ribosyltransferase, respectively. Such a biosynthetic system might represent the evolutionary precursor to fused *trpGD*, AS II-*P*:ribosyltransferase as in *E. coli*. A second more speculative possibility for monofunctional anthranilate synthase is that contiguous *trpG* and *trpD* genes are fused in *P. putida* and *A. calcoaceticus* and in *E. coli*, and the primary translation product is bifunctional AS II-*P*-ribosyltransferase. The bifunctional AS II-*P*-ribosyltransferase might be processed in *P. putida* and *A. calcoaceticus* to yield the two monofunctional proteins. This idea develops as a result of the Gly-Gly-Arg-Arg CO_2H-terminus of *P. putida*

AS II (Section 4.5). In eucaryotic cells processing of proproteins to functional proteins occurs by proteolytic cleavage at sites containing dibasic residues (68, 69). J. E. Dixon (personal communications) has isolated from rat liver and partially purified a protease with specificity for tandem dibasic residues but not a single lysine or arginine. This enzyme thus may function to process proproteins in rat liver. Attempts to detect a proteolytic enzyme in *P. putida* having specificity for tandem arginine residues have been unsuccessful. It is of interest that *E. coli* protease II selectively cleaves bifunctional aspartokinase I-homoserine dehydrogenase I into the two active monofunctional enzymes (70). The rat liver protease of Dixon does not cleave purified *E. coli* AS II:*P*-ribosyltransferase. The possibility of proteolytic processing of a hypothetical AS II:*P*-ribosyltransferase protein in *S. marcescens* is clearly excluded by the DNA sequence of the *trpG-trpD* intercistronic region (45). These two genes are separated by a nontranslated 19 nucleotide sequence.

A comparison of the DNA sequence of the *E. coli trpG-D* connecting region and the *S. marcescens trpG-trpD* intercistronic region has suggested to Miozzari and Yanofsky (45) a possible mechanism for gene fusion in this case. In *S. marcescens* the CO_2H-terminal AS II sequence is -Ala-Trp-Ala-Leu-Ala-Lys(CO_2H). Following the codon for Lys is a TAA stop codon that terminates translation of AS II. Sixteen nucleotides intervene between the TAA stop codon and the ATG initiator for *P*-ribosyltransferase. In *E. coli* the TAA translation stop is replaced by a CTA codon that allows translation to continue and give a hexapeptide "connector or hinge" that fuses AS II to *P*-ribosyltransferase. The extensive nucleotide and amino acid sequence homology of the CO_2H-terminus of AS II and the NH_2-terminus of *P*-ribosyltransferase from *S. marcescens* with the corresponding regions on each side of the connector region in *E. coli* clearly emphasizes the relatedness of the two genes for the monofunctional enzymes with the single gene for the bifunctional enzyme and suggests a common origin. Twenty of 23 amino acids in the *trpG* enzyme CO_2H-terminus and 16 of 22 amino acids at the start of *trpD* protein are identical. A hypothetical sequence of four steps was suggested to illustrate how the fusion might have occurred in the common ancestor that preceded *S. marcescens* and *E. coli*.

4.8 RELATIONSHIP OF ANTHRANILATE SYNTHASE TO OTHER GLUTAMINE AMIDOTRANSFERASES

Mechanism of Glutamine Utilization

Anthranilate synthase is one enzyme of a family of glutamine amidotransferases. It is therefore of interest to compare anthranilate synthase to other glutamine amidotransferases. Carbamyl-P synthase is the only other glutamine amidotransferase that is associated with other enzymes of a pathway and is multifunctional in eukaryotic organisms (71–73). On the other hand, anthranilate synthase and all other glutamine amidotransferases exhibit common properties, which suggests a similar mechanism for glutamine utilization. These properties include: (*a*) activity with NH_3 in addition to glutamine; (*b*) alkylation of a cysteine residue specifically inactivates the glutamine-dependent activity with lesser or no effect on NH_3-dependent activity; (*c*) glutaminase activity; (*d*) in some cases a covalent γ-glutamyl acyl enzyme intermediate has been isolated. Overall, these results suggest a structure-function relationship similar to that shown in Figure 4.4 (bottom) for anthranilate synthase. Important features are the distinct domains for amide transfer (glutaminase) and the reaction of NH_3.

Quaternary Structure

Glutamine amidotransferases exhibit two types of subunit structures: (*a*) monomeric or identical subunits and (*b*) nonidentical subunits. The subunit structures are therefore analogous to the anthranilate synthases from bacteria and *Euglena*. Carbamyl-*P* synthetase from *E. coli* (73) and *S. typhimurium* (75) contains nonidentical protein chains and a structure-function relationship identical to monofunctional AS I·AS II. *E. coli* CTP synthetase (76), GMP synthetase (77), amidophosphoribosyl transferase (L. J. Messenger and H. Zalkin, unpublished), and FGAR amidotransferase (78) contain glutamine and NH_3 sites on the same protein chain similar to *Euglena* anthranilate synthase (3). These latter proteins may be viewed as bifunctional: they contain two independent functions on a single protein chain: (*a*) NH_3-dependent amination, (*b*) glutamine hydrolysis and amide transfer. We have, however, not been able to disect by proteolysis the glutamine hydrolysis-amide transfer function from the NH_3-dependent activity in GMP synthetase and amidophosphoribosyltransferase.

Search for Structural Similarity

We have considered the possibility that similarities in the mechanism for glutamine utilization might result from structural similarity in a glutamine binding amide transfer domain. The initial results of protein sequencing and immunological experiments are inconclusive and do not yet support this hypothesis. The main results are as follows. The amino acid sequence of a tryptic peptide containing the essential cysteine residue of *E. coli* GMP synthetase has recently been completed (C. D. Truitt, M. A. Hermodson, and H. Zalkin, unpublished). There is essentially no sequence similarity between this peptide and the region of *P. putida* AS II containing the essential cysteine residue. A heptapeptide sequence containing two cysteine residues essential for glutamine reactivity of chicken liver formylglycinamide ribonucleotide amidotransferase (79) also bears no similarity to the GMP synthetase sequence but has a Gly-Val-Cys that is homologous with residues 78–80 at the active site of *P. putida* AS II. Unfortunately *E. coli* GMP synthetase is an exceptional amidotransferase in which true affinity labeling does not occur with 6-diazo-5-oxonorleucine and 2-amino-4-oxo-5-chloropentanoic acid (80). Alkylation of GMP synthetase by these analogs is a second order reaction and is not preceded by a reversible enzyme·inhibitor complex. This suggests a somewhat different glutamine binding site in GMP synthetase compared to other amidotransferases.

A sensitive radioimmune assay was used to search for immunological cross-reactivity between *E. coli* antianthranilate synthase or anti-AS II with proteins in an *E. coli* strain having no anthranilate synthase due to a deletion of all *trp* operon structural genes (62). The maximal extent of binding of extract proteins to anti-AS II was similar to the binding of AS II although the affinity for cross-reacting proteins was several orders of magnitude lower. This was interpreted as evidence for AS II-like determinants (structure) in other proteins. Glutamate synthase was identified as one of the cross-reacting proteins but several other glutamine amidotransferases, including amidophosphoribosyltransferase, carbamyl-*P* synthetase, CTP synthetase, formylglycinamide ribonucleotide amido-transferase, and GMP synthetase did not cross-react. It is thus apparent that further evidence is needed to evaluate the hypothesis of a common glutamine binding amide transfer domain in glutamine amidotransferases.

ACKNOWLEDGMENT

The author's research is supported by U.S. Public Health Service Grant GM 24658 from the National Institute of General Medical Sciences.

REFERENCES

1 H. Zalkin (1973), in A. Meister, ed., *Advances in Enzymology*, Vol. 38, John Wiley and Sons, New York, pp. 1–39.

2 F. Lingens, W. Goebel, and H. Vesseler (1967), *Eur. J. Biochem.* **2,** 442–447.

3 C. N. Hankins and S. E. Mills (1976), *J. Biol. Chem.* **251,** 7774–7778.

4 J. Ito and C. Yanofsky (1966), *J. Biol. Chem.* **241,** 4112–4114.

5 M. Largen and W. L. Belser (1975), *J. Bacteriol.* **121,** 239–249.

6 S. F. Queener and I. C. Gunsalus (1970), *Proc. Natl. Acad. Sci. USA* **67,** 1225–1232.

7 H. Zalkin and L. H. Hwang (1971), *J. Biol. Chem.* **246,** 6899–6907.

8 F. Robb, M. A. Hutchinson, and W. L. Belser (1971), *J. Biol. Chem.* **246,** 6908–6912.

9 M. Largen, S. E. Mills, J. Rowe, and C. Yanofsky (1976), *Eur. J. Biochem.* **67,** 31–36.

10 R. V. Sawula and I. P. Crawford (1973), *J. Biol. Chem.* **248,** 3573–3581.

11 R. Twarog and G. L. Liggins (1970), *J. Bacteriol.* **104,** 254–263.

12 S. W. Queener, S. F. Queener, J. R. Meeks, and I. C. Gunsalus (1973), *J. Biol. Chem.* **248,** 151–161.

13 M. Kawamura, P. S. Keim, Y. Goto, H. Zalkin, and R. L. Heinrikson (1978), *J. Biol. Chem.* **253,** 4659–4668.

14 J. F. Kane and R. A. Jensen (1970), *Biochem. Giophys. Res. Commun.* **41,** 328–333.

15 N. Patel, W. M. Holmes and J. F. Kane (1974), *J. Bacteriol.* **119,** 220–227.

16 S. O. Hoch and I. P. Crawford (1973), *J. Bacteriol.* **116,** 685–693.

17 A. Catena and R. D. DeMoss (1970), *J. Bacteriol.* **101,** 476–482.

18 J. Wegman and I. P. Crawford (1968), *J. Bacteriol.* **95,** 2325–2355.

19 A. R. Procter and W. E. Kloos (1973), *J. Bacteriol.* **114,** 169–177.

20 E. N. Baskerville and R. Twarog (1972), *J. Bacteriol.* **112,** 304–314.

21 E. Baskerville and R. Twarog (1974), *J. Bacteriol.* **117,** 1184–1194.

22 C. N. Hankins and S. E. Mills (1977), *J. Biol. Chem.* **252,** 235–239.

23 S-L. Li, J. Hanlon, and C. Yanofsky (1974), *Biochemistry* **13,** 1736–1744.

24 Y. Goto, H. Zalkin, P. S. Keim, and R. L. Heinrikson (1976), *J. Biol. Chem.* **251,** 941–949.

25 J. F. Kane, W. M. Holmes, K. L. Smiley, and R. A. Jensen (1973), *J. Bacteriol.* **113,** 224–232.

26 H. Zalkin and S. H. Chen (1972), *J. Biol. Chem.* **247,** 5996–6003.

27 H. Nagano and H. Zalkin (1970), *J. Biol. Chem.* **245,** 3097–3103.

28 E. J. Henderson and H. Zalkin (1971), *J. Biol. Chem.* **246,** 6891–6898.

29 A. F. Egan and F. Gibson (1972), *Biochem. J.* **130,** 847–859.

30 J. Ito and C. Yanofsky (1969), *J. Bacteriol.* **97,** 734–742.

31 R. Hütter and J. A. DeMoss (1967), *J. Bacteriol.* **94,** 1896–1907.

32 F. M. Hulett and J. A. DeMoss (1975), *J. Biol. Chem.* **250,** 6648–6652.

33 A. Arroyo-Begovich and J. A. DeMoss (1973), *J. Biol. Chem.* **248,** 1262–1267.

34 D. F. Mann and H. G. Floss (1977), *Lloydia* **40,** 136–145.

35 C. H. Doy and J. M. Cooper (1966), *Biochim. Biophys. Acta* **127,** 302–316.

36 E. J. Henderson, H. Nagano, H. Zalkin, and L. H. Hwang (1970), *J. Biol. Chem.* **245,** 1416–1423.

37 L. H. Hwang and H. Zalkin (1971), *J. Biol. Chem.* **246,** 2338–2345.

38 J. Ito, E. C. Cox, and C. Yanofsky (1969), *J. Bacteriol.* **97,** 725–734.

39 M. Grieshaber and R. Bauerle (1972), *Nat. New Biol.* **236,** 232–235.

40 M. Grieshaber and R. Bauerle (1974), *Biochemistry* **13,** 373–383.

41 E. N. Jackson and C. Yanofsky (1974), *J. Bacteriol.* **117,** 502–508.

42 C. Yanofsky, V. Horn, M. Bonner, and S. Stasiowski (1971), *Genetics* **69,** 409–433.

43 H. Zalkin (1973), in S. Prusiner and E. R. Stadtman, eds., *The Enzymes of Glutamine Metabolism*, Academic Press, New York, pp. 523–543.

44 H. Nagano, H. Zalkin, and E. J. Henderson (1970), *J. Biol. Chem.* **245,** 3810–3820.

45 G. F. Miozzari and C. Yanofsky (1979), *Nature,* **277,** 486–489.

46 R. L. Somerville and R. Elford (1967), *Biochem. Biophys. Res. Commun.* **28,** 437–444.

47 M. J. Pabst, J. C. Kuhn, and R. L. Somerville (1973), *J. Biol. Chem.* **248,** 901–914.

48 E. J. Henderson, H. Zalkin, and L. H. Hwang (1970), *J. Biol. Chem.* **245,** 1424–1431.

49 H. Zalkin and D. Kling (1968), *Biochemistry* **7,** 3566–3573.

50 Bauerle, R. H. and P. Margolin (1966), *Cold Spring Harbor Symp. Quant. Biol.* **31,** 203–214.

51 L. Huang, A. L. Montoya, and E. W. Nester (1974), *J. Biol. Chem.* **249,** 4473–4479.

52 W. M. Holmes and J. F. Kane (1975), *J. Biol. Chem.* **250,** 4462–4469.

53 M. Largen, S. E. Mills, J. Rowe, and C. Yanofsky (1978), *J. Biol. Chem.* **253,** 409–412.

54 J. F. Kane, W. M. Holmes, and R. A. Jensen (1972), *J. Biol. Chem.* **247,** 1587–1596.

55 J. F. Kane (1977), *J. Bacteriol.* **132,** 419–425.

56 G. R. Welch and F. H. Gaertner (1976), *Arch. Biochem. Biophys.* **172,** 476–489.

57 G. R. Welch and F. H. Gaertner (1975), *Proc. Natl. Acad. Sci. USA* **72,** 4218–4222.

58 G. L. E. Koch, D. C. Shaw, and F. Gibson (1972), *Biochim. Biophys. Acta* **258,** 719–730.

59 P. F. Leu and J. G. Kaplan (1970), *Biochim. Biophys. Acta* **220,** 365–372.

60 R. L. Cafferata and M. Freundlich (1970), *Biochim. Biophys. Acta* **222,** 671–674.

61 J. J. Reiners and H. Zalkin (1975), *J. Bacteriol.* **123,** 620–630.

62 J. J. Reiners, L. J. Messenger, and H. Zalkin (1978), *J. Biol. Chem.* **253,** 1226–1233.

63 D. M. Bonner, J. A. DeMoss, and S. E. Mills (1965), In V. Bryson and H. J. Vogel, eds., *Evolving Genes and Proteins*, Academic Press, New York, pp. 305–318.

64 J. Yourno, T. Kohno, and J. R. Roth (1970), *Nature* **228,** 820–824.

65 I. Ino, P. E. Hartman, Z. Hartman, and J. Yourno (1975), *J. Bacteriol.* **123,** 1254–1264.

66 T. J. Silhavy, H. A. Shuman, J. Beckwith, and M. Schwartz (1977), *Proc. Natl. Acad. Sci. USA* **74,** 5411–5415.

67 I. P. Crawford (1975), *Bacteriol. Rev.* **39,** 87–120.

68 D. F. Steiner, W. Kemmter, H. S. Tager, and J. D. Peterson (1974), *Fed. Proc.* **33,** 2105–2115.

69 J. F. Habener, H. T. Chang, and J. R. Potts (1977), *Biochemistry* **16,** 3910–3917.

70 M. Pacqud and C. Richaud (1975), *J. Biol. Chem.* **250,** 7771–7779.

71 P. F. Leu and J. G. Kaplan (1970), *Can. J. Biochem.* **48,** 155–159.

72 L. G. Williams, S. Bernhardt, and R. H. Davis (1970), *Biochemistry* **9,** 4329–4335.

73 P. F. Coleman, D. P. Suttle, and G. R. Stark (1977), *J. Biol. Chem.* **252,** 6379–6385.

74 P. P. Trotta, M. E. Burt, R. H. Haschemeger, and A. Meister (1971), *Proc. Natl. Acad. Sci. USA* **68,** 2599–2603.

75 A. T. H. Abdelal and J. L. Ingraham (1975), *J. Biol. Chem.* **250,** 4410–4417.

76 C. W. Long, A. Levitzki, and D. E. Koshland, Jr. (1970), *J. Biol. Chem.* **254,** 80–87.

77 N. Sakamoto, G. W. Hatfield, and H. S. Moyed (1972), *J. Biol. Chem.* **247,** 5880–5887.

78 J.-M. Frére, D. D. Schroeder, and J. M. Buchanan (1971), *J. Biol. Chem.* **246,** 4727–4730.

79 S. Ohnoki, B.-S. Hong, and J. M. Buchanan (1977), *Biochemistry* **16,** 1070–1076.

80 H. Zalkin and C. D. Truitt (1977), *J. Biol. Chem.* **252,** 5431–5436.

Chapter 5

GENE FUSIONS IN THE TRYPTOPHAN PATHWAY: TRYPTOPHAN SYNTHASE AND PHOSPHORIBOSYL-ANTHRANILATE ISOMERASE: INDOLEGLYCEROL-PHOSPHATE SYNTHASE

IRVING P. CRAWFORD

Department of Microbiology
University of Iowa, Iowa City

5.1 INTRODUCTION

Reactions of the Tryptophan Pathway

The pathway of synthesis of tryptophan from chorismic acid, the last common intermediate in the synthesis of aromatic amino acids and vitamins, is uniform in all organisms studied. It is presented in Figure 5.1. Gene fusions and multienzyme complexes involving the early enzymes of the pathway, anthranilate synthase (AS) and anthranilate phosphoribosyltransferase (PRT), are described in Chapter 4. Here the concern will be with multifunctional proteins and multienzyme complexes catalyzing the third and fourth reactions, phosphoribosyl-anthranilate isomerase (PRAI) and indoleglycerol-phosphate synthase (InGPS), and the more complex final reaction, tryptophan synthase (TS).

The third and fourth reactions are chemically uncomplicated. In most organisms they are catalyzed by separate polypeptides in the 25,000–35,000 mol wt range (1). Enteric bacteria are unique among procaryotes in having a single, bifunctional polypeptide that catalyzes both reactions (2).

The conversion of indoleglycerol phosphate to tryptophan is a more complicated process, best considered in two parts. The TS-A portion is the reversible aldolytic cleavage of indoleglycerol-phosphate to indole and D-glyceraldehyde-3-phosphate; the TS-B part is the pyridoxal-5′-phosphate (PLP) dependent addition of the alanyl portion of L-serine to indole to form L-tryptophan (3). In all organisms studied to date this two-part reaction is catalyzed by a complex enzyme containing more than one subunit. Usually there are two dissimilar polypeptide chains present, a smaller α chain concerned primarily with the TS-A reaction and a larger, PLP-containing β chain concerned primarily with the TS-B reaction. Often these two subunits associate noncovalently in an $\alpha_2\beta_2$ multimer which can dissociate into free α and β_2 portions having greatly diminished TS-A and TS-B activities, respectively (3, 4). Quite early (5) it was noted that the TS enzyme in ascomycetous fungi is different in nature from the bacterial and plant forms. Section 5.3 describes recent evidence that in many fungi the α- and β-chains have been fused into a single multifunctional polypeptide chain.

Although the reactions of the tryptophan pathway shown in Figure 5.1 are conserved, it is quite apparent that different organisms employ quite dif-

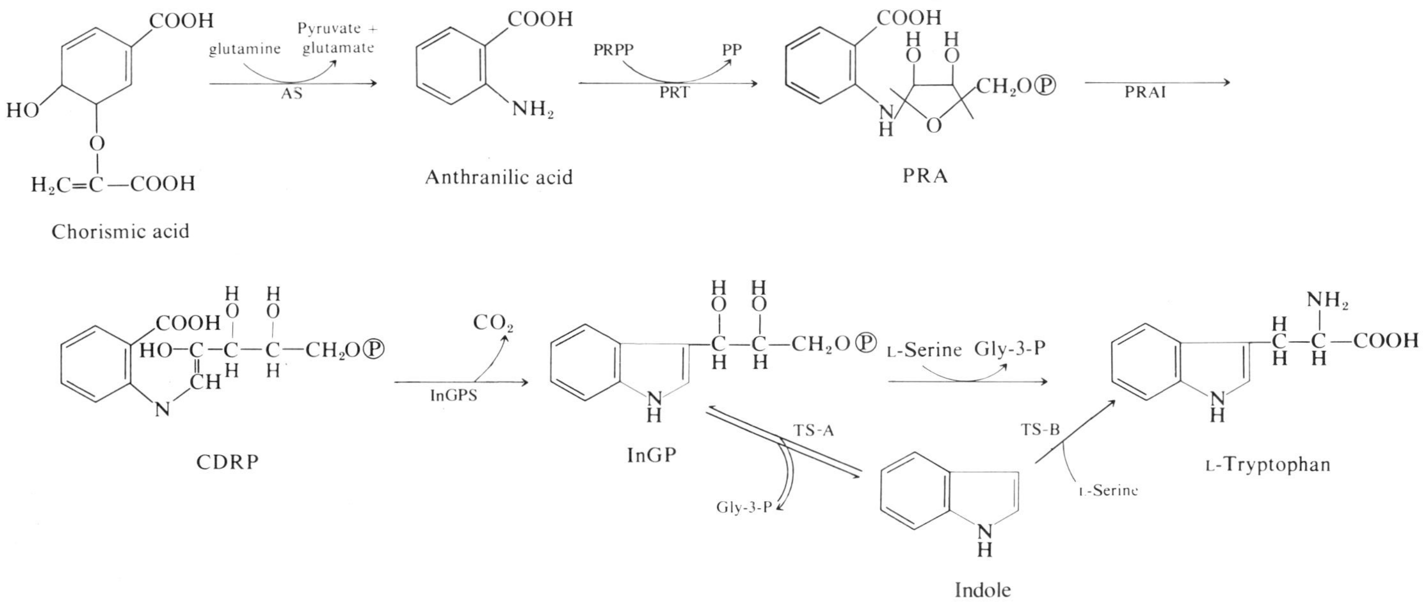

Fig. 5.1 Tryptophan biosynthetic pathway. Chorismic acid is the last intermediate common to the synthesis of aromatic amino acids and vitamins. Abbreviations: AS, anthranilate synthase; PRPP, 5-phosphoribosyl-pyrophosphate; PP, pyrophosphate; PRT, anthranilate phosphoribosyltransferase; PRA, *N*-phosphoribosyl-anthranilate; PRAI, phosphoribosyl-anthranilate isomerase; CDRP, 1-(o-carboxyphenylamino)-1-deoxyribulose-phosphate; InGP, indoleglycerol-phosphate; InGPS, indoleglycerol-phosphate synthase; Gly-3-P, D-glyceraldehyde-3-phosphate; TS, tryptophan synthase; TS-A and TS-B, the A and B half-reactions of tryptophan synthase.

ferent catalytic machinery to accomplish them. Examples of discrete, autonomous enzymes, multienzyme complexes, and multifunctional peptides are all available in this relatively short, unbranched pathway. Recently it was shown (6, 7) that the bacterial TS β chain catalyzing the ultimate step in the pathway, the TS-B reaction, itself exhibits two discrete domains that may be functionally as well as structurally specialized. The various genetic and biochemical tools used to uncover the unexpected variety in the apparatus used by different cells to execute the reactions of the tryptophan pathway are emphasized in the discussions that follow, for it is quite likely that when these tools are applied to the study of additional organisms we will learn that our present understanding of the variety of enzymes forms employed for tryptophan synthesis is still incomplete. In fact, we may presently recognize only a few tips of a complex iceberg.

Comparative Studies of the Tryptophan Genes and Enzymes

The apparatus for synthesizing tryptophan has probably been studied in more different organisms, both procaryotic and eucaryotic, than that of any other biosynthetic pathway. These comparative studies have been reviewed recently (1, 8), so they are not given in full detail here. An important lesson to be drawn from them is the importance of performing both genetic and biochemical investigations. When a distinction must be made between intimate association of products of separate genes in a multienzyme complex and posttranslational proteolysis of a single, multifunctional gene product, careful genetic analysis is invaluable. In a number of instances this analysis has been done on complexes in the tryptophan pathway with conclusive results.

The chromosomal distribution of the *trp* genes in some of the better studied bacterial groups is shown in Figure 5.2. Only in the case of the enteric bacteria are all the *trp* genes found in a single operon. Although enteric bacteria differ in the presence or absence of a fusion of *trpG* and *trpD*, as described in Chapter 4, all that have been studied show a fusion of *trpC*, the gene for InGPS, and *trpF*, the gene for PRAI. If the enteric group is defined broadly to include aeromonads and the marine luminous bacteria [there are sound taxonomic grounds for doing this (9)], then it can be stated that no other bacterial taxon examined to date shows an analogous concentration of all the genes for the tryptophan pathway in one operon, and none exhibits a fusion of the *trpC* and *trpF* genes. Interestingly enough, this *trpC-trpF* fusion reappears in the eucaryotic kingdom in the filamentous fungi (10, 11, 12), where it is accompanied by a fusion of the *trpA* and *trpB* genes. Yeasts, however, appear to have the fusion of the TS genes without the PRAI-InGPS fusion (13, 14). This is an example of the value of

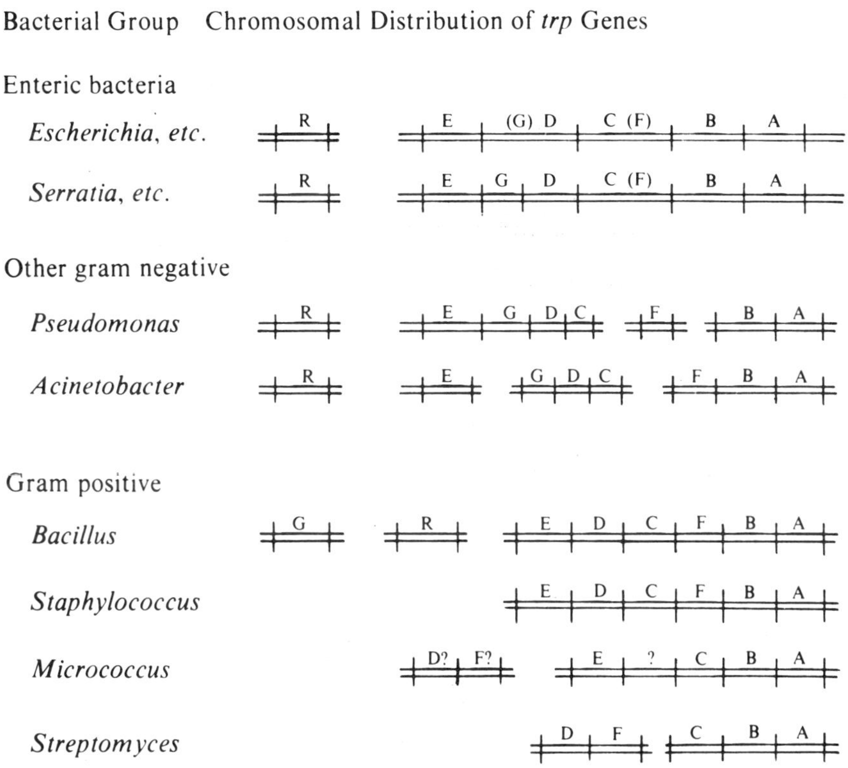

Fig. 5.2 Tryptophan gene distributions in bacteria. Analogous genes have the same letter designation throughout. The *trp* prefix for all gene symbols has been omitted for simplicity. Gene-enzyme key: *trpA*, TS α-chain; *trpB*, TS β-chain; *trpC*, InGPS; *trpD*, PRT; *trpE*, AS large subunit; *trpG*, AS small, glutamine amidotransferase subunit; *trpR*, tryptophan repressor. Gene fusions in the enteric bacteria are shown with functional components in the correct order; the component in parenthesis is not normally used in the gene designation. In *Acinetobacter* the gene order in the three gene clusters is not certain; the most probable order is shown. A question mark following a gene designation denotes doubt about its true location.

genetic analysis, for in the case of the filamentous fungi it has been shown that mutations occurring in specific regions within a single structural gene affect one or the other of the two catalytic activities of the polypeptide, or its ability to complex with the AS gene product and contribute to that activity (13, 15, 16). Genetic results obtained with the tryptophan loci of the better studied fungi are summarized in Table 5.1. Certain phycomycetes have been studied biochemically with evidence of less extensive aggregation of the enzymes (11), but in the absence of genetic analysis it cannot be decided unequivocally whether or not in their case multifunctional polypeptides are produced initially, then dissected by proteolysis.

Table 5.1 Tryptophan Gene-Enzyme Relationships in Fungi

Organisms	Enzyme				
	AS	InGPS	PRAI	PRT	TS
Neurospora crassa	tryp-2	tryp-1[a]		tryp-4	tryp-3
Linkage group[b]	VI	III		IV	II
Aspergillus nidulans	trypA	trypC[a]		trypD	trypB
Linkage group[b]	II	VIII		II	I
Coprinus lagopus	trp-3	trp-2[a]		trp-4	trp-1
Linkage group[b]	III	G		—[c]	III
Saccharomyces cerevisiae	trp-2	trp-3	trp-1	trp-4	trp-5
Linkage group[b]	V	XI	IV	IV	VII

[a] These genes encode a single polypeptide that has catalytic activity in both the PRAI and InGPS reactions. It also aggregates with and activates the AS polypeptide.

[b] Linkage group designations are not comparable between organisms. Data for *Neurospora*, *Aspergillus*, and *Saccharomyces* were compiled earlier (1). *Coprinus* results are from Tilby (12).

[c] Not linked to the other *trp* genes or to three additional linkage groups (12).

To return to the bacterial situations diagrammed in Figure 5.2, it is clear that in fluorescent pseudomonads and acinetobacters *trpC* and *trpF* are independent, unlinked genes. Other bacteria with separate *trpC* and *trpF* genes include *Pseudomonas acidovorans* (17), whose chromosomal *trp* gene distribution resembles acinetobacters more than fluorescent pseudomonads, *Rhizobium meliloti* (18) and *Streptomyces coelicolor* (19). In two gram-positive species, *Bacillus subtilis* and *Staphylococcus aureus*, the *trpC* and *trpF* genes are adjacent to each other in a cluster that includes most of the genes of the pathway. From position alone it cannot be determined in these cases whether the occurrence of PRAI and InGPS on separate polypeptides accurately reflects the genetic situation or results from proteolytic cleavage of a fused gene product. This question is considered again in Section 5.2.

Following early observations on comparative aspects of the tryptophan pathway (5) it was hypothesized that organisms higher in the phylogenetic scheme might exhibit more gene fusions and multifunctional polypeptides than more primitive ones. Although one eucaryote that has been studied biochemically, *Euglena gracilis*, conforms to the hypothesis by having all its tryptophan synthetic activities except AS catalyzed by a single polypeptide of about 155,000 mol wt (20), all higher plants studied so far have independent PRT, PRAI, InGPS, and TS enzymes (21). In the absence of genetic evidence it cannot be claimed that these biochemical results with plant enzymes rule out the possibility of fused genes counteracted by proteolytic cleavage, but the plain fact is that except for the usual association of the α and β chains of TS in a multienzyme complex (22) and some evidence for dimerization of InGPS in peas (21), there is no evidence of aggregate formation of any kind among the tryptophan enzymes of the higher plants investigated.

Regulation of the Tryptophan Genes

Considerable attention has been directed toward studies of the regulation of the rate of synthesis of tryptophan and the control of *trp* gene activity in microbes. Again a wide variety of patterns has emerged, from fixed levels of enzyme production to wide-ranging and complex controls of the amount of enzyme produced from the *trp* genes, which are not always regulated *en bloc* (1, 8). From the standpoint of multifunctional enzymes, however, an analysis of regulatory responses becomes important only if it reveals an underlying gene structure or clarifies an interaction between domains of a complex enzyme.

Regulation of the flow of metabolites down the tryptophan pathway normally involves feedback inhibition by the end product of the first enzyme, AS. Though this assumes importance in the study of enzyme aggregates involving AS (see Chapter 4), it is not pertinent otherwise. The use of inhibitors other than tryptophan to study the interdependence of catalytic centers will be mentioned later in connection with particular enzyme aggregates.

At first glance another form of regulation, repression or induction of gene expression, would seem more useful. Genes that are physically separate on the chromosome often show regulatory responses that are different in amount or in kind (1). Unfortunately, this is not an invariant rule. Some separate bacterial genes are regulated synchronously, and some neighboring genes show independent regulation. Therefore, regulatory studies normally fall short as substitutes for genetic analysis in establishing whether a given aggregate is synthesized as a multifunctional polypeptide or not. Establishment of noncoordination in the production of polypeptide components of a complex enzyme may be taken as fairly strong evidence that they are produced by separate genes (after ascertaining that changes in conditions do not affect the rate of destruction of one component rather than its rate of synthesis), but establishing coordination does not prove they are produced by a single gene. For example, the α and β chains of TS are regulated in unison in all organisms studied, yet the genetic evidence that they are the products of separate genes in bacteria is incontrovertible (3).

Summary of the Occurrence of Multifunctional Proteins and Multienzyme Complexes in the Latter Portion of the Tryptophan Pathway

Table 5.2 lists the multifunctional proteins and multienzyme complexes involving the enzymes PRAI, InGPS, and TS in the tryptophan pathway. For clarity all gene designations are made to conform with the uniform nomenclature used for the bacterial *trp* genes (see Figure 5.2), though the actual designations employed by those working with the fungi are quite different (Table 5.1). All instances where two or more enzyme activities are associated with a single polypeptide chain are listed as multifunctional proteins, whether studied genetically or not. Examples from the bacteria, fungi, and euglenoids are presently known.

Where two or more activities are represented in a complex protein, each having a different polypeptide chain primarily responsible for catalysis, the tendency has been to treat the aggregate as a multienzyme complex. Where there is appropriate genetic justification for doing so, they are so classified in Table 5.2. Where the genetic data are absent or inadequate they are listed as indeterminant cases. It can be seen that the two indeterminant cases, the PRAI-InGPS complex recently reported in *Brevibacterium flavum* (23) and TS in some bacteria and higher plants, have precedents among the genetically studied groups for both fused and separate genes. Each of the

Table 5.2 Enzyme Aggregates in the Latter Part of the Tryptophan Pathway

Type of Aggregate	Organism	*Trp* Genes Involved	Evidence	
			Genetic	Biochemical
Multifunctional protein	Enteric bacteria	*trpC·F*	+	+
	Ascomycetes	*trpC·F*	+	+
	Ascomycetes	*trpA·B*	+	+
	Euglenoids	*trpD·C·F·B·A*	−	+
Multienzyme complex	Many bacteria	*trpA–trpB*	+	+
	Neurospora, *Aspergillus*, *Coprinus*	*trpE–trpC·F*	+	+
	Saccharomyces	*trpE–trpC*	+	+
Indeterminate cases	*Brevibacterium*	*trpC–trpF*	−	+
	Some bacteria, higher plants	*trpA–trpB*	−	+

examples listed in Table 5.2 is discussed in more detail below. This chapter then concludes with a review of a technique that has recently become available to simplify acquisition of genetic data in questionable cases, followed by a reevaluation of the strength and nature of selective pressures favoring multifunctional proteins and/or multienzyme complexes in the tryptophan pathway.

5.2 PHOSPHORIBOSYL-ANTHRANILATE ISOMERASE: INDOLEGLYCEROL-PHOSPHATE SYNTHASE, A BIFUNCTIONAL PROTEIN IN ENTERIC BACTERIA

Biochemical Evidence

The first of the intermediates in the tryptophan pathway between anthranilate and indoleglycerol-phosphate to be described and characterized was 1-(o-carboxyphenylamino)-1-deoxyribulose-5-phosphate (CDRP). In their initial description Smith and Yanofsky (24) noted that one class of tryptophan auxotroph in *E. coli* and *S. typhimurium* was unable to synthesize this compound, while another was unable to convert it to indoleglycerol-phosphate and therefore excreted it into the medium in its dephosphorylated form (CDR). This dephosphorylated accumulation product had been described earlier by Doy and Gibson (25). Smith and Yanofsky noted that mixing extracts of the two kinds of mutants resulted in a preparation capable of converting anthranilate and 5-phosphoribosyl-1-pyrophosphate (PRPP) to indoleglycerol-phosphate (24). At the time this did not seem remarkable, but as it became clear that one kind of mutant unable to produce CDPR maps in the same gene as the CDR-accumulating mutants it became more interesting. The next section treats these genetic findings.

In 1966 Creighton and Yanofsky (26) described the purification from extracts of *E. coli* of the enzyme capable of converting either phosphoribosyl-anthranilate (PRA) or CDRP to indoleglycerol-phosphate. They determined that the two activities present, isomerization of PRA to CDRP (PRAI) and decarboxylation and ring closure of CDRP to form indoleglycerol-phosphate (InGPS) (Figure 5.1), were catalyzed by a 45,000 mol wt protein consisting of a single polypeptide chain. Later Creighton (27) examined the kinetics of the two reactions and concluded from his analysis that there are two distinct and nonoverlapping active sites on the molecule. Comparison of the enzymes from *Enterobacter aerogenes* and *S. typhimurium* (2), *Serratia marcescens* (28), and *Aeromonas formicans* (29) with the *E. coli* version revealed no significant differences in the size of the single polypeptide chain or its catalytic capabilities. Subsequently Largen and Belser (30) extended these observations to several species in the genera *Proteus* and *Erwinia* and to *Enterobacter liquifaciens* and *Serratia marinorubra*. Immunochemical cross-reactivity of the *trpC* gene product in many of these species has been described (31). Very recently the marine luminous bacterium, *Beneckea harveyi*, has been added to the group of bacteria possessing a bifunctional PRAI-InGPS enzyme (32).

As the *trpC* gene product is a conveniently obtained, stable bifunctional polypeptide catalyzing adjacent steps in a biosynthetic pathway, some effort has been devoted to searching for evidence of channeling of the intermediate metabolite or cooperativity between the two active sites. In addition to the kinetic experiments mentioned earlier (27), the binding of a fluorescent inhibitor, phosphoribityl-anthranilate, to both active sites on the molecule has been studied (33). Even with the most sensitive techniques no cooperativity or interference between the two active sites has been shown. The simplest explanation of the biochemical results obtained so far is that PRA binds at one site on the molecule, where it is converted to CDRP. CDRP then diffuses freely from this site to a second site on the molecule where it is converted to indoleglycerol-phosphate.

Genetic Evidence

About the time the bifunctional *trpC* polypeptide was being purified, genetic analysis of the *trp* operon in both *E. coli* and *S. typhimurium* had developed to the point where detailed analysis of the gene was feasible. Working with a large collection of mutants in *S. typhimurium*, Blume and Balbinder (34) clearly distinguished *trpD* mutants lacking PRT activity from *trpC* mutants and showed that the latter were of two types, those that accumulated CDRP and those that did not. They did not perform PRAI assays and did not detect InGPS activity in any of their *trpC* mutants, but they did observe instances of genetic complementation between different point mutants in the gene, implying that pairs where each member lacks one of the two activities can cooperate to restore prototrophy when both are present in the same cytoplasm. Smith (35) studied a smaller number of mutants in *E. coli*, but his collection included missense mutants retaining some activity in one or the other reaction as well as a nonsense mutant lacking both activities. Of the five PRAI-negative mutants studied, only one had InGPS activity approaching the level found in *trpD* mutant strains, and two of the remaining four had very low levels indeed, around 1% of normal. Smith showed that the level of immunologically detectable *trpC* product present correlated well with the InGPS activity remaining, leading to the conclusion that the missense mutation abolishing PRAI activity left the InGPS active site intact but resulted in an increased rate of degradation of the entire polypeptide. Missense mutations abolishing InGPS mapped at the end of *trpC* nearest to *trpD*, while those lacking PRAI mapped close to *trpB* (35). As the direction of translation of the *trp* operon is known, this orientation requires that the InGPS portion of the molecule be amino-terminal and the PRAI portion carboxy-terminal in the bifunctional polypeptide. It is noteworthy that in bacteria having separate genes for PRAI (*trpF*) and InGPS (*trpC*) the latter is always associated with *trpD*, and the former, if linked to any *trp* gene, is associated with *trpA* and *trpB* (Figure 5.2).

Subsequently a more detailed analysis of the *E. coli trpC* gene, focusing on frameshift and chain-terminating nonsense mutations was accomplished (36). This confirmed the earlier results and showed that frameshift and nonsense mutations lacking both activities occur almost randomly throughout the gene. (Even quite large nonsense fragments of most gene products are rapidly degraded in normal *E. coli* cells.) Complementation between pairs of missense mutants lacking one of the two activities was demonstrated using plasmid merodiploids in *recA*$^-$ cells; nonsense mutants were found to be noncomplementary in this highly sensitive system. Some deletions removing as much as the distal quarter of the polypeptide do appear to have low levels of InGPS activity (again the levels are low because of degradation of the shortened polypeptide), but the reciprocal circumstance, removal of part of the amino-terminal end of the molecule by deletion to see if PRAI activity remained, was not tested.

Intracellular proteolysis of gene products can have a profound effect on genetic experiments with a multifunctional protein. One study of the stability of the normal *trpC* gene product has been done (37). This showed that all of the enzymes of the *trp* pathway are stable during glucose starvation. But when the cell is starved for one of several amino acids, phosphate, or pyrimidines, as well as during inhibition by chloramphenicol, InGPS activity declines while all other *trp* activities, including PRAI, remain relatively constant. Starvation for cysteine or sulfate results in a relative loss of both InGPS and PRAI, however. The molecular explanation of these results is not yet at hand. The data indicate that the bifunctional PRAI-InGPS molecule is less stable than many other gene products under certain nongrowing conditions, and that one of the two activities can be lost preferentially, but it cannot be asserted that this is a consequence of bifunctionality or a general characteristic of bifunctional proteins.

Evidence for Domains

For other multifunctional proteins there is evidence that each of the component parts forms a compact domain, and these domains are linked together by "connecting segments" of the polypeptide. Because a connecting segment may be quite susceptible to proteolytic attack, these domains can sometimes be purified as independent entities following limited digestion with proteases. This has not been accomplished yet for the PRAI-InGPS enzyme from enteric bacteria. A genetic approach to the same result is to examine mutants for the occurrence of enzyme fragments capable of independent existence and catalytic function. Except for the low level of InGPS activity in some deletions removing the carboxy-terminal segment of the molecule (37), this approach has not succeeded either. Yet the kinetic and physical chemical data described early in Section 5.2 imply that the two halves of the molecule are not functionally interdependent, and the identical reactions are catalyzed in many other bacteria by entirely distinct and independent molecules. Perhaps if suitable crystals were available the technique of x-ray crystallographic analysis would show whether or not the molecular anatomy reveals an unusual three dimensional structural interdependence between these functionally independent PRAI and InGPS domains. The appearance of promising crystals of *E. coli* PRAI-InGPS in the laboratories of workers at the University of Basel (K. Kirschner, personal communication) gives hope that this powerful technique may soon be applicable to the problem.

Comparative View of Phosphoribosyl-Anthranilate Isomerase and Indoleglycerol-Phosphate Synthase Enzymes

Individual genes and proteins of the third and fourth step of the tryptophan pathway appear to be the rule for bacteria outside the enteric group (Table 5.3). One exception may be the cyanobacterium *Anabaena variabilis* (11). Although in *Bacillus subtilis* the genes for these two enzymes are adjacent to one another and flanked by *trpD* and *trpB* just as they are in the enteric operon (see Figure 5.2), in other cases they are totally separate and even independently regulated (1).

Fused and completely separate enzymes are not the only possibilities, however, and recently Sugimoto and Shiio (23), investigating the tryptophan enzymes of the coryneform organism *Brevibacterium flavum*, observed the occurrence of PRAI and InGPS in a multienzyme complex. The two activities coelute on a Sephadex G-200 column at a position indicating a molecular weight of around 125,000 daltons. When the same extract is applied to a DEAE cellulose column these activities appear in three peaks, InGPS alone, PRAI alone, and a complex of the two. The molecular weights of the separated components, 90,000 for PRAI and 70,000 for InGPS, do not add up to the molecular weight of the complex. Additional work to clarify the subunit structure of this enzyme will be awaited with interest.

Table 5.3 Molecular Weight of Bacterial PRAI and InGPS Enzymes

Organism	Molecular Weight				Purified	Reference
	Bifunctional PRAI-InGPS	Monofunctional PRAI	Monofunctional InGPS	Multienzyme Complex		
Escherichia coli	45,000				Yes	26
Salmonella typhimurium	45,000				Yes	2
Enterobacter aerogenes	45,000				Yes	2
Enterobacter-Citrobacter[a]	47,000				No[e]	30
Serratia[b] *spp.*	45,000				No[e]	28, 30
Aeromonas formicans	45,000				No[e]	29, 30
Proteus[c] *spp.*	53,000				No[e]	30
Erwinia[d] *spp.*	48,000				No[e]	30
Pseudomonas putida		39,000	32,000		No	67
Acinetobacter calcoaceticus		<30,000	<30,000		No	68
Bacillus subtilis		27,000	30,500		No	69
Bacillus pumilus		24,300	25,800		No	70
Bacillus alvei		26,400	26,400		No	70
Clostridium butyricum			28,000		No	71
Staphylococcus aureus		<30,000	<30,000		No	72
Brevibacterium flavum		90,000	70,000	125,000	No	22

[a] Includes *Enterobacter cloacae*, *Citrobacter ballerupensis*, *Citrobacter freundii*, and *Erwinia dissolvens*.
[b] Includes *Serratia marcescens*, *Serratia marinorubra*, and *Serratia liquifaciens*.
[c] Includes *Proteus vulgaris* and *Proteus morganii*.
[d] Includes *Erwinia carotovora* and *Enterobacter hafniae*.
[e] In addition to coincident elution from Sephadex G-200 columns, the PRAI and InGPS activities eluted together on DEAE cellulose chromatography (29).

Unfortunately no method exists to study the genetics of the tryptophan pathway in *Brevibacterium*. In their work Sugimoto and Shiio did investigate the regulation of gene expression by the end product (23) and by the inhibitor 5-methyltryptophan and mutations conferring resistance to it (38). Several conditions evoked noncoordinate synthesis of PRAI and InGPS, suggesting that the genes for these enzymes may be noncontiguous and independently regulated. Until this suggestion is confirmed by genetic means it must remain tentative, however, for the rates of degradation of the enzymes in question are not known and may differ as they do in *E. coli* (37). Nonetheless, it appears inescapable that *Brevibacterium* PRAI and InGPS are associated in a multienzyme complex, a circumstance unique among bacteria (1) and higher plants (21) studied so far.

In a complex such as that of *Brevibacterium*, characterized only by biochemical means, it is possible but unlikely that the PRAI and InGPS activities are first produced as a bifunctional polypeptide, then processed by specific proteolytic cleavage into separable components having affinity for each other. In this case the severed domains could remain tightly associated through noncovalent interactions. Similar interactions are weak if they exist at all in other bacterial PRAI and InGPS enzymes. Moreover, the rate of degradation of the two components would have to be different under various conditions, as mentioned earlier. A complete resolution of these uncertainties may not have to await the development of a system of genetic exchange and construction of a conventional chromosomal map for *B. flavum*, however. The powerful techniques of recombinant DNA methodology should be applicable to this sort of problem (8). The economic importance of this species, used for commercial production of amino acids, should spur attempts at genetic analysis along both conventional and unconventional lines.

Apparent gene fusions involving PRAI and InGPS in fungi and euglenoids are described in Section 5.4.

5.3 TRYPTOPHAN SYNTHASE, A BIFUNCTIONAL PROTEIN IN FUNGI

Biochemical Evidence

Although suggested quite early (15, 39) by genetic results to be described in the next section, biochemical proof that the TS of *Neurospora* contains only a single kind of subunit was slow in coming. Purification of the wild type enzyme was hampered because levels are quite low, but missense mutations in the TS gene resulting in indoleglycerol-phosphate accumulation show a tenfold increase in the amount of that gene product when tryptophan deprivation results in accumulation of the intermediate (40, 41). Matchett (42) subsequently showed that adding indoleacrylate to the medium causes even the wild type to accumulate indoleglycerol-phosphate, and this induces a 15–20-fold increase in TS levels.

Although the first attempts to purify *Neurospora* TS (43, 44) seem to have yielded a partially proteolyzed molecule, more recently two laboratories using stringent precautions against proteolysis obtained pure proteins with similar properties. In one case the level of the enzyme in the wild type strain of *N. crassa* was derepressed with indoleacrylate (45); in the other case the TS mutant td201 was used (46). In both cases the purified enzyme had a molecular weight of 150,000; in the presence of reducing agents and sodium dodecyl sulfate the enzyme dissociated into a monomer of 75,000 mol wt (45, 46). Both laboratories concluded from amino and carboxy-terminal analyses that the two monomers released per molecule were identical. Though the td201 enzyme retains only TS-B activity, in the case of the wild type both the TS-A and TS-B activities purified together (45). The complementation results in genetic experiments described in the next section provide strong support for the conclusion that the subunit structure of *Neurospora* TS is α_2 rather than $\alpha\beta$.

The TS from *Saccharomyces cerevisiae* has been purified to near homogeneity (47). This organism also contains proteases that can rapidly inactivate TS (described later in this section), so the fact that the purified preparation, having a molecular weight of 143,000, dissociated on denaturation into fragments of several sizes, the major one of 37,000 daltons (47), cannot at present be construed as proof that its subunit structure differs from that of the *Neurospora* enzyme. Genetic evidence strongly supports a single subunit catalyzing both TS activities in yeast as well as *Neurospora*.

Genetic Evidence

From the earliest investigations it was apparent that mutations of several types occur in the TS locus of *N. crassa* (5, 39) and *S. cerevisiae* (48). In each case some mutants retain some TS-A activity ("indole accumulators"), others some TS-B activity ("indole utilizers"), but most lack all TS activities. In both *N. crassa* (5, 49) and *S. cerevisiae* (50) these several mutant phenotypes are localized within the gene. Mutations affecting TS-A map at one end of the gene, separate from mutations affecting TS-B. Some of the mutations leading to inactivity in both reactions are missense (termed "profound" in *N. crassa*); these map in the central part of the *Neurospora* gene (5, 49). Others are chain terminating, nonsense mutations that may occur at many sites within the gene. In *S. cerevisiae* there is a region in the middle of the gene where nonsense mutations give rise to a smaller molecule possessing a low level of TS-A activity (50, 51). The existence of these active fragments implies that the direction of transcription is from the TS-A catalytic region to the TS-B catalytic region; it further suggests strongly that the TS-A catalytic region may fold into a domain that is at least somewhat stable and enzymatically active in its half-reaction (50).

Heterocaryons constructed from pairs of *N. crassa* mutants and heterozygous diploid *S. cerevisiae* strains allow analysis by complementation of mutations affecting TS. All indole accumulating mutants complement all indole utilizing mutants, but no wild type enzyme is present under these circumstances, and the mechanism obviously involves the appearance of indole as a free intermediate (8, 52). This differs from the complementation of bacterial *trpA* and *trpB* mutants, where at least some molecules are formed that are indistinguishable from wild type. Another type of complementation of missense mutants in these microorganisms involves heteropolymer formation; heteropolymers seen with two "profound" mutants or a "profound" mutant and an indole utilizing or indole accumulating mutant may exhibit a low level of tryptophan synthetic activity. This is easily distinguished from wild type enzyme by the low specific activity and aberrant activity ratios in the two half-reactions (8, 52).

Genetic results with large numbers of independent TS mutants of yeast and *N. crassa* are only understandable if the enzyme consists of a single polypeptide chain that forms a homomultimer, and for *Neurospora*, at least, the biochemical evidence is now in complete agreement with this conclusion. Both from the fact that in all other organisms examined (except *Euglena*) TS is a heteropolymer of α and β subunits and from the existence of fragments of the yeast enzyme showing weak TS-A activity, it may be surmised that the fungal TS polypeptide has two domains. Proteolytic attack on TS by endogenous proteases in crude and partially purified preparations of the fungal enzyme is a well established phenomenon, but despite numerous attempts it has not yet been possible to document the existence of domains of fungal TS corresponding to the α and β polypeptide chains of other organisms.

Tryptophan Synthase Inactivating Enzymes

As this subject has been reviewed recently (53) it is mentioned here only in passing. It was first noticed in yeast (54) but later observed in *Neurospora* as well that TS is labile in some extracts. As cultures approach the stationary phase they elaborate several intracellular proteases (and one inhibitor of these proteases) active against TS and certain other cellular enzymes.

Two of these proteases have been highly purified and studied in some detail. They correspond to previously described yeast proteases A and B (55). Their function appears to be the degradation of anabolic enzymes no longer required in large amounts by the resting cell. Presumably the amino acids derived thereby are made available for other cellular functions. TS molecules without their PLP cofactors are preferentially attacked.

As these "housekeeping" enzymes should carry their task to completion, it is not surprising that they have not been observed to release recognizable TS-A and TS-B catalytic domains from the intact molecule. The use of other, foreign proteases of restricted specificity should be employed for this purpose. An investigation of the products of partial proteolysis of a fused bifunctional enzyme along the lines of that recently done with the β_2 subunit and the $\alpha_2\beta_2$ complex of *E. coli* TS (6, 56) would be of great interest, as the most likely mechanism of origin of the fungal enzyme, head to tail fusion of a *trpB* and a *trpA* gene, might well result in a connecting region of the polypeptide between the two original gene products susceptible to selective cleavage.

Catalytic Properties

Most recent studies of the catalytic mechanism of the TS reaction(s) have employed the $\alpha_2\beta_2$ complex of *E. coli*. An excellent, up-to-date review of these studies is available (57). For nearly two decades it has been clear that indole formed in the TS-A reaction is *not* released from the enzyme before being converted to tryptophan in the physiological reaction (3, 57). This is also true for the fused fungal enzyme (58, 59, 60). In contrast to the enteric bacterial PRAI-InGPS, the fused fungal TS elements form a composite active site, or at least one in which the intermediate, indole, is "channeled" from the site of its formation to that of its utilization (59, 60). But this is also the case in all the multienzyme complex forms of TS studied (57). Presumably some ancestral organism had separate enzymes for the TS-A and TS-B reactions. This situation and any intermediate ones where the two active sites act independently but are present in a complex remain to be found. Though it cannot be proved it can be postulated that the ability to keep the unusually hydrophobic intermediate, indole, tightly enzyme-bound gives a potent selective advantage to organisms having the associated form.

It is possible to give the enzyme substrates for only one half-reaction (indoleglycerol-phosphate or indole and glyceraldehyde-phosphate for TS-A, indole and serine for TS-B) and study the effect chemical and mutational modifications directed at the other active site have on this activity. The conclusion for both the bacterial (57) and fungal (59) enzymes is that each active site has a effect on the other, but Matchett (59) has argued that the fungal enzyme is a more "constrained" system with less conformation flexibility.

5.4 LESS WELL STUDIED MULTIFUNCTIONAL PROTEINS IN THE TRYPTOPHAN PATHWAY

Fungal Phosphoribosyl-Anthranilate Isomerase: Indoleglycerol-Phosphate Synthase

As indicated in Table 5.2, AS is associated with PRAI and InGPS in *N. crassa* and with InGPS in *S. cerevisiae*. This complex enzyme has been purified from *N. crassa* (61) and has a molecular weight between 320,000 and 340,000. It dissociates into two polypeptides on sodium dodecyl sulfate acrylamide gel electrophoresis, one of 70,000 and the other of 94,000 mol wt (61). The smaller subunit possesses ammonia dependent AS activity. The larger one has both PRAI and InGPS activities, and, when associated in an $\alpha_2\beta_2$ complex with it, confers upon the smaller one the ability to use glutamine as an amino group donor in the AS reaction. Preparations of the intact enzyme are often unstable. The instability appears due to proteolytic enzymes that, attacking the 94,000 mol wt subunit first, give rise to several fragments capable of associating with AS and conferring glutamine utilization upon it (61). The smallest of these active fragments has a molecular weight of about 25,000 and resembles in size and activity the *trpG* gene product in bacteria having a monofunctional AS (see Chapter 4).

There is excellent genetic evidence that the 94,000 mol wt subunit of the complex is the product of a single gene (16). As this gene product exhibits three enzymatic functions often found on independent polypeptides in other organisms, InGPS, PRAI, and the glutamine amidotransferase part of AS, it is a trifunctional polypeptide that may represent the fusion of three genes. There is genetic evidence that mutations affecting InGPS and PRAI singly are located at opposite ends of the gene (16). Mutations affecting interaction with the AS gene product are in the center of the gene. The suggested structure of this locus, having the glutamine amidotransferase function sandwiched between the third and fourth enzyme functions, especially if viewed alongside the *S. cerevisiae* complex described below which lacks the PRAI function, leads to the supposition that in this case too, as with the enteric bacterial *trpFC* fusion, there will be no interaction between the PRAI and InGPS active sites and no "channeling" of the intermediate.

Other filamentous fungi have been less well investigated, but from the initial survey by Hütter and DeMoss (11) to more detailed biochemical studies with *Claviceps sp.* (62) there appear to be many occurrences of this trifunctional polypeptide among the basidiomycetes, filamentous ascomycetes, myxomycetes, and some phycomycetes. The situation in the genus *Coprinus* is still a little uncertain. A complex catalyzing the usual three activities has been observed (11, 63), but AS activity seems not to be dependent on the product of the PRAI-InGPS gene in *C. radiatus* (63) whereas the usual interdependence was shown in *C. lagopus* (12). One subdivision of phycomycetes, the *Oomycetes* group, differs quite substantially from other fungi in having its AS unassociated with other enzymes of the pathway. In this group the PRAI and InGPS cosediment at 3.5S (11). Whether this is a bifunctional polypeptide or not remains to be established by genetic or additional biochemical experimentation.

In *S. cerevisiae* and two other yeast genera AS is associated with InGPS but not with PRAI (11). Genetic evidence in *S. cerevisiae* (64) suggests that the InGPS and AS glutamine amidotransferase functions are encoded by a single gene, with missense mutations that affect one or the other function segregated at opposite ends of the gene. Biochemical study of this putative bifunctional polypeptide should be of considerable interest; it is not known whether the molecule is subject to protease attack.

The *Euglena gracilis* Multifunctional Protein

The enzymes of the tryptophan pathway have been examined in extracts of photosynthetically grown *E. gracilis* cells. To date there have been no tryptophan auxotrophs reported in this organism and no means of genetic analysis has been demonstrated. After determining that all of the pathway activities except AS are associated in a single multifunctional aggregate (65), Hankins and Mills purified this complex to ascertain its subunit structure (20). They determined that this approximately 325,000 mol wt aggregate could be dissociated only into two very similar or identical 155,000 ± 5000 mol wt subunits by any of the treatments tried. As this is the largest subunit observed for any multifunctional tryptophan pathway protein, it is tempting to believe that it represents a five gene fusion involving the functional elements for PRT, PRAI, InGPS, and the two TS components. As the AS of this organism contains a single polypeptide chain (see Chapter 4), the conclusion is that the trptophan pathway of *E. gracilis* requires but two structural genes.

Though *E. gracilis* exceeds even the fungi in the extent of gene fusion in the tryptophan pathway, the early speculation of Bonner et al. (5), that more highly evolved forms would show progressively more gene fusions, is not born out by studies of algae and higher plants (21). It is possible that the organisms showing the greater degree of gene fusion are in fact in an evolutionary *cul-de-sac* (20).

5.5 THEORETICAL CONSIDERATIONS

Does Proteolysis Confound Recognition of Multifunctional Proteins?

As mentioned in Section 5.1, genetic analysis is an invaluable aid in establishing whether a given enzyme aggregate is a multienzyme complex or a multifunctional protein. Much of the biochemical work described in Sections 5.3 and 5.4 has had to contend with proteolytic enzymes that sometimes copurify with the enzymes in question. There is evidence that these proteases serve a physiological function (53), and they may begin their action long before the cell wall is broken. Until a method to forestall proteolytic artifacts is developed, the purely biochemical approach can hardly be relied on to disprove the possibility of multifunctionality.

Recently developments in genetic analysis may prove very useful for solving this kind of problem. With recombinant DNA technology it is possible to insert chromosomal segments carrying genes with selectable functions into suitable vehicles and continue their analysis in an organism such as *E. coli* where highly sophisticated genetic analysis can be done. Of course, the mere fact that two functional units are near each other on the chromosome and can be cloned together does not prove they are elements of a fused gene pair. Subsequent strategies to establish the facts are diverse and are not considered here. Although there are pitfalls in this approach and the number of successful trials of the method is still small (8), the prospect of extending the power of genetic analysis to organisms lacking a means of genetic exchange is a happy one indeed.

Selective Advantages of Multienzyme Complexes and Multifunctional Proteins

Certain of the theoretical advantages of aggregating the enzymes of a given pathway are exemplified in the well studied complexes in the tryptophan pathway. Both the bacterial and fungal forms of TS illustrate the concept of "channeling" intermediates, whose advantages were pointed out some time ago by Davis (66). The bifunctional protein in *E. coli* catalyzing the third and fourth steps in the pathway does not channel the intermediate, however, and the two active sites do not seem to interact in any significant fashion. Some fungal multienzyme complexes omit an intervening enzyme, which also suggests there may be some significant advantage to aggregation other than channeling.

Another possible selective factor is that fusion of elements in the same pathway causes them to be regulated in unison. Unfortunately for this hypothesis, the range of regulation observed in the fungi and euglenoids, where fusions are rampant, is much less than in many bacteria where no gene fusions are seen at all. In fungi, and perhaps eucaryotes in general, proteolytic degradation also appears to play a regulatory role.

It must be admitted that the selective advantage for the fusion of genes in the tryptophan pathway still eludes definition. Perhaps the obvious gaps in our knowledge of the frequency and kind of gene fusions selected, and perhaps discarded, during evolution prevent identification of the real selective forces at work. Perhaps the aggregates seen once the cell is disrupted are only the strongest remnants of a larger structure that exists in the normal cellular environment. In order to be accepted, gene fusions would have to be compatible with the spatial constraints of such a superstructure. One additional advantage present and future students of the problem will enjoy is the ability to construct gene fusions *in vitro*, then to introduce them into suitable cells to ascertain their efficiency.

REFERENCES

1 I. P. Crawford (1975), *Bacteriol. Rev.* **39,** 87.

2 J. F. McQuade, III and T. E. Creighton (1970), *Eur. J. Biochem.* **16,** 199.

3 C. Yanofsky and I. P. Crawford (1972), in P. D. Boyer, ed., *The Enzymes*, Vol. 7, 3rd ed., Academic Press, New York, p. 1.

4 I. P. Crawford and C. Yanofsky (1968), *Proc. Natl. Acad. Sci. U.S.A.* **44,** 1161.

5 D. M. Bonner, J. A. DeMoss, and S. E. Mills (1965), in V. Bryson and H. J. Vogel, eds., *Evolving Genes and Proteins*, Academic Press, New York, p. 305.

6 A. Högberg-Raibaud and M. E. Goldberg (1977), *Proc. Natl. Acad. Sci. U.S.A.* **74,** 442.

7 A. Högberg-Raibaud and M. E. Goldberg (1977), *Biochemistry* **16,** 4014.

8 I. P. Crawford (1979), *Crit. Rev. Biochem.*, in press.

9 L. Baumann and P. Baumann (1976), *Microbios Lett.* **3,** 11.

10 A. Arroyo-Begovich and J. A. DeMoss (1973), *J. Biol. Chem.* **248,** 1262.

11 R. Hütter and J. A. DeMoss (1967), *J. Bacteriol.* **94,** 1896.

12 M. J. Tilby (1976), *J. Gen. Microbiol.* **93,** 126.

13 J. A. DeMoss (1965), *Biochem. Biophys. Res. Comm.* **18,** 850.

14 C. H. Doy and J. M. Cooper (1966), *Biochem. Biophys. Acta* **127,** 302.

15 R. Hütter (1973), in Z. Vanek, Z. Hostalek and J. Cudlin, eds., *Genetics of Industrial Microorganisms*, Academia, Prague, p. 109.

16 J. A. DeMoss, R. W. Jackson, and J. H. Chalmers, Jr. (1967), *Genetics* **56,** 413.

17 W. E. Buvinger and H. E. Heath (1979), *J. Bacteriol.*, in press.

18 A. W. B. Johnston, M. J. Bibb, and J. E. Beringer (1978), *Mol. Gen. Genet.*, **165,** 323.

19 C. M. Smithers and P. P. Engel (1974), *Genetics* **78,** 799.

20 C. N. Hankins and S. E. Mills (1977), *J. Biol. Chem.*, **252,** 235.

21 C. N. Hankins, M. T. Largens, and S. E. Mills (1976), *Plant Physiol.* **57,** 101.

22 D. P. Delmer and S. E. Mills (1968), *Biochem. Biophys. Acta* **167,** 431.

23 S. Sugimoto and I. Shiio (1977), *J. Biochem. Tokyo* **81,** 823.

24 O. H. Smith and C. Yanofsky (1960), *J. Biol. Chem.* **235,** 2051.

25 C. H. Doy and F. Gibson (1959), *Biochem. J.* **72,** 586.

26 T. E. Creighton and C. Yanofsky (1966), *J. Biol. Chem.* **241,** 4625.

27 T. E. Creighton (1970), *Biochem. J.* **120,** 699.

28 M. A. Hutchinson and W. L. Belser (1969), *J. Bacteriol.* **98,** 109.

29 I. P. Crawford, S. Sikes, and D. K. Melhorn (1967), *Arch. Mikrobiol.* **38,** 72.

30 M. T. Largen and W. L. Belser (1975), *J. Bacteriol.* **121,** 239.

31 G. R. Reyes and V. Rocha (1977), *J. Bacteriol.* **129,** 1448.

32 C. D. Bieger and I. P. Crawford, manuscript in preparation.

33 K. Kirschner, W. Cohn, and H. Bisswanger (1978), in C. Trojanovsky, ed., *Proteins and Ions Involved in Fast Dynamic Phenomena*, Elsevier, Amsterdam.

34 A. J. Blume and E. Balbinder (1966), *Genetics* **53,** 577.

35 O. H. Smith (1967), *Genetics* **57,** 95.

36 C. Yanofsky, V. Horn, M. Bonner, and S. Stasiowski (1971), *Genetics* **69,** 409.

37 R. D. Mosteller, K. R. Nishimoto and R. V. Goldstein (1977), *J. Bacteriol* **131,** 153.

38 S. Sugimoto and I. Shiio (1978), *J. Biochem. Tokyo* **83,** 879.

39 Y. Suyama, A. M. Lacy, and D. M. Bonner (1964), *Genetics* **49,** 135.

40 W. H. Matchett and J. A. DeMoss (1962), *J. Bacteriol.* **83,** 1294.

41 J. R. Turner and W. H. Matchett (1968), *J. Bacteriol.* **95,** 1608.

42 W. H. Matchett (1972), *J. Bacteriol.* **110,** 146.

43 S. Ensign, S. Kaplan, and D. M. Bonner (1964), *Biochem. Biophys. Acta* **81,** 357.

44 M. Carsiotis, E. Apella, P. Provost, J. Germershausen, and S. R. Suskind (1965), *Biochem. Biophys. Res. Comm.* **18,** 877.

45 W. H. Matchett and J. A. DeMoss (1975), *J. Biol. Chem.* **250,** 2941.

46 H. Tsai, C. Y. Yang, and J. H. J. Tsai (1974), *Biochem. Biophys. Res. Comm.* **61,** 1332.

47 D. H. Wolf and M. Hoffmann (1974), *Eur. J. Biochem.* **45,** 269.

48 T. R. Manney (1964), *Genetics* **50,** 109.

49 A. M. Lacy (1965), *Biochem. Biophys. Res. Comm.* **18,** 812.

50 T. R. Manney, W. Duntze, N. Janosko, and J. Salazar (1969), *J. Bacteriol.* **99,** 590.

51 T. R. Manney (1968), *Genetics* **60,** 719.

52 W. W. Duntze and T. R. Manney (1968), *J. Bacteriol.* **96,** 2085.

53 H. Holzer, T. Katsunuma, E. Schött, A. R. Ferguson, A. Hasilik, and H. Betz (1973), *Adv. Enzyme Regul.* **11,** 53.

54 T. R. Manney (1968), *J. Bacteriol.* **96** 403.

55 T. Saheki and H. Holzer (1974), *Eur. J. Biochem.* **42,** 621.

56 E. W. Miles and W. Higgins (1978), *J. Biol. Chem.* **253,** 6266.

57 E. W. Miles (1979), *Adv. Enzymol.* **49,** 127.

58 C. Yanofsky and M. Rachmeler (1958), *Biochem. Biophys. Acta* **28,** 640.

59 W. H. Matchett (1974), *J. Biol. Chem.* **249,** 4041.

60 J. A. DeMoss (1962), *Biochem. Biophys. Acta* **62,** 279.

61 F. M. Hulett and J. A. DeMoss (1975), *J. Biol. Chem.* **250,** 6648.

62 D. F. Mann and H. G. Floss (1977), *Lloydia* **40,** 136.

63 H. Henke, J. L. Guerdoux, and R. Hütter (1973), *C.R. Acad. Sci. Paris* **276,** 2001.

64 M. Landolt-Sydler, Y. Schürch-Rathgeb, and R. Hütter (1973), *Path. Microbiol.* **39,** 40.

65 J. C. Lara and S. E. Mills (1972), *J. Bacteriol.* **110,** 1100.

66 R. H. Davis (1967), in H. J. Vogel, J. O. Lampen, and V. Bryson, eds., *Organizational Biosynthesis*, Academic Press, New York, p. 303.

67 T. Enatsu and I. P. Crawford (1968), *J. Bacteriol.* **95,** 107.

68 R. Twarog and G. L. Liggins (1970), *J. Bacteriol.* **104,** 254.

69 S. O. Hoch, C. Anagnostopoulos, and I. P. Crawford (1969), *Biochem. Biophys. Res. Comm.* **35,** 838.

70 S. O. Hoch and I. P. Crawford (1973), *J. Bacteriol.* **116,** 685.

71 E. N. Baskerville and R. Twarog (1972), *J. Bacteriol* **112,** 304.

72 A. R. Proctor and W. E. Kloos (1973), *J. Bacteriol* **114,** 169.

Chapter 6

DNA POLYMERASE I OF *ESCHERICHIA COLI*

HANS KLENOW

Biokemisk Institut B
Panum Instituttet
University of Copenhagen, Denmark

6.1 INTRODUCTION

DNA polymerase I of *E. coli* was the first enzyme of its kind to be recognized, and it has been studied more thoroughly than other DNA polymerases. In fact the main body of knowledge of the principles of enzymatic synthesis and repair of DNA is due to investigations of this enzyme. It has been studied extensively in A. Kornberg's laboratory. A number of excellent review papers on the enzyme have been published (1, 2, 3). The present review is written with special emphasis on the multifunctional nature of the enzyme molecule. For a discussion of the function of this enzyme the reader is referred to the review by Lehman and Uyemura (3).

The enzyme, which consists of a single polypeptide chain, catalyzes the following reactions: extension of polydeoxyribonucleotide chains in the 5′ →

3′ direction (polymerase reaction); exonucleolytic degradation of double stranded DNA in the 5′ → 3′ direction; exonucleolytic degradation of single stranded DNA in the 3′ → 5′ direction; pyrophosphorolytic degradation of DNA and exchange of pyrophosphate with deoxyribonucleoside triphosphates. The latter two reactions are probably features of the activity responsible for DNA synthesis, and they are therefore not dealt with in detail in this review.

Of the two exonuclease activities associated with the enzyme the 5′ → 3′ exonuclease is confined to a domain of the molecule that may be physically separated from the polymerase activity and the 3′ → 5′ exonuclease activity. A similar physical separation of the 3′ → 5′ exonuclease activity from the polymerase activity has not been obtained. Several lines of evidence do, however, indicate that the polymerase and 3′ → 5′ exonuclease activities are catalyzed by two independent catalytically active sites. These two sites may be confined to two different domains of the enzyme molecule.

Since the enzyme consists of a single polypeptide chain harboring two or possibly three different domains catalyzing separate enzyme reactions, DNA polymerase I of *E. coli* may be considered to be a true multifunctional protein.

6.2 PHYSICAL AND CHEMICAL PROPERTIES

The sedimentation constant ($s^{\circ}_{20,w}$) of the enzyme is 5.6, and optical properties indicate the presence of right-handed β-helix structure. The molecular weight of the native enzyme has been found both by sedimentation equilibrium and by sedimentation velocity to be about 1.09×10^5. Similar values have been obtained by these methods under denaturing conditions indicating the presence of only one polypeptide chain. Electrophoresis under denaturing conditions also shows the presence of only one type of polypeptide chain, and the molecular weight is as found for the native enzyme. Polymerase activity and 5′ → 3′ exonuclease activity are both almost completely lost in the presence of guanidine and mercaptoethanol. The major part of both activities may, however, be recovered upon removal of the denaturing agents indicating a partial refolding of the chain into its native conformation (4). The enzyme contains one g-atom of zinc/mole, and removal of zinc results in almost complete loss of polymerase activity. This activity is restored by adding zinc ions back. Co^{2+} and Mn^{2+} partially restores the activity of the zinc free enzyme (5). The acylating reagent *N*-carboxymethylisatoic anhydride (NCMIA) reacts with DNA polymerase forming a highly fluorescent derivative (6). The acylation has a differential effect on the three enzyme activities (7) (see Section 6.3).

The enzyme contains about 1000 amino acid residues, and the amino acid composition shows the presence of only one sulfhydryl group and one disulfide bond (4) (see Table 6.1). The sulfhydryl group is probably nonfunctional since it may react with either iodoacetate or mercuric ions without affecting the polymerase activity or the 5′ → 3′ exonuclease activity (6). It is, however, not known whether the 3′ → 5′ exonuclease activity is affected. The N-terminal amino acid sequence is Met-Val-Glx-Ile-Pro-Glx-Leu-Asx but in a small number of enzyme molecules the N-terminal Met is missing (8). When DNA polymerase is treated with proteolytic enzymes the 5′ → 3′ exonuclease activity disappears while the polymerase activity is retained. According to gelfiltration and electrophoresis under denaturing conditions the polymerase activity is now present in a polypeptide with a molecular weight of about 75000. This peptide has also 3′ → 5′ exonuclease activity (9, 10). If treatment with the proteolytic enzyme subtilisin is performed in the presence of DNA the 5′ → 3′ exonuclease activity is also preserved. This activity may now be isolated separate from the two other activities for

Table 6.1 Amino Acid Composition of DNA Polymerase I of *E. coli*

Amino Acid	Residues/109,000 g Protein
Lysine	61
Histidine	19
Arginine	48
Aspartic acid, asparagine	88
Threonine	53
Serine	40
Glutamic acid, glutamine	126
Proline	53
Glycine	63
Alanine	102
Half-cystine	3
Valine	61
Methionine	24
Isoleucine	55
Leucine	112
Tyrosine	33
Phenylalanine	25
Tryptophan	9
(Ammonia)	(58)
Total	975

instance by chromatography on hydroxyl apatite (8, 11). Both gel filtration and electrophoresis show that the 5′ → 3′ exonuclease activity is associated with a peptide with a molecular weight of about 35000 (11, 12), its sedimentation constant is 2.8 (13). Cleavage of DNA polymerase I by limited proteolysis into two catalytically active fragments thus indicates, therefore, that the enzyme in the presence of DNA is folded in two separate domains that are themselves quite resistant to proteolysis. The two domains may be connected by a short flexible peptide particularly susceptible to proteolytic attack. The two fragments do not seem to have any affinity for one another since they sediment independently in a sucrose gradient (14). The amino acid composition of the two fragments is very similar except that the large fragment contains all three sulphur atoms present in the native enzyme (8, 15). The N-terminal amino acid sequence has been determined for the small fragment and found to be Met-Val-Glx-Ile, identical with that of the native enzyme. For the large fragment the N-terminal sequence is Val-Ile-Met, that is different from that of the native enzyme (8). This shows that DNA polymerase may be cleaved by limited proteolysis into a small 35,000 mol wt N-terminal fragment with 5′ → 3′ exonuclease activity and a large 75,000 mol wt C-terminal fragment with both 3′ → 5′ exonuclease activity and polymerase activity. This arrangement of the activities has also been deduced from experiments with the *pol* Al mutant (13). This mutant has very low levels of polymerase activity but contains nearly normal 5′ → 3′ exonuclease activity. This latter activity is associated with a protein molecule that seems to be identical with the small fragment of the wild type enzyme. The *pol* Al mutant is an *amber* mutant, and it has been concluded that the 5′ → 3′ exonuclease activity is associated with an *amber* peptide that has been converted to the small fragment by intracellular proteolysis. This suggests that the 5′ → 3′ exonuclease activity is situated in the amino-terminal part of DNA polymerase I molecule. The large fragment of the wild type enzyme has also been obtained as a contaminant of the intact enzyme. It is probably formed by limited proteolysis during the isolation procedures (10).

The studies of the chemical and physical properties of DNA polymerase I have been hampered by a scarcity of enzyme. The recent cloning of the gene for DNA polymerase I results in an enhancement of the cellular content of the enzyme by a factor of more than 75 and thus should change this situation radically (16).

6.3 CATALYTIC PROPERTIES

The catalytic activities of the intact enzyme have been thoroughly studied since it was first isolated some 20 years ago. It was, however, rather late in the history of the enzyme that it was found that it may be cleaved by limited proteolysis into two fragments with different enzymic activities, and less information is, therefore, available on the properties of the fragments than on the intact enzyme. On the other hand, it was only after the two fragments had been isolated separately that a clear distinction between the two exonucleolytic functions of the intact enzyme became possible. For the evaluation of the effect of having all the three catalytic activities of the enzyme in one polypeptide chain rather than having two activities in one molecule and the third activity in another molecule a fairly limited number of comparative studies are available. In this respect the properties of a number of mutant enzymes that have been more or less altered in specific parts of the molecule have contributed to our understanding of the function of each separate enzyme activity and of the concerted effect of all three activities.

Outline of the Properties of the Individual Enzyme Activities

The polymerization reaction is an elongation of polydeoxyribonucleotide chains in the 5′ → 3′ direction. The substrates are deoxyribonucleoside triphosphates (or probably rather their complexes with Mg^{2+}) and the 3′-hydroxyl end group of a polydeoxyribonucleotide chain, that is the terminus of a primer chain. The reaction is a nucleophilic attack of the 3′-hydroxyl end group of the primer terminus on the α-phosphorus atom of the reacting deoxyribonucleoside triphosphate. This results in the extension of the primer chain by one deoxyribonucleotide residue through formation of a new 3′ → 5′ phosphodiester bond and the simultaneous release of a pyrophosphate molecule.

$$(\mathrm{DNA})_{n\ \mathrm{residues}} + \mathrm{dNTP} \rightleftharpoons (\mathrm{DNA})_{n+1\ \mathrm{residues}} + \mathrm{pyrophosphate}$$

The 3′-hydroxyl end group of the primer molecule is indispensable for the reaction and chains with other terminal groups like 3′-phosphate, 3′-deoxy, 3′-fluoro, or 3′-arabinosylresidues do not support synthesis (17). Besides the substrates the reaction also requires a polydeoxyribonucleotide template strand forming a double stranded structure with the primer chain. Its function is to select the reacting deoxyribonucleoside triphosphate. The latter is selected to form a Watson-Crick base pair at the position next to the primer terminus. The DNA required for synthesis is often termed a template-primer molecule and should have the structure of a double stranded molecule with a protruding 5′ end group. The polymerase reaction extending the 3′ end group may proceed until a complete duplex structure has been formed, and the reaction will stop because of the lack of exposed template.

If the 3′ terminal nucleotide of the primer chain is not base paired with the template strand no synthesis occurs. On the other hand, duplex DNA molecules with gaps or nicks with 3′-hydroxyl groups may support synthesis. In the former case synthesis is a gap filling reaction and in the latter case it requires the simultaneous removal of the strand in front of synthesis. This may be accomplished either by strand displacement by unwinding the strand in front of synthesis or by exonucleolytic degradation of the strand in the 5′ → 3′ direction giving rise to a process termed nick translation (Figure 6.1).

The substrate for the 5′ → 3′ exonuclease activity is DNA with double stranded 5′ end groups, that is double stranded DNA, which contains either nicks or a protruding 3′ end group (18, 19). The products are primarily deoxyribonucleoside 5′-monophosphates but di- and oligodeoxyribonucleotides up to eight residues in length may also be excised. The formation of the latter products can not be attributed to endonucleolytic activity but is due to an inherent property of the exonuclease activity.

Single stranded DNA is the substrate for the 3′ → 5′ exonuclease activity, and the product is deoxyribonucleoside 5′-monophosphates. Also single stranded 3′ end groups that protrude from a double stranded DNA structure will be degraded by the 3′ → 5′ exonuclease activity.

The Individual Enzyme Activities

The Polymerase Activity In the first phase of the polymerase reaction the enzyme binds the substrates, that is the primer template molecule and a

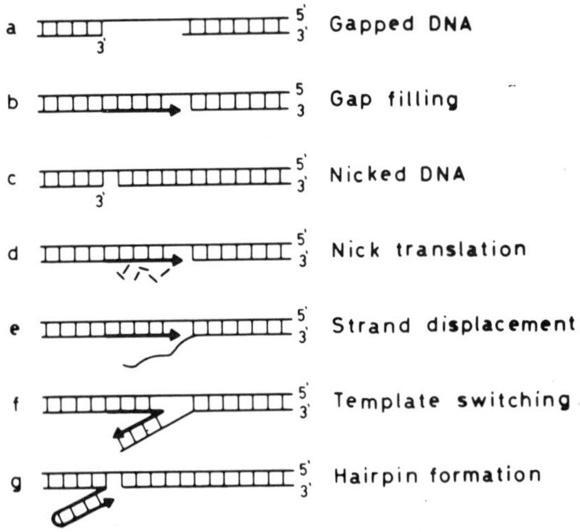

Fig. 6.1 Polymerase action at gaps and nicks. Model for formation of hairpin structures.

deoxyribonucleoside triphosphate. Binding of these compounds can occur independently of each other, and during the polymerase reaction their binding may possibly be in a random order. Binding studies to DNA have been performed only with the intact enzyme. While it does not bind to closed circular duplex DNA molecules it is bound after the introduction of nicks in such a molecule. One enzyme molecule is bound at each nick whether it is a 3′-hydroxyl nick or a 3′-phosphate nick. The binding does not require Mg^{2+}. One enzyme molecule also binds to a oligo d(A-T) · oligo d(A-T) molecule. This binding is independent of Mg^{2+} and is prevented by high concentration of phosphate. The intact enzyme also binds to single stranded DNA, and the number of enzyme molecules bound is proportional to the length of the DNA molecule (20). This may reflect a binding to the template site on the enzyme and may, therefore, be due to the part that corresponds to the large fragment. While this kind of binding is also independent of Mg^{2+} the presence of this ion will cause both the intact enzyme and the large fragment to bind as well to 3′ hydroxyl end groups of single-stranded DNA. This is because the 3′ → 5′ exonuclease activity which requires Mg^{2+} is present in both enzyme molecules (21, 22). The affinity of the enzyme both to the single stranded template and to the primer terminus may be important for the correct binding and orientation of the enzyme molecule for its function in the gap filling synthesis, nick translation, and strand displacement reactions.

The binding constants for deoxyribonucleoside triphosphates have been determined by equilibrium dialysis experiments with the intact enzyme, and competition studies have shown that a single site on the enzyme binds all four deoxyribonucleoside triphosphates. Binding requires Mg^{2+}, probably because it is the Mg^{2+}-triphosphate complex that is bound. A number of analogs of the normal deoxyribonucleoside triphosphates compete with the binding of, for example, dATP or dTTP. These analogs include rATP, 2′ 3′-dideoxyATP, and α,β-dTTP methylene diphosphonate but not 2′-deoxyribonucleoside monophosphates. Binding is inhibited by high ionic strength (23).

Similar studies have not been performed with the large fragment, but some of the above mentioned findings have been confirmed by experiments with the acylating amino group reagent NCMIA with both the intact enzyme and the proteolytically cleaved enzyme (7). NCMIA reacts with the intact enzyme, and it may cause an almost complete loss of polymerase activity while the 3′ → 5′ exonuclease activity is unaffected, and the 5′ → 3′ exonuclease activity increases by a factor of 4 to 5. The effect of NCMIA on the polymerase activity may be completely prevented by the simultaneous presence of Mg^{2+} and a deoxyribonucleoside triphosphate like dATP or ATP, whereas Mg^{2+} alone or triphosphates alone or Mg^{2+} plus dAMP are not able to protect against the effect of NCMIA. When polymerase activity is protected by Mg^{2+} and triphosphates, NCMIA still causes a considerable increase in 5′ → 3′ exonuclease activity. These findings are in agreement with the presence of a site that binds Mg^{2+} complexes of triphosphates whether they are of the ribo- or the deoxyribonucleoside type, while the site does not bind deoxyribonucleoside monophosphates. NCMIA probably reacts with amino groups near or at the binding site for triphosphate-Mg^{2+} complexes, resulting in loss of polymerase activity due to prevention of binding of triphosphates. On the other hand, the presence of triphosphates and Mg^{2+} protects against the effect of NCMIA possibly by induction of conformational changes around the triphosphate binding site in such a way that otherwise reactive amino groups near or at the site are no longer exposed to the action of NCMIA (Figure 6.2). The increase of the 5′ → 3′ exonuclease activity observed also when the polymerase reaction is protected by the dATP and Mg^{2+} may, therefore, be ascribed to the reaction of NCMIA with amino groups outside the triphosphate binding site.

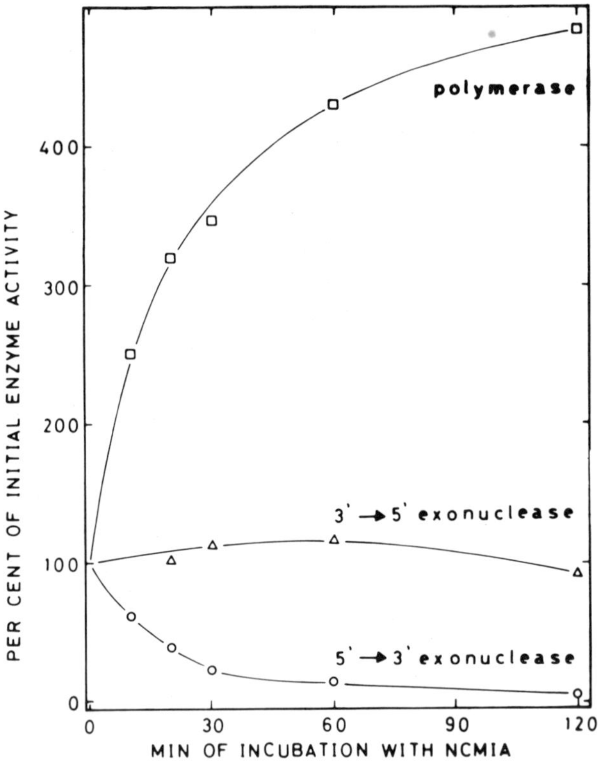

Fig. 6.2 Effect of reaction of *N*-carboxymethylisatoic anhydryde (NCMIA) on the three enzyme activities of the intact enzyme.

After cleavage of the enzyme by treatment with subtilisin NCMIA affects the polymerase activity and the 3′ → 5′ exonuclease activity similarly to that seen with the intact enzyme. In this case complete protection of the polymerase activity requires, however, higher concentration of triphosphates than with the intact enzyme. With the cleaved enzyme NCMIA causes, however, a decrease in the 5′ → 3′ exonuclease activity of the small fragment as compared to the increase observed with the intact enzyme. This indicates that there is an influence on the 5′ → 3′ exonuclease domain by the part of the intact enzyme that corresponds to the large fragment, causing an increase in activity upon treatment with NCMIA whether the triphosphate binding site is protected by the presence of triphosphate-Mg^{2+} complexes or not. After proteolytic cleavage of the enzyme this influence is lost, and an important amino group becomes exposed to react with NCMIA.

The polymerase activity is greatly influenced by monovalent cations. In glycine and phosphate buffers the reaction is probably completely dependent on, for example, K^+ or Na^+. In amine buffers like Tris-HCl or collidine-HCl the reaction is stimulated up to about 15-fold by K^+, Rb^+, Cs^+, or NH_4^+. Na^+ and Li^+ have little or not stimulating effect under these conditions, and the two ions counteract the stimulating effect of the four former ions. The activation constant for K^+ decrease with increasing concentration of amine buffers and with increasing pH. It is remarkable that the activity in the pH range 7.5 to 10.5 is virtually unchanged in amine buffers, provided that the concentration of KCl is optimal at the individual pH values (24).

The reaction is inhibited by high ionic strength. Inhibition is often observed at ionic strength higher than 0.2 and is dependent on pH and on the type of monovalent cations present (24). This inhibitory effect is probably due to a decreased binding of the enzyme molecule to the primer template molecule. It appears that optimal ionic conditions for the intact enzyme and the cleaved enzyme are different. With activated calf thymus DNA as template primer the optimal potassium phosphate concentration for polymerization is about 80 m*M* for the native enzyme while it is about 40 m*M* for the cleaved enzyme, and the activity of the cleaved enzyme at 40 m*M* is about twice that of the intact enzyme at 80 m*M*. This phenomenon makes it possible to monitor the proteolytic cleavage of the intact enzyme as an increase by a factor of more than 2 of the polymerase activity under appropriate ionic conditions (12) (Figure 6.3).

A number of base analogs may substitute for the naturally accuring bases in deoxyribonucleoside triphosphate in the polymerase reaction. The main requirement seems to be the ability of the base analogs to participate in formation of Watson-Crick base pairs. Thus, thymine in dTTP may be replaced by uracil, 5-bromouracil, or 5-fluorouracil, and cytosine in dCTP

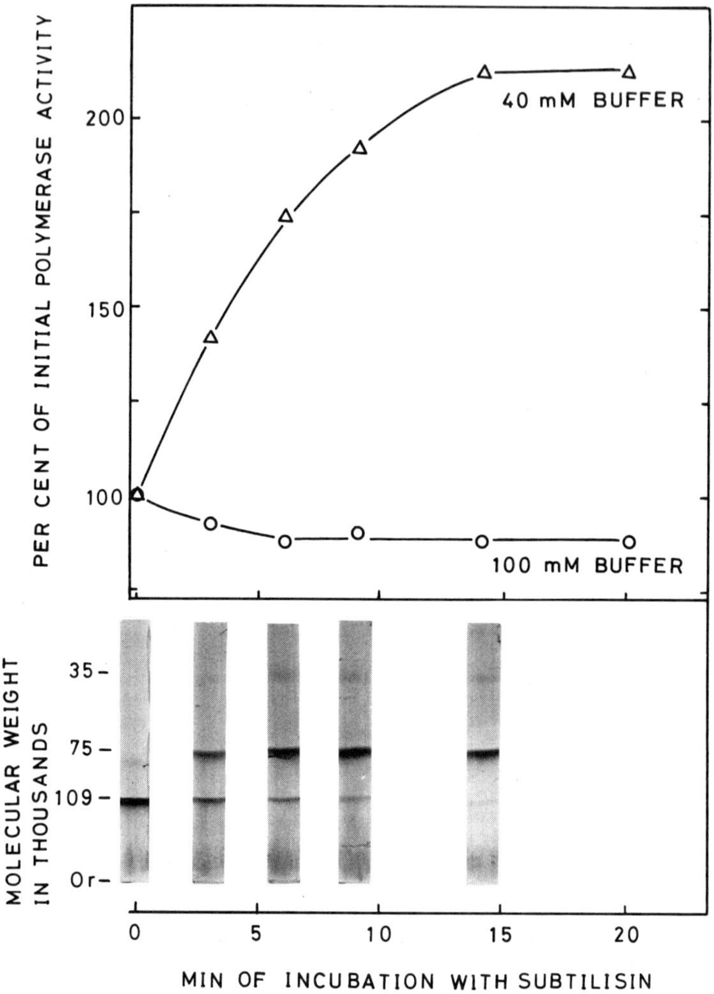

Fig. 6.3 Effect on the concentration of phosphate buffer in the assay mixture for the change in polymerase activity during proteolytic cleavage of DNA polymerase I. The template primer was activated calf thymus DNA. Samples were analyzed for both polymerase activity and for peptide chains by polyacrylamide gel electrophoresis in the presence of sodium dodecyl sulphate.

may be replaced by 5-methyl cytosin or 5-bromocytosin; and guanine in dGTP may be replaced by hypoxanthine (25).

The polymerase reaction is highly specific for deoxyribonucleoside triphosphates when Mg^{2+} is present in the reaction mixture. If, however, Mg^{2+} is replaced by Mn^{2+} and a deoxyribonucleoside triphosphate is replaced by the corresponding ribose compound, synthesis takes place, and the ribonucleoside monophosphate is incorporated into the polymer. The rate of incorporation of rCMP is almost as high as that of dCMP under normal conditions. For the three other ribose compounds the rate is considerably lower, that of rUMP being barely detectable (26). Divalent cations have, therefore, a remarkable effect on the polymerase reaction since a ribose compound like rATP is efficiently bound to the triphosphate binding site in the presence of Mg^{2+}, but it serves as a substrate only when Mg^{2+} is replaced by Mn^{2+}. The incorporation of ribonucleoside monophosphates into polymer structure also means that the enzyme under appropriate conditions is able to utilize a primer with a 3′ terminal ribonucleotide as substrate. This has also been directly demonstrated with synthetic primer molecules containing 3′ terminal ribonucleotides (27). Also the large fragment of the enzyme is able to utilize a primer molecule with a 3′ terminal ribonucleotide (28).

As already mentioned, a template-primer molecule may contain a 3′-OH gap or a 3′-OH nick. In the latter case the intact enzyme primarily catalyzes the coordinated extension of the primer chain with the concurrent 5′ → 3′ exonucleolytic hydrolysis of the strand in front of synthesis in such a way that the nick is moved along the helix without a net change in the amount of DNA (nick translation) (29) (Figure 6.4). The rate of incorporation of deoxyribonucleoside monophosphates in polymer structure is slower in this case than that obtained in a gap filling reaction (30). With the large fragment of the enzyme, elongation of the primer chain of a nicked DNA molecule takes place with concommittant displacement of the strand in front of synthesis (22, 31). This may result in a net synthesis of DNA without any simultaneous degradation (32). The structure of DNA obtained after a template-primer molecule has been replicated many times in the presence of DNA polymerase I differs from that of natural DNA in two ways. Branched molecular structures may be observed in the electron microscope. These molecules renature rapidly after treatment at high pH or high temperature (33). These features have been interpreted to have come about by a template switch during replication at a nick in a duplex DNA molecule. The template switch probably results from a displaced strand that has been adopted as a template for the polymerase reaction. Subsequently the displaced strand of the template-primer molecule may be restored to its original duplex state by a process termed branch migration. This permits the product strand to anneal to itself to form a hairpin structure (2, 34). This hairpin structure will renature rapidly after denaturation (see Figure 6.1). These structure analyses were performed in 1965 with a preparation of the intact enzyme. In view of the later finding of the nick translating effect of the intact enzyme and the strand displacing ability of the large

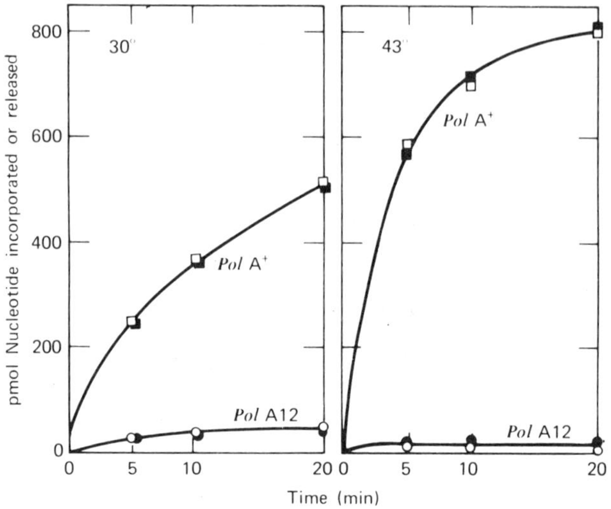

Fig. 6.4 Nick translation of DNA. In the presence of deoxyribonucleoside triphosphates the intact wild type enzyme (Pol A$^+$) catalize a coordinated incorpoation of nucleotides by polymerase action (□) and release of nucleotides by 5′ → 3′ exonuclease action (■) at two temperatures. Under the conditions of nick translation the mutant enzyme (Pol A12) has very low polymerase activity (○) and 5′ → 3′ exonuclease activity (●) at 30°C. The two activities are almost absent at 43°C. Reproduced from ref. 44 with permission by the American Society of Biological Chemists.

proteolytic fragment it seems likely that the branched, undenaturable product referred to above may have been formed due to the presence of the large fragment of the enzyme in the preparation used.

The intact enzyme is able to utilize RNA as template, in fact the synthetic homopolymer poly A·poly dT has been found to be a very efficient template-primer molecule giving rise to a rapid synthesis of poly d(T) (35). All-ribonucleotide template-primer molecules like poly(A)·oligo(U), poly(C)·oligo(I) and poly(I)·oligo(C) also support synthesis efficiently (36) as do several natural RNA molecules with secondary structures (37, 38).

The question of processivity of the polymerase reaction has been investigated with the intact enzyme. At one extreme the polymerizing enzyme may be released from the template-primer molecule after the primer strand has been extended by one nucleotide (nonprocessive mechanism or processivity of one) and at the other extreme the enzyme may be released only after extension of the primer by a large number (n) of nucleotides (processivity of n). It appears that for DNA polymerase I the degree of processivity is greatly influenzed by a number of factors like temperature and ionic strength. Although the enzyme primarily acts in a highly processive manner with a processivity of up to more than 2000 (39) it may under certain conditions be almost nonprocessive (40).

The 5′ → 3′ Exonuclease Activity This activity has been studied both with the intact wild type enzyme and its small fragment as well as with several mutant enzymes.

With nicked double stranded DNA as a substrate both the intact wild type enzyme and the small fragment give rise to products consisting of 70 to 80% mononucleotides, 16 to 18% dinucleotides, 3 to 4% trinucleotides, and less than 1% larger oligomers. Also thymine dimers introduced by ultraviolet light are excised by both enzymes. The nature of the reaction giving rise to formation of oligonucleotides has been studied with the intact enzyme using $d(A)_{4000}$ annealed to $pC(pC)_3(pT)_{200}$ as substrate. Although the 5′ end of the latter polymer does not form hydrogen bonds with $d(A)_{4000}$ it is nevertheless released by the 5′ → 3′ exonuclease activity as $(pC)_4$ pT, $(pC)_4(pT)_2$, and $(pC)_4(pT)_3$ (40). This shows the specificity of the enzyme for double stranded DNA and indicates that the enzyme activity is able to hydrolyze double stranded DNA in the duplex region beyond a set of nonhydrogen-bonded residues, excising oligonucleotides containing these residues, that is the enzyme is able to excise mismatched sequences at the 5′ end of double stranded DNA. This holds also for thymine dimers, which do not form effective hydrogen bonds with the complementary strand. The finding that di- and trinucleotides are excised by both the intact enzyme and the small fragment from a completely double stranded substrate may mean that the enzyme is able to bring about a fraying of a few nucleotide at the 5′ end. By hydrolysis beyond this frayed structure short oligonucleotides are formed. In the presence of nicked DNA the 5′ → 3′ exonucleolytic activity of the intact enzyme is much stimulated by the the four deoxyribonucleoside triphosphates leading to simultaneous synthesis. Under these conditions the rate of synthesis matches the rate of 5′ → 3′ exonuclease activity. The net result of this concerted polymerase and 5′ → 3′ exonuclease action is a nick translation propagating the nick along the DNA molecule (see Figure 6.4). Under these conditions a substantial amount of the products of the 5′ → 3′ exonuclease action is oligonucleotides four to eight nucleotides long, products that are almost lacking in the absence of simultaneous synthesis. Simultaneous synthesis apparently causes a more pronounced fraying of the 5′ end of the strand in front of synthesis giving rise to longer oligonucleotides and increased rate of 5′ → 3′ exonuclease activity (41). These findings probably mean that the domain responsible for polymerase activity and the domain responsible for 5′ → 3′ exonuclease activity are spacially arranged in a way that gives rise to a coordination of the two activities.

In the absence of DNA the two proteolytic fragments of the enzyme do not show any physical affinity to each other as they sediment independently in a sucrose gradient. In the presence of DNA containing a number of nicks which are, however, fewer than the number of the two fragment molecules, the same coordinated effect of the polymerase and the 5′ → 3′ exonuclease activity is observed as with the intact enzyme. That is, in the presence of the large fragment and triphosphates the 5′ → 3′ exonuclease activity is substantially increased, the polymerase and the exonuclease activities match, and four to eight nucleotides long oligomers are formed in substantial amounts. If the molar amounts of the two enzyme fragments are not in excess over the molar amounts of nicks in the DNA no coordinated action of the polymerase and the 5′ → 3′ exonuclease activity is seen. This indicates that under these conditions the two fragments are probably distributed randomly at the nicks. It also suggests lack of mutual binding between the two fragments in the presence of nicked DNA (41). These findings show that the coordinated action of the two domains in the enzyme does not normally require that they are part of the same polypeptide chain as long as they are located next to each other at a nick in a DNA molecule. The stimulating effect of the large fragment is, however, not always dependent on simultaneous synthesis. An example of this is the excision of thymine dimers from nicked DNA that has been irradiated with ultraviolet light. These dimers appear to be excised by the small fragment at a rate that is only about one-tenth of that obtained with the intact enzyme. However, in the presence of the large fragment the rate of dimer excision increases by a factor of about 4, and if both the large fragment and the four deoxyribonucleoside triphosphates are present a further increase in 5′ → 3′ exonuclease activity is obtained. In contrast, there is no difference in 5′ → 3′ exonuclease activity with nonirradiated DNA when the intact enzyme in the absence of triphosphates and the small fragment is compared. This suggests that at a nick close to thymine dimers some interaction between the two fragments takes place with the result of a more rapid excision of thymidine dimers even in the absence of simultaneous synthesis (42).

It has already been mentioned that reaction with NCMIA causes a pronounced change in the 5′ → 3′ exonuclease activity. Reaction with the intact enzyme results in a four- to fivefold increase in activity, while reaction with the cleaved enzyme results in a rapid loss of this activity. This indicates an interaction between the two domains of the intact enzyme molecule. Another example of such an interaction is the change in susceptibility to ionic strength of the 5′ → 3′ exonuclease upon proteolytic cleavage of the intact enzyme. The cleavage is accompanied by appearance at low salt concentrations of an ionic strength optimum peak with an activity that over a broad range of ionic strength is considerably higher than that of the intact enzyme. The effect is especially pronounced (a sixfold increase) when

DNase treated poly d(A-T) is a substrate. Under the conditions used the two enzyme fragments would be randomly distributed at the nicks in the substrate molecule. The increase in 5′ → 3′ exonuclease activity has been used to follow the progress of the proteolytic cleavage (12).

A RNA strand in hybrid structure with a DNA strand as in poly (dT)·oligo (rA) may be degraded by the 5′ → 3′ exonuclease activity of the intact enzyme (28). It is probably of great physiological importance that the primer RNA of the bacteriophage M13 single stranded DNA may thus be removed by 5′ → 3′ exonuclease activity of DNA polymerase I (43).

A number enzymes from mutants of *E. coli* have been isolated and found to have altered 5′ → 3′ exonuclease activity. In the enzyme isolated from the *pol* A12 mutant the coordinated action of the polymerase and the 5′ → 3′ exonuclease activities has been lost. The separate specific activity of the polymerase as measured in a gap filling reaction, of the 5′ → 3′ exonuclease, and of the 3′ → 5′ exonuclease are nearly normal. However, the 5′ → 3′ exonuclease activity is not stimulated by the presence of the deoxyribonucleoside triphosphate, and nick translation is abnormally low at 30°C and almost absent at 43°C indicating a substantial defect in the coordinated action of the two domains of the enzyme. This can probably be ascribed to a structural perturbation of the mutant enzyme molecule (see Figure 6.4). It has a much less rigid and compact structure than that of the wild type enzyme since it has a slower electrophoretic mobility and a slower sedimentation rate, although the molecular weight is unchanged. Also, this more loose structure probably is the reason for its increased temperature sensitivity, especially at low salt concentrations. Stability of polymerase, 3′ → 5′ exonuclease, and 5′ → 3′ exonuclease is affected and both domains of the enzyme molecule are, therefore, affected by the mutation. It seems, therefore, that a more compact rigid structure of the molecule is essential for obtaining the optimal spacial arrangement necessary for a coordinated effect of the activities of the two domains (30). Two enzymes, from the mutants *pol* Aexl and *pol* A107, respectively, have normal polymerase activity but abnormally low or absent 5′ → 3′ exonuclease activity. *Pol* Aexl is a temperature sensitive conditional lethal mutant, and the DNA polymerase present in this mutant has not obviously been altered structurally. The polymerase and the 3′ → 5′ exonuclease activities are not grossly different from those of the wild type enzyme at either 30°C or 43°C. The 5′ → 3′ exonuclease activity is, however, reduced at both temperatures. The presence of the four deoxyribonucleoside triphosphates stimulates the 5′ → 3′ exonuclease activity at 30°C but it almost completely abolishes this activity at 43°C, that is the 5′ → 3′ exonuclease activity is temperature sensitive only in the presence of triphosphates. At elevated temperature the capacity to promote nick translation is, therefore, lost, and polymerization proceeds consequently exclusively with simultaneous strand displacement (44). The enzyme from the polA107 mutant has normal polymerase and 3′ → 5′ exonuclease activity, while the 5′ → 3′ exonuclease activity is much lower than that of the wild type enzyme, and also in this case DNA synthesis on nicked molecules proceeds mainly with simultaneous strand displacement and branched molecules are often formed (45). Like the wild type enzyme this mutant enzyme is also cleaved into a 35,000 and 75,000 mol wt fragment, and the cleavage is under appropriate assay condition likewise accompanied by a substantial increase in polymerase activity. The 5′ → 3′ exonuclease activity is at the same low level before and after cleavage (46). These findings indicate that in the mutant enzyme the domain corresponding to the small fragment has the same influence on the polymerase activity as found in the wild type enzyme although the former enzyme almost completely lacks 5′ → 3′ exonuclease activity. The *pol* A107 mutation is therefore probably located in the region of the *pol* A gene coding for the small fragment and is a point mutation.

The 3′ → 5′ Exonuclease Activity This activity has been studied with both the intact enzyme and its large proteolytic fragment. The activity requires single stranded DNA as substrate, and it catalyzes the hydrolytic release of mononucleotides from the 3′ end of the DNA strand. The enzyme activity requires the presence of K^+ (22). The very low activity that is obtained with double stranded DNA as substrate is probably due to a partial local melting at the 3′ terminus of the molecule (8, 18).

Using [^{3}H] poly dT_{50} as a substrate it has been found that the 3′ → 5′ exonuclease activity is inhibited by deoxyribonucleoside-5′-phosphates. Whereas the polymerase function is unaffected, the 3′ → 5′ exonuclease activity decreases with increasing concentrations of, for example, dTMP. The kinetics of the effect is consistent with a competition of dTMP with poly dT_{50} for the active site of the exonuclease. All four deoxyribonucleoside monophosphates are inhibitors, although with different inhibitor constants (47). Independent experiments have shown that the intact DNA polymerase I molecule contains a single binding site for deoxyribonucleoside monophosphates (48). The dissociation constants obtained in the binding studies and the inhibitor constants obtained in the exonuclease experiments are in excellent agreement. Furthermore, the structural requirements for inhibitors of the 3′ → 5′ exonuclease function are the same as those for binding of nucleotides to the intact enzyme. Only nucleotides containing a 5′-phosphoryl group and a 3′-hydroxyl group whether they are of the ribonucleotide or of the 2′ deoxyribonucleotide type are active. These findings strongly suggest that the binding site on the enzyme for nucleoside monophosphates is the inhibitor site of the 3′ → 5′ exonuclease activity.

Not only single stranded DNA is degraded by the 3′ → 5′ exonuclease activity but also unpaired or mispaired 3′-terminal nucleotides in template primer molecules are efficiently removed by this activity; even a single 3′-unpaired terminal nucleotide is removed. This is the case both when the unpaired 3′-terminus nucleotide is a deoxyribonucleotide and when it is a ribonucleotide. In both cases Mg^{2+} is required (see Figure 6.5) (21, 27). When two unpaired terminal nucleotides are ribonucleotides they are both removed by the 3′ → 5′ exonuclease although at a reduced rate (27). Chain extension of primer molecules with unpaired termini takes place only after such termini have been removed and a base paired terminus is reached. Under conditions for synthesis the first base paired terminus is conserved, whether it is a deoxyribonucleotide or a ribonucleotide. These observations have been obtained when either the 3′ → 5′ exonuclease activity is a part of the intact enzyme or the large fragment (21, 27). These findings show that mismatched bases incorporated by erroneous reactions at the 3′-termini of extended DNA chains are excised by the 3′ → 5′ exonuclease activity. This proofreading effect of the exonuclease, therefore, contributes to the fidelity of template copying. In agreement with this it has been found (49) that inhibition of the 3′ → 5′ exonuclease activity of DNA polymerase by nucleoside 5′-monophosphates increases error frequency during DNA synthesis *in vitro* (49).

The presence of pyrimidine dimers formed by ultraviolet irradiation of the template strand appears to block the progression of the DNA polymerase reaction. It has been found with the large fragment of DNA polymerase I that when synthesis is blocked in this way there is a rapid conversion of deoxyribonucleoside triphosphates to monophosphates in contrast to what is seen during DNA synthesis with unirridiated DNA template. These results have been interpreted in the following way: The photoproducts in the template strand do not prevent incorporation of nucleotides in the primer strand, but bases incorporated opposite the photoproducts are registered as being mismatched, and the terminal nucleotides are, therefore, excised by the 3′ → 5′ exonuclease activity. This results in prevention of chain growth and in hydrolysis of triphosphates to monophosphates and pyrophosphate, giving rise to idling of the DNA polymerase reaction (50). It would be interesting to learn if inhibition of the 3′ → 5′ exonuclease activity with, for example, nucleoside monophosphates would allow chain growth to occur opposite pyrimidine dimers in the template strand.

It has already been mentioned that by treatment with the amino group reagent NCMIA it is possible to suppress the polymerase activity of both the intact enzyme and its large fragment almost completely without changing the 3′ → 5′ exonuclease activity (7).

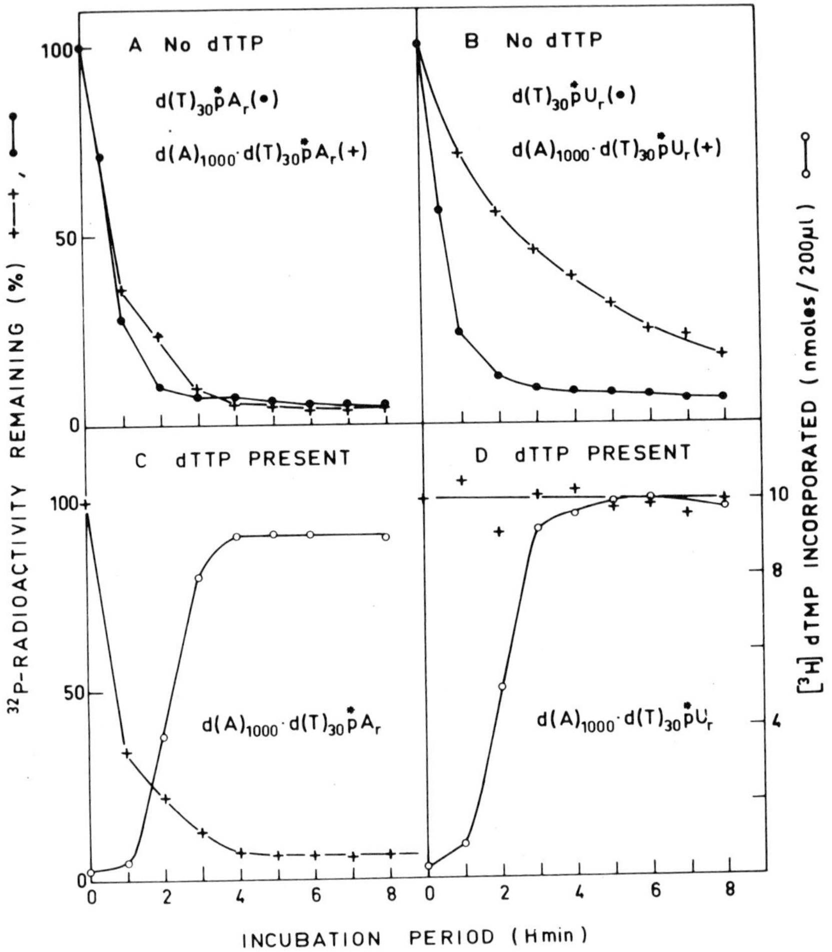

Fig. 6.5 Specificity of the 3′ → 5′ exonuclease activity of the large fragment of DNA polymerase I with primers containing a single ^{32}P-labeled ribonucleotide residue at the 3′ end. the primers were d(T) $_{30}$pAr (in *A* and *C*) or d(T) $_{30}$pUr (in *B* and *C*). The absence of dTTP (*A*) the terminal rAMP residue was removed rapidly both in the absence (O) and in the presence (+) of an excess of a template strand d(A) $_{1000}$. The terminal rUMP residue (*B*) was, however, removed much slower in the presence (+) than in the absence (O) of template. In the presence of dTTP (10 nm/200 μl) and an excess of template (in *C* and *D*) the terminal rAMP residue was removed rapidly (*C*, +) while the terminal rUMP residue was not removed at all (D, +). dTMP was rapidly incorporated (O) in both cases. Reproduced from ref. 27 with permission of the American Society of Biological Chemists.

It has not been possible to separate the polymerase activity from the 3′ → 5′ exonuclease activity by limited proteolysis of the large fragment, and it can not be totally excluded that the same active site is responsible for the polymerase and the 3′ → 5′ exonuclease reaction, the former reaction being the formation of a phosphodiester bond between the 3′-end group of a template primer molecule and the β-phosphate of a reacting deoxyribonucleoside triphosphate and the latter reaction being the hydrolytic cleavage of a phosphodiester bond at the 3′-end of a single stranded DNA molecule. The following properties of the two activities do, however, support the notion that they are confined to two different active sites of the large fragment, perhaps harbored in different domains of the molecule. This would imply that the primer terminus site for hydrolysis is distinct from the primer terminus site for polymerization. The 3′ → 5′ exonuclease reaction may be strongly inhibited by nucleoside monophosphates without any effect on the polymerase reaction. On the other hand, the polymerase reaction may be almost completely abolished by reaction of some amino groups in the enzyme molecule with NCMIA under conditions where the 3′ → 5′ exonuclease activity is unaffected. Ribonucleotides at the 3′-terminus may be removed by the 3′ → 5′ exonuclease activity in the presence of Mg^{2+}, while the incorporation of ribonucleotides into a DNA primer strand requires Mn^{2+}. The polymerase function is active in amine buffers like Tris buffer, although the rate of the reaction is increased by the presence of monovalent cations like K^+; in contrast the 3′ → 5′ exonuclease is inactive in Tris buffer in the absence of K^+.

6.4 REMARKS ON THE FUNCTION AND EVOLUTION OF DNA POLYMERASE I

The nick translating property of DNA polymerase I of *E. coli* makes it excellently suited to play an essential role in repair processes like those taking place *in vivo* during excision of pyrimidine dimers introduced by ultraviolet irradiation or excision of mismatched bases. The first step in such processes is probably an incision by an exonuclease that recognizes distortion of the DNA double helix. If the nick introduced by such an endonuclease creates a 3′OH terminus at the 5′ side of the distorted region the following steps of repair consisting of excision of the damaged region and replacement of the excised region may be catalyzed by the nick translating effect of DNA polymerase I. The final step of the repair process may be catalyzed by polynucleotide joining enzyme (DNA ligase), which would intervene to close the nick. The finding that mutants deficient in the polymerase activity (51) or the 5′ → 3′ exonuclease activity (45) of DNA polymerase I are sensitive to ultraviolet irradiation supports the notion that the enzyme has essential repair functions.

The nick translating property of DNA polymerase I may also play an important role in DNA replication. As already mentioned, the DNA polymerase reaction requires the presence of a primer strand with a 3′OH group. This poses a problem in connection with replication of DNA since the two newly synthesized DNA strands at a replication fork have opposite orientation and since the polymerization as catalyzed by all known DNA polymerases is uniquely 5′ → 3′. The problem of the synthesis of the strand whose overall direction of synthesis is 3′ → 5′ is probably solved by the finding that DNA replication may occur by synthesis of small fragments joined by DNA ligase. The 5′ → 3′ synthesis of short fragments could, therefore, proceed in the overall 3′ → 5′ direction. The problem connected with such discontinuous DNA synthesis is how the synthesis of each fragment is initiated. There is compelling evidence that RNA serves as a primer in discontinuous DNA synthesis and that this primer is excised prior to incorporation of DNA fragments into the chromosome. In *in vitro* experiments on the conversion of single stranded circular M 13 bacteriophage DNA to fully double stranded circular structure it has been shown that DNA polymerase I is required for the removal of a short RNA primer fragment and for filling the corresponding gap (43). This process could very well be the result of the nick translating activity of the enzyme. The involvement of DNA polymerase I *in vivo* in a process of this type is supported by the finding that mutants low in DNA polymerase I are characterized by a greatly retarded joining of nascent DNA fragments. This has been ascribed to the decreased nick translating activity (3).

The two proteolytic fragments together of DNA polymerase I have nick translating activity only when they are present in excess over the number of nicks. It thus seems to be a distinct advantage for the organism that harbors this enzyme that the polymerizing function and the 5′ → 3′ exonuclease function are present in the same polypeptide chain and confined to different domains. It is probable that the enzyme has evolved by the fusion of separate genes for the two functions. Before fusion the two structural genes may have been in a neighboring position and they may have fused by two sequential mutational events near the intercistronic region. These events may have been frame shift mutations allowing the message punctuation signals originally separating the two structural genes to be translated in the reading frame. The signals would therefore not be recorded and the two genes translated as one polypeptide chain. The region that has undergone a frame shift mutation may create a vulnerable point for proteolytic cleavage of the translation product. The two domains would be formed sequentially during the translation beginning with the one corresponding to the small fragment containing the 5′ → 3′ exonuclease activity.

REFERENCES

1 A. Kornberg (1962), *Enzymatic Synthesis of DNA*, John Wiley, New York.

2 A. Kornberg (1974), *DNA Synthesis*, W. H. Freeman, San Francisco.

3 I. R. Lehman and D. G. Uyemura (1976), *Science* **193,** 963–969.

4 T. M. Jovin, P. T. Englund, and L. L. Bertsch (1969), *J. Biol. Chem.* **244,** 2996–3008.

5 C. F. Springgate, A. S. Mildvan, R. Abramson, J. L. Engle, and L. A. Loeb (1973), *J. Biol. Chem.* **248,** 5987–5993.

6 T. M. Jovin, P. T. Englund and A. Kornberg (1969), *J. Biol. Chem.* **244,** 3009–3018.

7 H. Klenow and K. Overgaard-Hansen, unpublished results.

8 H. Jacobsen, H. Klenow, and K. Overgaard-Hansen (1974), *Eur. J. Biochem.* **45,** 623–627.

9 D. L. Brutlag, M. R. Atkinson, P. Setlow, and A. Kornberg (1969), *Biochem. Biophys. Res. Commun.* **37,** 982–989.

10 H. Klenow and I. Henningsen, (1970), *Proc. Nat. Acad. Sci. USA* **65,** 168–175.

11 H. Klenow and K. Overgaard-Hansen (1970), *FEBS Letters* **6,** 25–27.

12 H. Klenow, K. Overgaard-Hansen, and S. Patkar (1971), *Eur. J. Biochem.* **22,** 371–381.

13 I. R. Lehman and J. R. Chien (1973), *J. Biol. Chem.* **248,** 7717–7723.

14 P. Setlow and A. Kornberg (1972), *J. Biol. Chem.* **247,** 232–240.

15 P. Setlow, D. Brutlag, and A. Kornberg (1972), *J. Biol. Chem.* **247,** 224–231.

16 W. S. Kelley, K. Chalmers, and N. Murray (1977), *Proc. Nat. Acad. Sci. USA* **74,** 5632–5636.

17 A. Kornberg, (1969), *Science* **163,** 1410–1418.

18 R. P. Klett, A. Cerami, and E. Reich (1968), *Proc. Nat. Acad. Sci. USA* **60,** 943–950.

19 M. P. Deutscher and A. Kornberg (1969), *J. Biol. Chem.* **244,** 3029.

20 P. T. Englund, R. B. Kelly, and A. Kornberg (1969), *J. Biol. Chem.* **244,** 3045–3052.

21 D. Brutlag and A. Kornberg (1972), *J. Biol. Chem.* **247,** 241–248.

22 H. Klenow and K. Overgaard-Hansen (1972), in R. D. Wells and R. B. Inman, eds., *DNA Synthesis in vitro*, University Park Press, Baltimore, p. 13–25.

23 P. T. Englund, J. A. Huberman, T. M. Jovin, and A. Kornberg (1969), *J. Biol. Chem.* **244,** 3038–3044.

24 H. Klenow and I. Henningsen (1969), *Eur. J. Biochem.* **9,** 133–141.

25 M. J. Bessman, I. R. Lehman, J. Adler, S. B. Zimmerman, E. S. Simms, and A. Kornberg (1958), *Proc. Nat. Acad. Sci. USA* **44,** 633–639.

26 P. Berg, H. Fancher, and M. Chamberlin (1963), in H. J. Vogel, V. Bryson, and J. O. Lampen, eds., *Informational Macromolecules*, Academic Press, New York, p. 467.

27 H. Kössel and R. Roychoudhury (1974), *J. Biol. Chem.* **249,** 4094–4099.

28 R. Roychoudhury and H. Kössel (1973), *Biochem. Biophys. Res. Commun.* **50,** 259–265.

29 R. B. Kelly, N. R. Cozzarelli, M. P. Deutscher, I. R. Lehman, and A. Kornberg (1970), *J. Biol. Chem.* **245,** 39–45.

30 D. Uyemura and I. R. Lehman (1976), *J. Biol. Chem.* **251,** 4078–4084.

31 Y. Masamune and C. C. Richardson (1971), *J. Biol. Chem.* **246,** 2692–2701.

32 D. Brutlag and A. Kornberg (1970), *Fed. Proc.* **29,** 406.

33 R. B. Inman, C. L. Schilkraut, and A. Kornberg (1965), *J. Mol. Biol.* **11,** 285–292.

34 C. S. Lee, R. W. Davis, and N. Davidson (1970), *J. Mol. Biol.* **48,** 1–22.

35 J. D. Karkas (1972), *Proc. Nat. Acad. Sci. USA* **69,** 2288–2291.

36 J. D. Karkas (1973), *Proc. Nat. Acad. Sci. USA* **70,** 3834–3838.

37 E. C. Travaglini and L. A. Loeb (1974), *Biochemistry* **13,** 3010–3017.

38 S. C. Gulati, D. L. Kacian, and S. Spiegelman (1974), *Proc. Nat. Acad. Sci. USA* **71,** 1035–1039.

39 D. Uyemura, R. Bambara, and I. R. Lehman (1975), *J. Biol. Chem.* **250,** 8577–8584.

40. W. R. McClure and T. M. Jovin (1975), *J. Biol. Chem.* **250,** 4073–4080.

41 P. Setlow and A. Kornberg (1972), *J. Biol. Chem.* **247,**232–240.

42 E. C. Friedberg and I. R. Lehman (1974), *Biochem. Biophys. Res. Commun.* **58,** 132–139.

43 K. Geider and A. Kornberg (1974), *J. Biol. Chem. 249,* 3999–4005.

44 D. Uyemura, D. C. Eichler, and I. R. Lehman (1976), *J. Biol. Chem.* **251,** 4085–4089.

45 H. L. Heijneker, D. J. Ellens, R. H. Tjeerde, B. W. Glickman, B. van Dorp, and P. H. Pouwels (1973), *Molec. Gen. Genet.* **124,** 83.

46 H. L. Heijneker and H. Klenow (1975), in R. B. Setlow and P. C. Hanawalt, eds., *Molecular Mechanisms for the Repair of DNA*, Plenum, p. 219–223.

47 B. G. Que, K. M. Downey, and A. G. So (1978), *Biochemistry* **17,** 1603–1606.

48 J. Huberman and A. Kornberg (1970), *J. Biol. Chem.*, **245,** 5326–5334.

49 J. J. Byrnes, K. M. Downey, B. G. Que, M. Y. W. Lee, V. L. Black, and A. G. So (1977), *Biochemistry* **16,** 3740–3746.

50 G. Villani, S. Boiteux, and M. Radman (1978), *Proc. Natl. Acad. Sci. USA* **75,** 3037–3041.

51 T. Kato and S. Kondo (1970), *J. Bacteriol.* **104,** 570–881.

Chapter 7

FATTY ACID SYNTHETASE COMPLEXES

ECKHART SCHWEIZER

Lehrstuhl für Biochemie
Institut für Mikrobiologie und Biochemie
Universität Erlangen-Nürnberg
Erlangen, Germany

The main purpose of this article is to show how it was possible, principally as a result of extensive genetic analyses, to explain the molecular properties of a multienzyme complex such as the yeast fatty acid synthetase. These genetic data, together with biochemical evidence, finally led to the concept of multifunctional proteins being the constituent parts of this enzyme complex. Since other aspects of the fatty acid synthetase system, like the mechanism and control of the reaction process or the functional and structural characteristics of the various known fatty acid synthetases, have been treated in several recent and extensive reviews (1–6), these will be mentioned in less detail here.

7.1 THE FATTY ACID SYNTHETASE MULTIENZYME SYSTEM

The biosynthesis of long chain fatty acids is one of the basic capabilities of living organisms with only a very few exceptions known (7–9). As it is outlined in Figure 7.1, this entire biosynthetic process is the result of the combined action of a series of different enzyme systems. Among these the fatty acid synthetase multienzyme system occupies a central position since it catalyzes the *de novo* formation of long chain saturated fatty acids from acetate and malonate. The fundamental studies performed in Lynen's laboratory on the fatty acid synthetase of yeast have revealed that the overall process of

$$\text{acetyl-CoA} + 7\ \text{malonyl-CoA} + 14\ \text{NADPH} + 14\ H^+ \rightarrow \\ \text{palmityl-CoA} + 7\ \text{CoASH} + 7\ CO_2 + 14\ \text{NADP}^+ + 14\ H_2O$$

may be essentially resolved into seven repetitive cycles of five different partial reactions, each cycle elongating the growing fatty acyl chain by another unit of two carbon atoms (10). For the enzyme system of yeast, as for all other multienzyme complex type fatty acid synthetases (see below), one cycle of these elongation steps is thought to proceed according to the scheme shown in Figure 7.2. During the elongation process the substrates malonate and acetate, as well as all acyl intermediates formed, are covalently attached to specific binding sites on the enzyme protein. According to their chemical nature, their substrate specificities, and characteristic inhibitor sensitivities three different types of acyl binding sites are present in the fatty acid synthetases of yeast (11–14) and of other organisms where it has been looked for (15–17): a serine OH-group, a cysteinyl-, and a phosphopantetheine-SH group. As it has been first demonstrated by Vagelos and co-workers (18) for the *E. coli* fatty acid synthetase, 4′-phosphopantetheine is attached by phosphodiester linkage to a specific component protein of the enzyme system ("acyl carrier protein"). The cysteinyl SH-group is considered as part of the condensing enzyme, the OH-group(s) as active site of the acetyl-, malonyl-, and palmityl-transferase component enzymes. In Figure 7.2, —SH_c designates the "central" or pantetheine SH-group and —SH_p the "peripheral" or cysteinyl SH-group. The —SH_p group binds first the priming substrate, acetate, and subsequently also accepts its higher homologues, which are transiently formed during the elongation process. Thereby new malonate molecules are allowed to enter the central acyl binding site in order to condense, subsequently, with acetate or its higher homologues. Besides the five elongation reactions indicated in Figure 7.2 and in accordance with the above mentioned view of the fatty acid synthetase reaction mechanism two additional component enzymes have been identified: one of them transacylating acetate from coenzyme A to its enzyme binding site, the other catalyzing the transfer of palmitate from the enzyme to either coenzyme A (yeast, bacteria) or water (animals). Actually each of the three fatty acid synthetase-catalyzed acyl transfers is a several

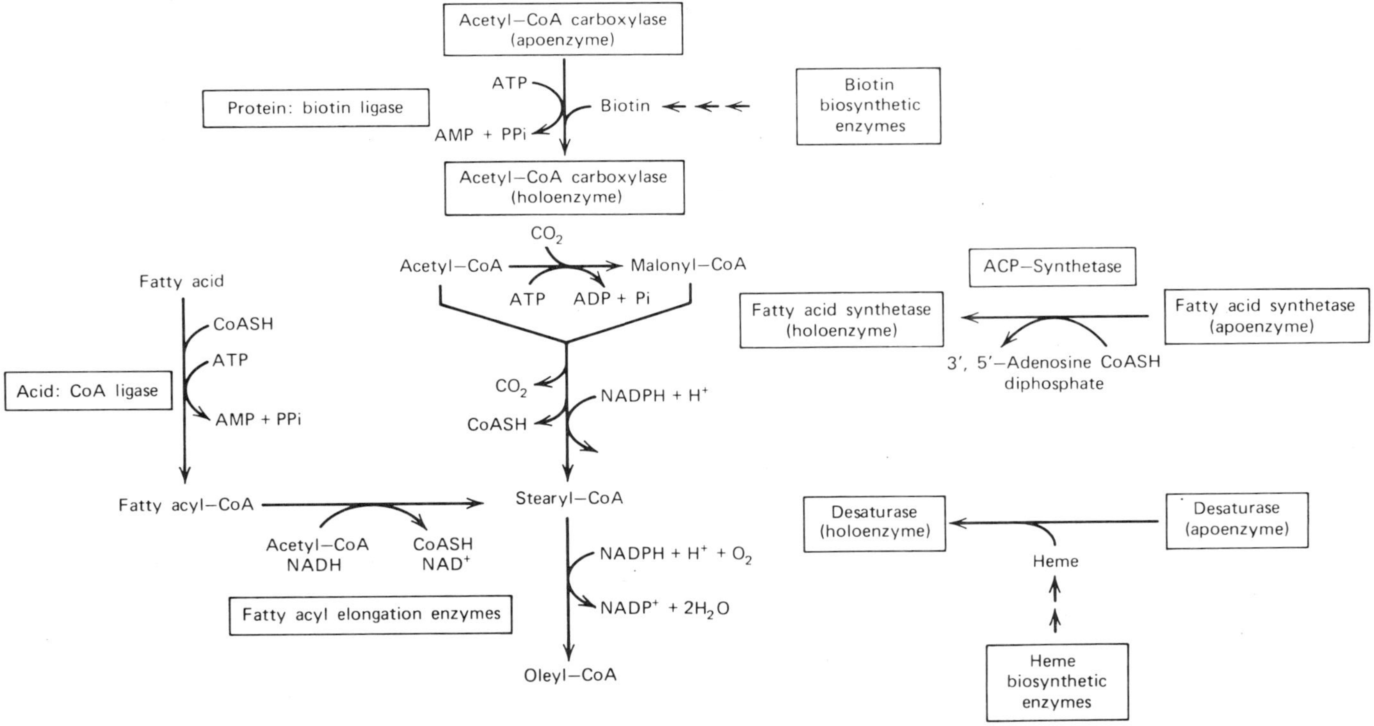

Fig. 7.1 Pathways of eucaryotic fatty acyl CoA biosynthesis.

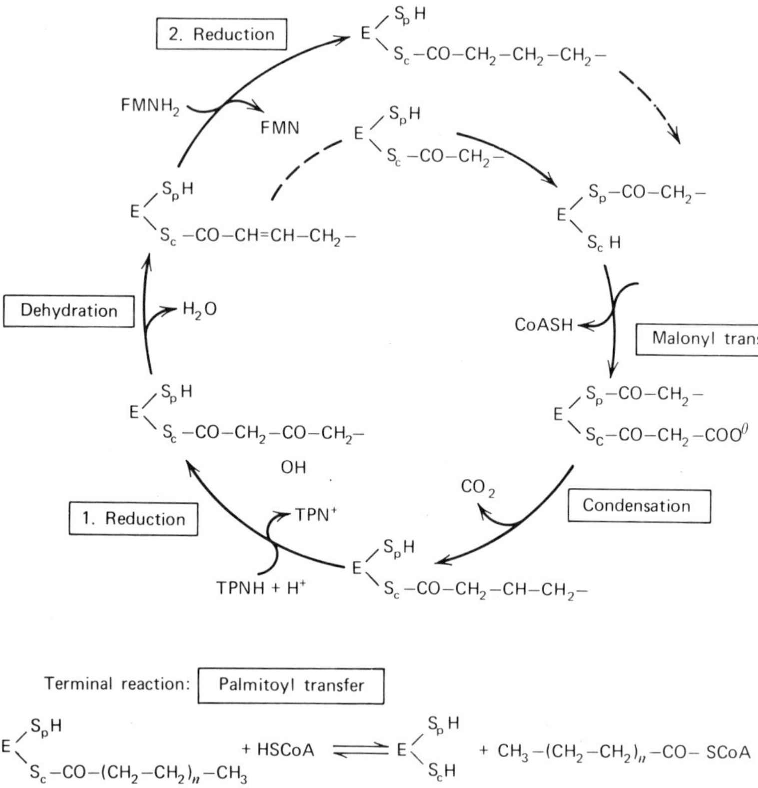

Fig. 7.2 Fatty acyl chain elongation reactions in the yeast fatty acid synthetase complex (1).

step reaction by itself. Starting with acyl coenzyme A, the acyl group is transferred through an intermediate involving the OH-group of the respective transferase to phosphopantetheine and, as in the case of acetate, from there even further to cysteine.

Apparently the chemical mechanism of fatty acid synthesis is essentially the same in all biological systems so far studied. However, this uniformity ends when the molecular organization of the constituent component enzymes is considered. According to their macromolecular structure the known fatty acid synthetases may be generally divided into two different groups: the so-called type I fatty acid synthetases, which constitute polyfunctional proteins with all known component enzyme activities contained in only two, or even one, polypeptide chain; and the type II fatty acid synthetases, which are nonaggregated and consist of a series of independent and monofunctional enzyme entities. The fatty acid synthetase multienzyme complexes are found in animal tissues, in eucaryotic microorganisms, and in some bacteria, while the nonassociated and possibly more primitive synthetases occur in most procaryotic systems like eubacteria, bluegreen algae, and plant chloroplasts (for a review see 5, 6). Nevertheless, some experimental evidence suggests that *in vivo* the type II fatty acid synthetase component enzymes may be organized in a loose manner, too. Radioautographic data obtained with acyl carrier protein labeled *E. coli* cells indicating a preferential association of radioactivity with the cell membrane lend some support to this view (19).

Among the type I fatty acid synthetases the multienzyme complex of yeast has been especially well characterized by both biochemical and genetic methods. According to Lynen (10) the intact complex has a molecular weight of 2.3 $\times$ 10^6 daltons. By both SDS-polyacrylamide gel electrophoresis of the SDS-denatured enzyme (20) and by disc-gel electrophoresis in the presence of urea of a chemically modified fatty acid synthetase (21) the yeast complex is shown to contain two nonidentical subunits α and β (Figure 7.3). Both subunits appear to be present in about equal amounts and have molecular weights of 185,000 (α) and 180,000 (β) daltons, according to their mobility in SDS gels. This finding, together with earlier determinations of the flavin (FMN) and pantetheine content of the intact yeast fatty acid synthetase complex (22, 23), suggests an $\alpha_6\beta_6$ molecular structure of this multienzyme complex. Genetic studies performed in our own laboratory indicated that the subunits α and β are

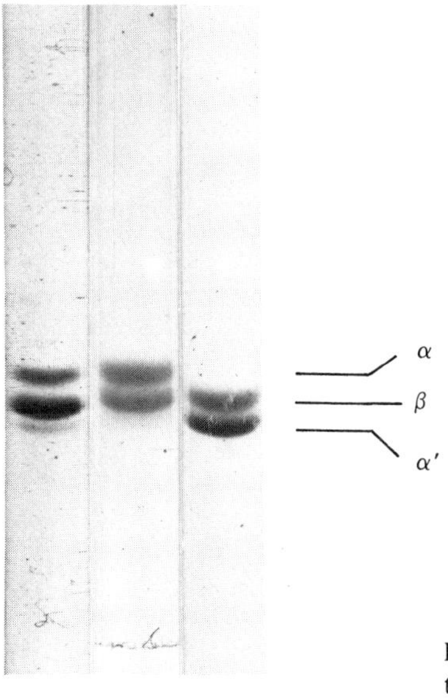

a b c

Fig. 7.3 SDS polyacrylamide gel electrophoresis patterns of purified wild type (*a*), *fas* 2-123 (*b*) and *fas* 2-46 (*c*) fatty acid synthetases.

genetically encoded by two different and unlinked loci on the yeast genome, *fas 2* and *fas 1*, respectively (24). Accordingly, each subunit was assumed to represent a multifunctional protein accomodating several of the various different fatty acid synthetase component functions within a single polypeptide chain.

7.2 FATTY ACID SYNTHETASE MISSENSE MUTATIONS IN YEAST

During the last several years a great number of fatty acid synthetase-deficient *Saccharomyces cerevisiae* mutants (*fas* mutants) have been isolated in this laboratory. All *fas* mutants so far studied map at either *fas 1* or *fas 2*. The inability of *fas* mutants for the *de novo* fatty acid synthesis has been established *in vivo* by the use of odd chain length supplementary fatty acids (25). The cellular lipids synthesized under these conditions contain exclusively odd chain length fatty acids, whereas in correspondingly grown wild type cells the majority of fatty acids are of the normal even chain length type. *In vitro* the mutational inactivation of the fatty acid synthetase complex in *fas 1* and *fas 2* mutants has been demonstrated in the crude cell homogenates and with the purified mutants enzyme complexes. While overall fatty acid synthetase activity was lost in all cases, all except one of the various component enzyme activities were usually retained. Thus mutants specifically affected with any one of the known fatty acid synthetase component enzyme activities could be identified. As it is indicated in Table 7.1, besides the primary enzymatic lesion, some variability of the residual partial activities compared to their respective wild type levels is often observed. For instance, condensing enzyme activity considerably increases with all enoyl reductase mutations studied, although both enzymes are encoded by two different *fas* loci. On the other hand, a decrease of enoyl-reductase activity accompanies the mutational loss of many other *fas 1* encoded functions. The complete inactivation of one to two additional partial enzymes occurs with most palmityl- (group Vb) and acetyl- (group Va) transferase mutations. As it is evident from the data listed in Table 7.1, among both groups of mutants many also lack malonyl transferase or condensing enzyme activity or both. This interference of a single mutational event with several component enzyme activities at the same time possibly reflects protein-protein interactions between different domains of the same multifunctional polypeptide chain or even between the two subunits α and β.

Based on the biochemical and genetic data listed in Table 7.1, a tentative model of the two fatty acid synthetase subunits α and β has been developed. In this model, as it is depicted in Figure 7.4, both subunits are folded into a sequence of several specific domains, each representing a different component enzyme active site. As an exception, the three acyl transferase active sites are depicted as overlapping domains on the basis of the enzymatic characteristics of transferase mutants as shown in Table 7.1. Since intragenic mapping data regarding the relative positions of different *fas 1* and *fas 2* allelic sites are not yet available, the various active centers in Figure 7.4 have been tentatively arranged according to the sequence of individual steps in the overall reaction process. Furthermore, the mapping studies will have to clarify whether the location of a specific component enzyme active site corresponds to a distinct domain along the polypeptide chain or whether several domains contribute to the functional topology of a single active site at the same time. The yeast fatty acid synthetase genetic system appears to be especially well suited for an experimental proof of the various concepts of protein folding and function interrelationships found in the literature (26–29).

7.3 CRM-NEGATIVE FATTY ACID SYNTHETASE MUTANTS

Only about half of the more than 3000 different *fas 1* and *fas 2* mutants so far isolated in this laboratory contain a fatty acid synthetase-like, though mutant, cross-reacting material (CRM). This is observed when the crude cell homogenate is challenged with a specific fatty acid synthetase antiserum. Quite generally it is found that from all CRM-positive mutants, but never from CRM-negative ones, a mutant fatty acid synthetase complex may be isolated (30). These mutant enzyme complexes are essentially indistinguishable from the wild type enzyme and from each other according to the criteria of molecular weight, electrophoretic mobility of α and β subunits in SDS gels, subunit patterns, and Ouchterlony immuno diffusion analysis. Nevertheless, some minor differences apparently do exist regarding the proteolytic stability of subunit α. This protein apparently contains a site which is especially sensitive against proteolytic attack. Even in wild type cells a certain proportion of α is usually converted into α', which is about 9000 daltons smaller than α (20). In accordance with this in different *fas 2*-mutants the relative amount of α' with respect to α may vary greatly between 0 and 100. (see Figure 7.3). This observation probably reflects subtle conformational changes of the α-tertiary structure as a result of the mutational event. Nevertheless, more drastic conformational differences between the various mutant fatty acid synthetases of CRM-positive strains appear unlikely since the entire β chain, and the α chain at least up to the length of the α' chain, remain completely stable in these mutants.

The CRM-positive *fas* mutants are opposed by an almost equally large number of completely CRM-negative *fas* mutants (30). Both *fas 1* and *fas 2*

Table 7.1 Fatty Acid Synthetase Component Enzyme Activities in *fas* 1 and *fas* 2 Mutants Representative for Various Different Complementation Groups

fas Mutant	Complementation Pattern	Fatty Acid Synthetase	Malonyl Transferase	Palmityl Transferase	Acetyl Transferase	Condensing Enzyme	Enzyme-Bond Pantetheine	1.Reductase	Dehydratase	2.Enoyl-Reductase
		% wild type activity								
1- 246	II		205	210	67	350	present	120	500	
1-1610	II	<1	90	38	200	900	”	111	145	11
1-2224	II	<1	139	48	297	733	”	129	294	4
1-1923	II	<1	77	46	77	300	”	100	132	
1-2465	Va	37	31	66		250	present	107	143	133
1-1924	Va	30	3	75		167	”	100	175	175
1-1320	Va	6	21	49	5	4	”	121	572	56
1-2670	Va	9	30	19	<1		”	100	825	10
1- 83	Va	2	38	74		112	”	100	172	225
2586	Va	<1		12			”	117	22	35
264	Va	30	4	6	11	63	”	69	27	11

185	Va/Vb	2	21	11		199		100	250	176
1-2329	Vb	<1	125	<1	168	125	present	80	120	34
1-1883	Vb	<1	272	<1	490		”	100	320	23
1-1447	Vb	3	80	<1	303	72	”	100	145	53
1-1099	Vb	<1	18	<1	93		”	100	82	230
2726	Vb	<1	53	9	105		”	135	202	164
518	Vb	<1		<1	78	38	”	120	74	160
1813	Vb	<1			203		”	130	138	96
1343	Vb	<1	33	12	124		”	62	138	26
1- 120	Vb				100		”	120	100	165
1-2439	Vc	<1	29	25	33	40	present	68		7
1-1467	Vc	<1	282	198	365	189	”	117		77
1-2550	Vc	<1	443	18	328	315	”	130		284
2449	Vc	<1	204	32	151	519	”	110		78
1- 535	Vc		38	50	15	60	”	220		290
1218	Vc/Vb	4	90	83	311	40	present	130		59
2653	Vc/II/Vb	<1	105	32	315	93	present	115		13
1928	Vc/II/Vb	<1	300	20	101	665	present	91		15
2- 15	VI		180	100	100		present	100	140	100
2- 153	VII		66	200	40			25	130	52
2- 90	VIII		42	220	140	41	present		150	220

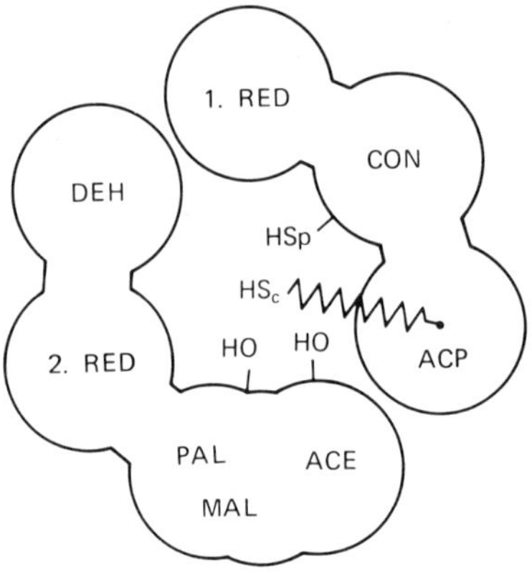

Fig. 7.4 Active sites on the multifunctional subunits α and β of yeast fatty acid synthetase. β-Subunit: 1. RED = β-ketoacyl reductase; CON = condensing enzyme; ACP = acyl-carrier-protein. β-subunit: DEH = dehydratase; 2. RED = enoyl reductase; PAL, MAL, ACE = palmityl-, malonyl-, and acetyl-transacylase.

alleles exhibit this phenotype to about the same extent. Since the antiserum used in these experiments had been prepared against the holoenzyme, and since it generally reacts with the two subunits of wild type fatty acid synthetase even after their dissociation by sodium dodecyl sulfate (30), it was concluded that in CRM-negative *fas* mutants both the α and the β subunits were absent, although only one of them was mutated. Consequently, it proved to be impossible to purify any mutant fatty acid synthetase protein from these mutants in analogy to the CRM-positive strains (30).

From basic studies on the occurrence and characteristics of genetic suppression in yeast it is clear that the frequency of induction of chain terminating nonsense mutations is in the range of 25% (31), that is definitely less than 50%. Frameshift mutations are usually even less frequent. Although mutants of this type are, in an operon like system of clustered genes, most likely to exhibit pleiotropically negative phenotypes, they apparently cannot account for all the CRM-negative *fas* mutants obtained. In agreement with this consideration was the finding that among 43 CRM-negative *fas 1* mutations studied only 27 were suppressible by known amber and ochre suppressors (32). The rest most probably were amino acid replacement missense mutations. Furthermore, the CRM-negative characteristics were very common also among temperature sensitive *fas* mutants. It therefore was concluded that not only nonsense and frameshift mutations but also many amino acid replacement (missense) mutations of the fatty acid synthetase α or β subunits could lead to the complete absence of both within the mutant cell.

In order to explain the instability of the fatty acid synthetase protein in certain *fas* mutants it was proposed that the mutant α or β subunits produced in these cells were either very unstable themselves or led to the formation of an unstable $\alpha_6\beta_6$ mutant complex. In both cases α and β were rapidly degraded by the yeast intracellular proteolytic system as it is characteristically observed with many mutant or otherwise abnormal proteins. If one makes the additional assumption that both subunits, even in their wild type conformation, are unstable unless properly associated with each other, then not only all association deficient α or β mutants but also all those mutants in which either one of the two different subunits or the entire mutant fatty acid synthetase complex are proteolytically unstable would exhibit a CRM-negative phenotype. In other systems similar reasonings have been verified experimentally (33, 34) Maybe this is also the reason why no mutants with a mutationally altered quarternary structure of the complex have been isolated thus far. These mutants, though principally existing, should be expected to belong to the CRM-negative group. Furthermore, it is expected that the CRM-negative phenotype is characteristic for many mutations mapping at various different positions of the α or β genes. Actually this has been demonstrated by Tauro et al. (32) for a rather large collection of CRM-negative *fas 1* missense and nonsense mutations.

Another explanation for the CRM-negative character of many *fas* mutants would be the mutual positive control of *fas 1* over *fas 2* and *vice versa*. Experimentally this mechanism has not been disproved, yet. Therefore experiments are presently under way in this laboratory to investigate this problem further.

7.4 COMPLEMENTATION AMONG DIFFERENT *FAS* MUTANTS

One of the basic techniques in eucaryotic mutant characterization is their analysis for mutual complementation in heteroallelic diploid cells. Usually, the number of different gene loci involved may be identified by this method. The results of intergenic complementation may sometimes be obscured by the superposition of intragenic complementation eventually occurring with mutants of homomultimeric proteins. Usually, however, both types of complementation may be distinguished from each other according to their typical complementation patterns: From intergenic complementation data a linear complementation map may be constructed, while intragenic complementation usually results in a nonlinear, complex map (35).

In the fatty acid synthetase system, complementation analysis has proved a very potent tool in the genetic and biochemical characterization of *fas* mutants. Based on this method the two distinct fatty acid synthetase gene loci *fas 1* and *fas 2* have been identified (36): By definition, every allele of the one *fas* locus complements with every allele of the other. On the other hand, the opposite is true for different alleles within the same locus. Theoretically, all crosses between homologous mutants of either *fas 1* and *fas 2* should yield noncomplementing, that is, fatty acid dependent diploid cells. Actually, noncomplementation is observed in about 85% of all crosses between allelic *fas 1* or *fas 2* mutants. About one eighth of all possible homologous crosses yield complementing diploids with essentially wild type growth characteristics. This proportion correlates with the finding that about half of all *fas 1* and *fas 2*mutants are incapable of any allelic complementation, while the rest exhibit distinct interallelic complementation characteristics when crossed with each other.

In Figures 7.5 and 7.6 (see back pocket at the end of the book) representative interallelic complementation patterns of various different *fas 1* and *fas 2* mutants are shown. All noncomplementing *fas 1* mutants and most of the corresponding *fas 2* alleles have been omitted from these maps. All other complementation patterns observed are listed with their representative relative frequencies. Obviously, besides a rather large number of mutants, particularly in *fas 2*, with irregular and individually different complementation characteristics, several major groups each with a distinct and consistent pattern of interallelic complementation can be defined. So far, essentially all of the *fas 1* patterns that are designated in Figure 7.5 by Roman numbers, have been characterized biochemically and, thus, were attributed to specific lesions of one of the component enzyme activities of the complex. The results of this study are contained in Table 7.1 Similarly, some of the *fas 2* patterns have also been characterized as being specific condensing enzyme (group VI), β-ketoacyl-reductase (group VIII), or pantetheine-less acyl carrier protein (group VII) mutations. Regarding the last group, studies currently under way will have to show whether they actually represent acyl carrier protein-, that is, pantetheine binding-deficient mutants, or whether their pantetheine deficiency results from an inactivated phospho-pantetheine acyl-carrier-protein ligase. It is evident from Figure 7.6 that besides the three *fas* complementation groups mentioned, there exist numerous additional patterns still awaiting biochemical characterization. From these studies it will be seen which of the various different *fas 2* complementation patterns are based on different component enzyme deficiencies and which ones on allelic complementation between isofunctional mutants. Therefore these studies are expected to reveal further insight into the fatty acid synthetase reaction mechanism and the molecular structure of the complex.

As it is obvious from Table 7.1, each of the four major *fas 1* complementation groups exhibits a distinct and specific component enzyme deficiency pattern. By this means they may be integrated into a linear complementation map. The three minor patterns, Va/Vb, Vc/Vb, and Vc/II/Vb, may be attributed either to allelic complementation of isofunctional mutants or to double mutational events. On the other hand, no linear complementation map may be derived from the *fas 2* complementation data of Figure 7.6. Quite obviously *fas 1* and *fas 2* complementation patterns are clearly different: the *fas 2* patterns look much more complex than *fas 1*. While there are rather sharp and consistent boundaries between the individual *fas 1* groups with little if any interallelic complementation within the same group, such a boundary is difficult to define between some of the *fas 2* complementation groups. Therefore, only the *fas 1* complementation map looks essentially like one that would be predicted for a system of several independent genes. Nevertheless, the already mentioned minor *fas 1* patterns as well as most of the *fas 2* complementation characteristics exhibit several features incompatible with intergenic complementation between several distinct, though clustered and coordinately transcribed, genes: first, the existence and frequency of occurrence of a complex complementation pattern not capable of being integrated into a linear map; second, the missense (that is temperature sensitive) character of completely noncomplementing *fas 1* or *fas 2* mutants; third, the antipolar character of many noncomplementing *fas* mutants, that is their randomly scattered location throughout the entire gene locus, as it has been demonstrated by Tauro et al. (32) for a large sample of *fas 1* mutants. These considerations originally led us to postulate that *fas 1* and *fas 2* each coded for a multifunctional rather than for a series of monofunctional proteins.

Assuming that the fatty acid synthetase subunits α and β were two polyfunctional proteins encoded by *fas 2* and *fas 1*, respectively, it is obvious that only in *fas 2/fas 1* heterozygous diploid cells a wild type fatty acid synthetase complex is synthesized. On the other hand, all *fas 1/fas 1* or *fas 2/fas 2* heteroallelic diploids should synthesize, even in the case of positive complementation, a hybrid hexameric complex containing two different mutant species of the same subunit within any single complex molecule. In principle restoration of enzymatic activity in such a hybrid complex may be brought about by either one of two different mechanisms: by the correction of mutationally induced misfolding due to protein-protein interactions (interallelic complementation); or by a cross-feeding of intermediates within the same complex molecule, but between different protomeric entities of the oligomer. If intracomplex cross-feeding of intermediates really occurs, this would indicate that the $\alpha\,\beta$-protomeric subunit does not represent an independent unit within the complex but communicates with at least one other protomer. Thus the minimal functional subunit of yeast fatty acid synthetase would be an $\alpha_2\beta_2$ dimer rather than the $\alpha\,\beta$ monomer. Data of

Lynen's laboratory indicating certain half-of-the-sites characteristics of the intact yeast fatty acid synthetase complex would be in support of this view (37, 38).

Interpreting the complementation data listed in Figures 7.5 and 7.6 it appears that probably both mechanisms mentioned above contribute to the complementation characteristics of allelic *fas 1* and *fas 2* mutants. Obviously, heterofunctional *fas* mutants, even if they are allelic, with very few exceptions always complement each other. In addition, many isofunctional mutants complement each other too. From this it appears reasonable to assume that the usually oserved restoration of function in a heterofunctionally affected hybrid complex is based on the appropriate cross-feeding of intermediates. Since in such a complex the two types of mutated but homologous subunits are each affected at a different component enzyme activity, by cross-feeding of intermediates between their residual and mutationally unaffected sites the inactivated enzymatic centers may be bypassed. Consequently this restoration mechanism for overall fatty acid synthetase function would yield predictable and consistent complementation characteristics for each pair of mutants solely depending on their respective enzymatic lesions. On the other hand, interallelic complementation, though frequently observed in proteins with two or more identical subunits, has remained an unpredictable event (39, 35). Therefore, complementation patterns based on this phenomenon are unlikely to exhibit the consistency and lesion-specific characteristics outlined above. The superposition of both effects, intracomplex cross-feeding of intermediates and interallelic complementation by protein-protein interactions therefore explains best the various experimental data mentioned above, such as the consistently positive complementation between heterofunctional mutants and the complex, nonlinear patterns of some mutants. Finally, the strict correlation between CRM-negativity and noncomplementation, as it is experimentally observed, may also be considered as a direct consequence of the cross-feeding principle. According to this every heterofunctional pair of alleles will complement and, hence, noncomplementation will be observed only between alleles that are either isofunctionel or of which at least one leads to the biosynthesis of a proteolytically labile gene product.

Based on these considerations, the much more regular complementation patterns observed in crosses between different *fas 1* mutants compared to allelic *fas 2* crosses may indicate that interallelic complementation is considerably less frequent between *fas 1* mutants than between *fas 2* mutants, while the cross-feeding reactions occur to a similar extent with both. This means that there are apparently much more intensive protein-protein interactions between β subunits than between α subunits. Immunological and electron microscopic data of Wieland and Lynen (40) are in good agreement with such a theory. Based on these data, they proposed a model of the complex where the α subunits are concentrated in the center plane of the complex, whereas the β subunits extend on both sides to its periphery (21).

7.5 YEAST FATTY ACID SYNTHETASE MOLECULAR STRUCTURE

Very recently the $\alpha_6\beta_6$ molecular structure of the yeast fatty acid synthetase complex, which was originally proposed on the grounds of mostly genetic evidence together with some protein-chemical data, has been considerably supported by the very elegant biochemical work of Wieland (21) and Engeser (41, 14) in Lynen's laboratory. By a controlled chemical modification procedure the intact enzyme complex could be completely and reversibly dissociated into its two constituent components α and β. Subsequently both subunits could be separated from each other in preparative scale by sucrose density gradient centrifugation. The biochemical characterization of the isolated subunits revealed that the two α-specific activities of the β-ketoacyl synthetase and β-ketoacyl reductase component enzymes were completely lost upon dissociation and could not be identified, neither on the α nor on the β subunit. On the other hand, the acetyl transferase, enoyl reductase, dehydratase, and malonyl transferase component enzymes, which had been allocated to subunit β genetically, were apparently much less dependent on a higher level of structural organization since they exhibited essentially wild type activities after removal of the chemical substituents initially introduced to the β subunit. Again, the α subunit was free of these activities. Thus, genetic and biochemical results are in excellent agreement regarding the molecular structure of the yeast fatty acid synthetase complex as well as the allocation of its various different component enzyme activities to the two subunits α and β.

Beyond this there is unfortunately very little experimental evidence, at the moment, regarding the topological arrangements of the various different active sites within the $\alpha\beta$ protomer and of the six pairs of subunits relative to each other within the intact complex. As already mentioned, Wieland and Lynen (21, 40) have developed a model of the hexameric complex based on immunological and electron microscopic data that allocates the α subunits to a central equatorial plane and the β subunits more to the peripheral part of the complex. There is at least some, genetic and biochemical, evidence that the various $\alpha\beta$-protomers in the complex do not function independently of each other but cooperate within dimeric or higher oligomeric minimal functional substructures. Many more questions still remain to be answered regarding the details of the reaction mechanisms and the topology of fatty acid elongation on the $\alpha\beta$-protomeric substructure and within the intact hexameric complex. Again, *in vitro* dissociation/re-association experiments may be helpful in contributing to this question.

7.6 OTHER FATTY ACID SYNTHETASE COMPLEXES

Available data on the fatty acid synthetase complexes of a large variety of different organisms indicate that, regardless of the source and size of the intact complex, they are all composed of 180–270,000 daltons mol wt subunits (5, 6). Compared to the yeast enzyme, however, the molecular weight of the intact complex is in most cases considerably smaller. All the animal fatty acid synthetase complexes, for instance, have molecular weights between 4.0 and 5.6×10^5 daltons. The rabbit mammary gland fatty acid synthetase, which originally was reported to have a much higher molecular weight (42), now seems to fall into this scheme, too (6). Consequently, these complexes may be considered to have a two subunit molecular structure. While it has been reliably demonstrated for the yeast enzyme that it consists of two different, nonidentical types of subunits, similarly conclusive evidence regarding the animal fatty acid synthetase structure is still missing. The synthetases isolated from goose uropygial gland (43) and from chicken or rat liver (44) are reported to contain, respectively, 2.07 and 1.6–2.2 M of 4′-phosphopantetheine per mole of enzyme. According to this result either an α_2 molecular structure is suggested, or an $\alpha\beta$ structure where, other than in the yeast enzyme, both subunits α and β carry the pantetheine prosthetic group. On the other hand, only 1 molecule of pantetheine per mole of synthetase have been identified in the synthetases of rat mammary gland (45) and pigeon liver (46). In these cases an $\alpha\beta$ structure similar to the protomer of the yeast enzyme seems to be present. Other microbiol fatty acid synthetases, like those present in etiolated *Euglena gracilis* cells (47), *Penicillium patulum* (1), *Neurospora* (48), *Corynebacterium diphtheriae* (49), and *Mycobacterium smegmatis* (50) have molecular weights between 1.7 and 2.6×10^6 daltons thereby resembling much more the yeast than the animal fatty acid synthetases. Whereas the Neurospora enzyme seems to be closely related to the yeast fatty acid synthetase also in all other respects so far studied (48), the *Mycobacterium smegmatis* fatty acid synthetase seems to be organized differently. Its six to eight subunits, which are considerably bigger than the yeast α and β subunits, appear to be identical according to a series of different criteria like size, charge, pantetheine and FMN content, or fatty acid synthesis stoichiometry (50).

A method for the reversible dissociation of animal and yeast fatty acid synthetases into two nonidentical subunits of molecular weight

225,000–250,000 has been reported by Porter's group (51, 52). However, their data differ in several respects from our results and from those obtained in Lynen's laboratory: the various partial activities were, even with the yeast enzyme, differently assigned to the subunits I and II compared to the yeast subunits α and β. Subunit I of dissociated rat, human, chicken liver, and yeast fatty acid synthetase contained β-ketoacyl and enoyl-reductase activities, while subunit II contained the acetyl- and malonyl-transferase. Dehydratase and palmityl-transferase (deacylase) were found on both subunits. In addition the dissociation and separate isolation of a low molecular weight pantetheine-labeled acyl carrier protein from eucaryotic fatty acid synthetase complexes has been occasionally reported, too (53, 54). Since, at least in yeast, the genetic data on fatty acid synthetase biosynthesis are rather straightforward and not compatible with the above mentioned subunit functions, an interference of proteolytic artefact formation (43, 55) with complex dissociation may possibly be an explanation for the discrepancies mentioned.

The kinetic and regulatory advantages connected with the combination of several sequential enzymatic activities within a single multienzyme complex are quite obvious. On the other hand, with a larger number of different subunits involved, the biosynthesis of a multienzyme complex should become increasingly complicated. Unfavorable association kinetics and the need for sufficiently high subunit concentrations and for sophisticated control mechanisms to ensure a coordinated biosynthesis of the subunits may tend to counteract the above mentioned catalytic advantages of a complex. Obviously these problems do not exist if a multifunctional enzyme protein rather than a multienzyme complex is synthesized. In this sense the apparent tendency in higher organisms to develop cluster genes (56) out of gene clusters may be explained. It therefore may not be surprising to find systems where the two multifunctional gene loci *fas 1* and *fas 2*, which are still unlinked in yeast, have been fused to an even more complex cluster gene coding for a single protein containing all fatty acid synthetase active sites within one polypeptide chain.

REFERENCES

1 F. Lynen (1972), in J. Ganguly and R. M. S. Smellie, eds., *Current Trends in the Biochemistry of Lipids*, Academic Press, New York, pp. 5–26.

2 J. J. Volpe and P. R. Vagelos (1973), *Ann. Rev. Biochem.* **42,** 21–60.

3 S. Numa and S. Yamashita (1974), *Curr. Top. Cell Regul.* **5,** 198–246.

4 K. Bloch (1977), *Adv. Enzymol.* **45,** 1–84.

5 K. Bloch and D. Vance (1977), *Ann. Rev. Biochem.* **46,** 263–298.

6 S. Smith (1977), in M. R. J. Salton, *Immunochemistry of Enzymes and their Antibodies*, Wiley, New York, pp. 125–146.

7 M. Shrifine and A. G. Marr (1963), *J. Gen. Microbiol.* **32,** 263–270.

8 S. Kanegasaki and S. Numa (1976), *Biochem. Biophys. Acta* **202,** 436–446.

9 M. Kates, B. Palameta, C. N. Joo, D. J. Kushner, and N. E. Gibbons (1966), *Biochemistry* **5,** 4092–4099.

10 F. Lynen (1969), *Methods Enzymol.* **14,** 17–33.

11 E. Schweizer, F. Piccinini, C. Duba, E. Ritter, and F. Lynen (1970), *Eur. J. Biochem.* **15,** 483–499.

12 J. Ziegenhorn, R. Niedermeier, C. Nüssler, and F. Lynen (1972), *Eur. J. Biochem.* **30,** 285–300.

13 G. B. Kresze, L. Steber, D. Oesterhelt, and F. Lynen (1977), *Eur. J. Biochem.* **79,** 181–190.

14 H. Engeser, F. Wieland, and F. Lynen (1977), *FEBS Lett.* **82,** 139–142.

15 G. T. Philipps, J. E. Nixon, J. A. Dorsey, P. H. W. Butterworth, C. J. Chesterton, and J. W. Porter (1970), *Arch. Biochem. Biophys.* **138,** 380–391.

16 C. A. Plate, V. C. Joshi, and S. J. Wakil (1970), *J. Biol. Chem.* **245,** 2868–2875.

17 P. R. Vagelos (1973), in P. D. Boyer, ed., *The Enzymes*, Academic Press, New York, pp. 155–199.

18 P. W. Majerus, A. W. Alberts, and P. R. Vagelos (1965), *J. Biol. Chem.* **240,** 4723–4726.

19 H. van den Bosch, J. R. Williamson, and P. R. Vagelos (1970), *Nature* **228,** 338–341.

20 E. Schweizer, B. Kniep, H. Castorph, and U. Holzner (1973), *Eur. J. Biochem.* **39,** 353–362.

21 F. Wieland (1978), doctoral thesis, University of Munich.

22 E. Schweizer, K. Willecke, W. Winnewisser, and F. Lynen (1970), *Vitam. Horm.* **28,** 329–343.

23 D. Oesterhelt, H. Bauer, and F. Lynen (1969), *Proc. Natl. Acad. Sci.* **63,** 1377–1382.

24 L. Kühn, H. Castorph, and E. Schweizer (1972), *Eur. J. Biochem.* **24,** 492–497.

25 T. W. Orme, J. Mc Intyre, F. Lynen, L. Kühn, and E. Schweizer (1972), *Eur. J. Biochem.* **24,** 407–415.

26 C. C. F. Blake (1976), *FEBS Lett.* **62,** E30–E36.

27 O. B. Ptitsyn (1978), *FEBS Lett.* **93,** 1–4.

28 B. W. Matthews (1977), in H. Neurath and R. L. Hill, eds., *The Proteins*, Vol. 3, Academic Press, New York, pp. 404–590.

29 D. A. Wetlaufer (1973), *Proc. Natl. Acad. Sci USA* **70,** 697–701.

30 G. Dietlein and E. Schweizer (1975), *Eur. J. Biochem.* **58,** 177–184.

31 D. C. Hawthorne and U. Leupold (1974), *Curr. Top. Microbiol. Immunol.* **64,** 1–47.

32 P. Tauro, U. Holzner, H. Castorph, F. Hill, and E. Schweizer (1974), *Mol. Gen. Genet.* **129,** 131–148.

33 A. L. Goldberg and J. F. Dice (1974), *Ann. Rev. Biochem.* **43,** 835–869. For a review see.

34 S. Lin and J. Zabin (1972), *J. Biol. Chem.* **247,** 2205–2211.

35 J. R. S. Fincham and P. R. Day (1965), *Fungal Genetics*, F. A. Davis Publ., Philadelphia.

36 E. Schweizer, L. Kühn, and H. Castorph (1971), *Hoppe-Seyler's Z. Physiol. Chem.* **352,** 377–384.

37 D. Oesterhelt, H. Bauer, G. B. Kresze, L. Steber, and F. Lynen (1977), *Eur. J. Biochem.* **79,** 173–180.

38 G. B. Kresze (1976), doctoral thesis, University of Munich.

39 F. H. C. Crick and L. E. Orgel (1964), *J. Mol. Biol.* **8,** 161–165.

40 F. Wieland, M. Stürzer, and F. Lynen (1977), *Proc. 11th FEBS Meet.*, Copenhagen, A5-1-756-1/2.

41 H. Engeser (1978), doctoral thesis, University of Munich.

42 E. M. Carey and R. Dils (1970), *Biochem. Biophys. Acta* **210,** 371–387.

43 J. S. Buckner and P. E. Kolattukudy (1976), *Biochemistry* **15,** 1948–1957.

44 M. J. Arslanian, J. K. Stoops, Y. H. Oh, and S. J. Wakil (1976), *J. Biol. Chem.* **251,** 3194–3196.

45 S. Smith and S. Abraham (1970), *J. Biol. Chem.* **245,** 3209–3217.

46 F. A. Lornitzo, A. A. Clureshi, and J. W. Porter (1974), *J. Biol. Chem.* **249,** 1654–1656.

47 J. Goldberg and K. Bloch (1972), *J. Biol. Chem.* **247,** 7349–7357.

48 J. Elovson (1975), *J. Bacteriol.* **124,** 524–533.

49 H. W. Knoche and K. E. Koths (1973), *J. Biol. Chem.* **248,** 3517–3519.

50 W. I. Wood, D. O. Peterson, and K. Bloch (1977), *J. Biol. Chem.* **253,** 2650–2656.

51 A. A. Qureshi, F. Lornitzo, R. A. Jenik, and J. W. Porter (1976), *Arch. Biochem. Biophys.* **177,** 364–378.

52 A. A. Qureshi, F. Lornitzo, R. Y. Hsu, and J. W. Porter (1976), *Arch. Biochem. Biophys.* **177,** 379–393.

53 A. A. Qureshi, F. Lornitzo, and J. W. Porter (1974), *Biochem. Biophys. Res. Comm.* **60,** 158–165.

54 D. A. K. Roncari (1974), *J. Biol. Chem.* **249,** 7035–7037.

55 J. K. Stoops, M. J. Arslanian, Y. H. Oh, K. C. Aune, T. C. Vanaman, and S. J. Wakil (1975), *Proc. Natl. Acad. Sci. USA* **72,** 1940–1944.

56 K. Kirschner and H. Bisswanger (1976), *Ann. Rev. Biochem.* **45,** 143–166.

Chapter 8

GRAMICIDIN S-SYNTHETASE

HORST KLEINKAUF and HANS KOISCHWITZ

Max-Volmer-Institut für Physikalische Chemie und Molekularbiologie
Abteilung Biochemie, Technische Universität Berlin
Berlin, West Germany

8.1 INTRODUCTION

In the early history of the study of protein biosynthesis, Lipmann (1) proposed a linear sequence of activating enzymes, where the sequence of "activating spots" determined the sequence of the synthesized peptide (Figure 8.1). This model of protein biosynthesis, which historically linked the polyenzyme and the template hypothesis, left many open questions. It did not explain the reproduction of proteins or the origin of a protein template. With the discovery of peptide synthesizing multienzyme systems in the late 1960s in the laboratories of Lipmann (18), Laland (19), and Kurahashi (16), [as has been reviewed by Lipmann (2)], the existence of protein templates has been verified. It has been established that complicated enzyme systems function as templates in the synthesis of peptides like gramicidin S, tyrocidine, linear gramicidin, bacitracine, edeine, or alamethicin (3). In 1971 the proposal was made that these enzymes were aggregates of amino acid activating subunits arranged in a specified

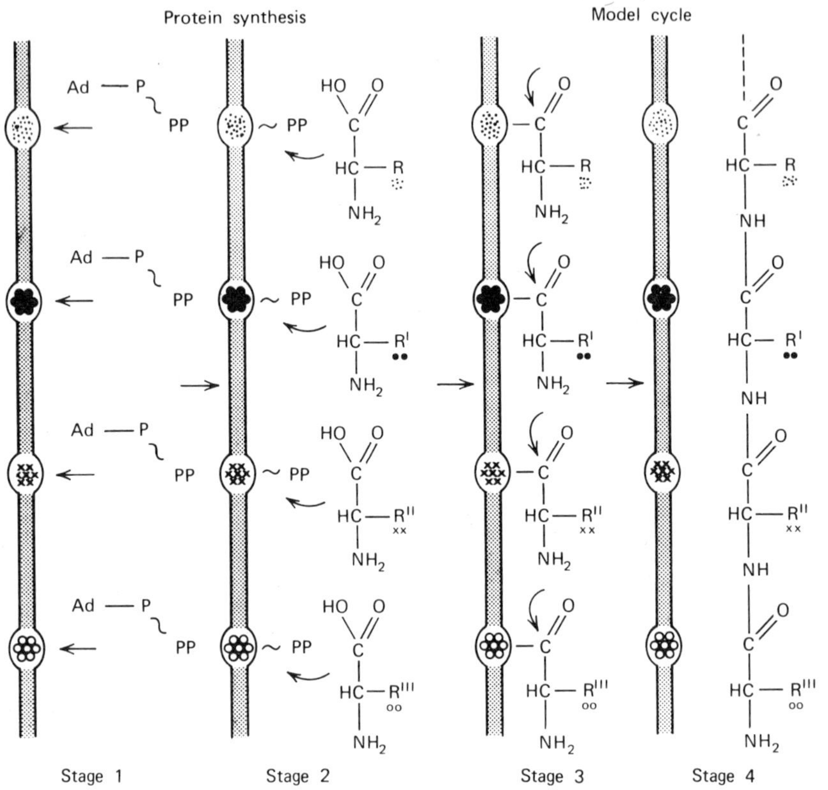

Fig. 8.1 Early model of protein biosynthesis by Lipmann. Amino acid specific activation spots are lined up in sequence. After pyrophosphorylation of the template (Stage 1), amino acids are bound (Stage 2), polymerized (Stage 3), and the peptide dissociates (Stage 4). For the structure of gramicidin S-synthetase a similar model of covalently bound activating subunits has been proposed. Reproduced, by permission, from F. Lipman, in W. D. McElroy and B. Glass, eds., *The Mechanics of Enzyme Action*. Copyright 1954 by the Johns Hopkins University Press.

sequence around a protein with the function of an intermediate substrate carrier containing the cofactor 4′-phosphopantetheine (4). Since then this proposed subunit structure has been under investigation in the enzyme systems forming gramicidin S and tyrocidine. In the case of gramicidin S-synthetase the structure proved to be resistent to protein dissociation agents, while tyrocidine synthetase enzymes could be dissociated into the proposed subunits by an enzymatic reaction not yet characterized (5, 6). This dissociation of tyrocidine multienzymes occurs in crude extracts or by incubation with a triton extractable membrane factor.

Although the existence of single multifunctional polypeptide chains, at least for gramicidin S-synthetase, has not yet been proven, it appears to be a fair interpretation of experimental data. Work is in progress to eliminate alternative structures involving noncovalent interactions, branched peptides, or covalent linkages other than peptide or disulfide bonds.

8.2 FUNCTIONS OF GRAMICIDIN S-SYNTHETASE

Gramicidin S-synthetase, which is produced by Bacillus brevis ATCC 9999 at the end of the logarithmic growth phase, consists of two enzymes of 100,000 daltons (GS 1) and 280,000 daltons (GS 2) which complement each other in peptide synthesis. The reaction sequence leading to the cyclic decapeptide gramicidin S is shown in Figure 8.2.

The activation of the starter amino acid phenylalanine (unless otherwise indicated, amino acids are always in the L-form) proceeds on the "racemase" (E.C.5.1.1.11.) or GS 1; this enzyme has been classified as a racemase (7) although it also catalyzes reactions 1, 2, and 6 besides the racemization of activated phenylalanine (reaction 3).

(1) $$[\text{GS 1}] + \text{Phe} + \text{ATP} \rightleftarrows \underset{\text{(I)}}{[\text{GS 1}](\text{Phe-AMP})} + \text{PP}_i$$

The phenylalanyladenylate is then accepted by a specific enzyme thiol group (S^1):

(2) $$[\text{GS 1}]\,(\text{Phe-AMP}) \rightarrow \underset{\text{(II)}}{[\text{GS 1}] - \text{S}^1\text{-Phe}} + \text{AMP}$$

The active ester intermediate (II) is then epimerized:

(3) $$[\text{GS 1}] - \text{S}^1\text{-Phe} \rightleftarrows [\text{GS 1}] - \text{S}^1\text{-D-Phe}$$

and presumably only the D-Phe intermediate is a substrate in the initiation reaction.

An absolute requirement for initiation to take place is the charging of the multienzyme GS 2 with proline in the two-step activation sequence already shown for Phe (9). Note that the specific thiol group accepting the second amino acid proline has been marked "S^2."

(4) $$[\text{GS 2}] + \text{ATP} + \text{Pro} \rightleftarrows \underset{\text{(III)}}{[\text{GS 2}]\,(\text{Pro-AMP})} + \text{PP}_i$$

(5) $$[\text{GS 2}]\,(\text{Pro-AMP}) \rightarrow \underset{\text{(IV)}}{[\text{GS 2}] - \text{S}^2\text{-Pro}} + \text{AMP}$$

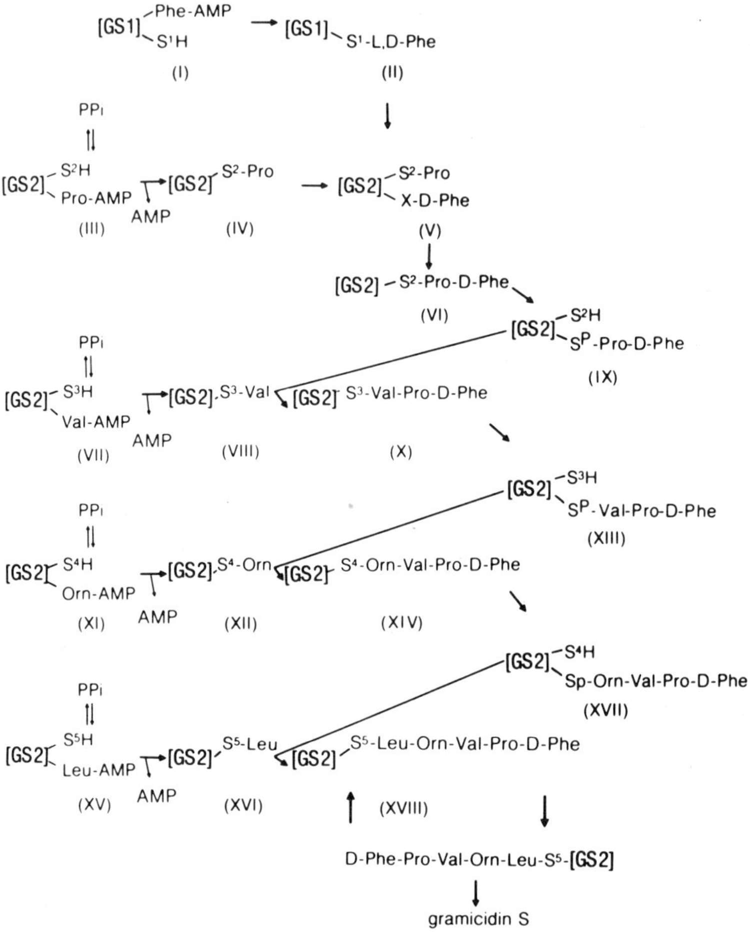

Fig. 8.2 Sequence of reactions in biosynthesis of gramicidin S (for explanation see text).

The charged enzyme (IV) interacts with the charged GS 1 (II) to transfer D-Phe as an active ester to GS 2. No stable "initiation" complex has yet been isolated of GS 1 and GS 2. So far no extensive search has been carried out for conditions that promote a possible ligand induced association. The interaction of both enzymes may be rate limiting in the overall process of peptide biosynthesis, since the rate is dependent on protein concentration (10). The experimental data fit a rate equation derived by Hamilton and Wang (11) for a process of dipeptide formation requiring two enzymes that can associate and dissociate. Details of affinity chromatography data obtained by Pass et al. (12, 13) that led to several proposals on interaction of GS 1 and GS 2 have been questioned (3, 14, 15) and are under reinvestigation.

The transfer reaction (reaction 6) preceding the initiation reaction (reaction 7) involves the postulated acceptor XH for D-Phe on GS 2, presumably also a thiol group (ref. 12).

(6) 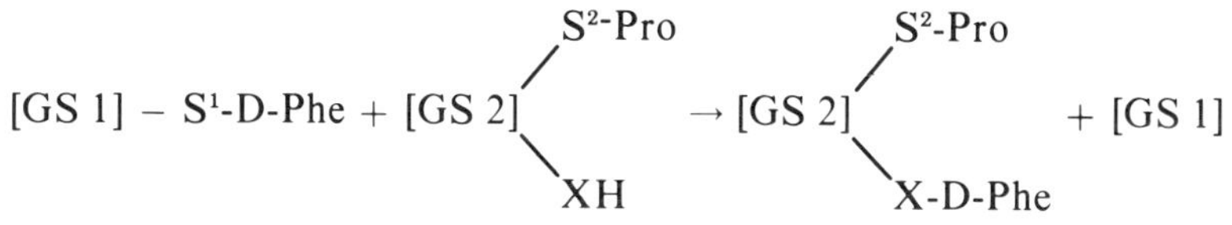

(V)

Initiation of peptide synthesis yields an active dipeptide intermediate:

(7) 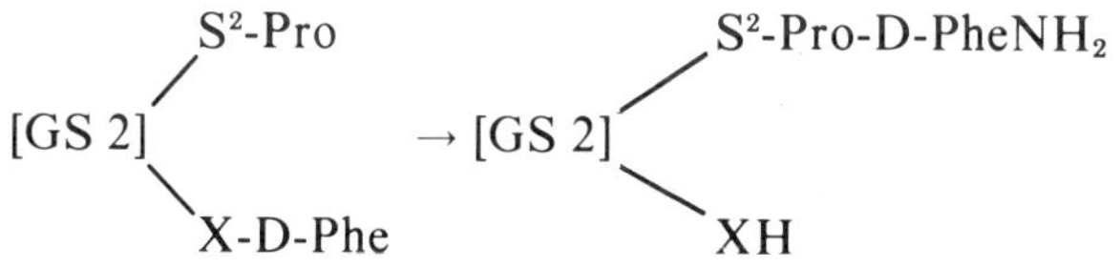

(VI)

By definition the active dipeptide is the *donor peptide*, which is elongated by reacting with the amino group of the following activated amino acid (acceptor). GS 2 has to be charged with the following amino acid valine:

(8) $$[GS\ 2] + Val + ATP \rightleftarrows [GS\ 2]\,(Val\text{-}AMP) + PPi$$
(VII)

(9) $$[GS\ 2](Val\text{-}AMP) \rightarrow [GS\ 2]\ S^3\text{-}Val + AMP$$
(VIII)

Again the notation "S^3" is used for the specific thiol group accepting valyl adenylate.

To explain the elongation process, the thiol of 4′-phosphopantetheine (SPH) is believed to transport peptides *specifically* to their elongation sites:

(10) 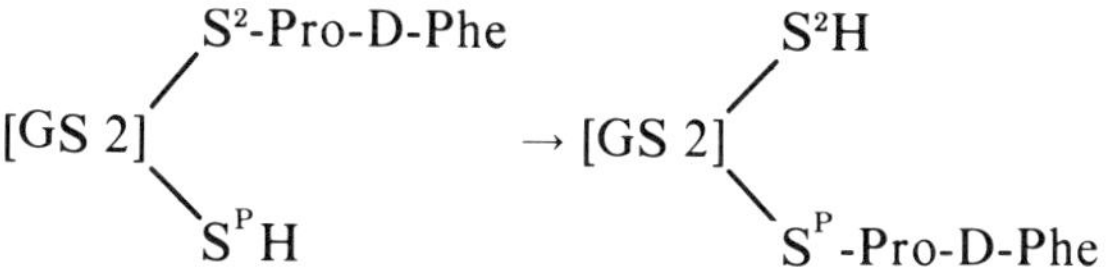

(VI) (IX)

(11)
$$[\text{GS 2}]\begin{array}{l}\diagup\ \text{S}^{\text{P}}\text{-Pro-D-Phe}\\ \diagdown\ \text{S}^{3}\text{-Val}\end{array} \rightarrow [\text{GS 2}]\begin{array}{l}\diagup\ \text{S}^{\text{P}}\text{H}\\ \diagdown\ \text{S}^{3}\text{-Val-Pro-D-Phe}\\ \quad(\text{X})\end{array}$$

If this sequence is repeated for the amino acids ornithine and leucine, we obtain enzymes charged with nascent peptides

(12)
$$[\text{GS 2}] + \text{Orn} + \text{ATP} \rightleftarrows [\text{GS 2}]\,(\text{Orn-AMP}) + \text{PPi}$$
(XI)

(13)
$$[\text{GS 2}]\,(\text{Orn-AMP}) \rightarrow [\text{GS 2}] - \text{S}^{4}\text{-Orn} + \text{AMP}$$
(XII)

(14)
$$[\text{GS 2}]\begin{array}{l}\diagup\ \text{S}^{3}\text{-Val-Pro-D-Phe}\\ \diagdown\ \text{S}^{\text{P}}\text{H}\\ \quad(\text{X})\end{array} \rightarrow [\text{GS 2}]\begin{array}{l}\diagup\ \text{S}^{3}\text{H}\\ \diagdown\ \text{S}^{\text{P}}\text{-Val-Pro-D-Phe}\\ \quad(\text{XIII})\end{array}$$

(15)
$$[\text{GS 2}]\begin{array}{l}\diagup\ \text{S}^{4}\text{-Orn}\\ \diagdown\ \text{S}^{\text{P}}\text{-Val-Pro-D-Phe}\end{array} \rightarrow [\text{GS 2}]\begin{array}{l}\diagup\ \text{S}^{4}\text{-Orn-Val-Pro-D-Phe}\\ \diagdown\ \text{S}^{\text{P}}\text{H}\\ \quad(\text{XIV})\end{array}$$

(16)
$$[\text{GS 2}] + \text{Leu} + \text{ATP} \rightleftarrows [\text{GS 2}](\text{Leu} - \text{AMP}) + \text{PPi}$$
(XV)

(17)
$$[\text{GS 2}](\text{Leu} - \text{AMP}) \rightarrow [\text{GS 2}] - \text{S}^{5}\text{-Leu} + \text{AMP}$$
(XVI)

(18)
$$[\text{GS 2}]\begin{array}{l}\diagup\ \text{S}^{4}\text{-Orn-Val-Pro-D-Phe}\\ \diagdown\ \text{S}^{\text{P}}\text{H}\end{array} \rightarrow [\text{GS 2}]\begin{array}{l}\diagup\ \text{S}^{4}\text{H}\\ \diagdown\ \text{S}^{\text{P}}\text{-Orn-Val-Pro-D-Phe}\\ \quad(\text{XVII})\end{array}$$

(19) $[GS\ 2]\begin{cases} S^{P}\text{-Orn-Val-Pro-D-Phe} \\ S^{5}\text{-Leu} \end{cases} \rightarrow [GS\ 2]\begin{cases} S^{P}H \\ S^{5}\text{-Leu-Orn-Val-Pro-D-Phe} \end{cases}$ (XVIII)

The initial step of the cyclization reaction of two pentapeptides may be intramolecular (on one enzyme charged with two peptides) or intermolecular (two enzymes); the mechanism, however, has not been convincingly established, as has been reviewed by Kurahashi (16) (Figure 8.3).

The accumulation of the enzyme bound intermediate peptides D-Phe-Pro, D-Phe-Pro-Val, and D-Phe-Pro-Val-Orn has been observed in the absence of Leu, the last constituent amino acid of gramicidin S. Upon addition of Leu, however, only very small amounts of the pentapeptides can be detected because of rapid gramicidin S-formation (17). It has been proposed that all possible intermediates may be present on one multienzyme simultaneously (18, 19). Based on this assumption Stoll et al. (20) charged the multienzyme GS 2 with either Phe, ^{3}H-Pro, Val, and Orn or the same set of amino acids where Pro was replaced by ^{14}C-azetidine-2-carboxylic acid. The isolated peptidyl enzymes were mixed and peptide synthesis was completed in the presence of unlabeled amino acids and $MgATP^{2-}$. Most of the resulting gramicidin S would be a hybrid if the termination reaction were intermolecular. However, the predominant species obtained were labeled either ^{3}H or ^{14}C. For the interpretation of a possible intramolecular cyclization another thiol group as a "pentapeptide waiting position" was postulated. A different conclusion has to be reached from an experiment of Roskoski et al. (4) who charged GS 1 and GS 2 exclusively with thiolesters of the constituent amino acids, making use of the high salt lability of aminoacyl adenylates. Gramicidin S was formed from the thiolesters, and the conclusion was drawn, based on the assumption of single amino acid binding sites (S^2, S^3, S^4, S^5), that the condensation is intermolecular. It has been pointed out that both mechanisms may operate, that the rate of reac-

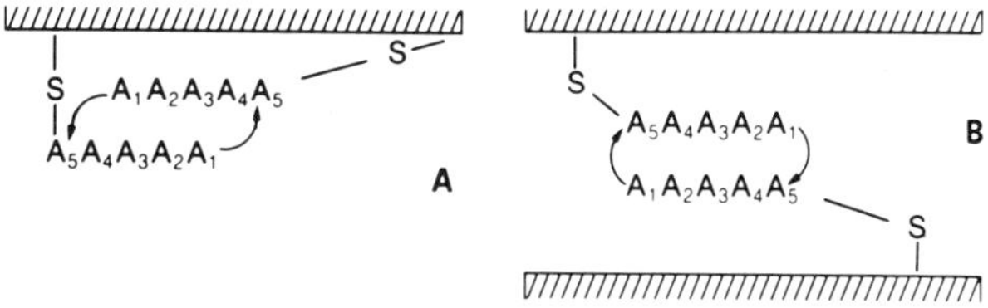

Fig. 8.3 Mechanisms of intramolecular (*A*) and intermolecular (*B*) cyclization.

tion is important, and that the possible occurrence of dimers of GS 2 has to be considered (14).

In summary we observe a sequence of reactions assembled on an enzyme system where intermediates are covalently transported and only the final product is released. To explain the control of the sequence of events a model has been proposed based on Lynen's fatty acid synthetase model (Figure 8.4) (3, 14). In this model the specificity of elongation steps is controlled by the structure of the donor peptide, which is "checked" or bound at the acceptor amino acid site. Thus each elongation step is treated as occurring on a specific peptide synthetase. A possible recognition signal could be the free N-terminal of the donor peptide, which by interaction with the protein surface would be protected from side reactions with activated carboxyls of acceptor acid or donor peptide.

The transport of peptides is thought to proceed by a "swing-arm" type of motion of 4′-phosphopantetheine, so that all activated amino acids have to be in reach of the thiol of the cofactor. This has also been concluded from the observation that no more than 6 amino acid activation sites have been found assembled in one enzyme (3). This model of enzyme structure (Figure 8.4) does not reflect the actual structure of the supermultienzyme yeast fatty acid synthetase, from which it has been derived. It remains useful, however, in interpretation of the reaction sequence. Structural studies of multienzymes containing a single pantetheine, like gramicidin S-synthetase

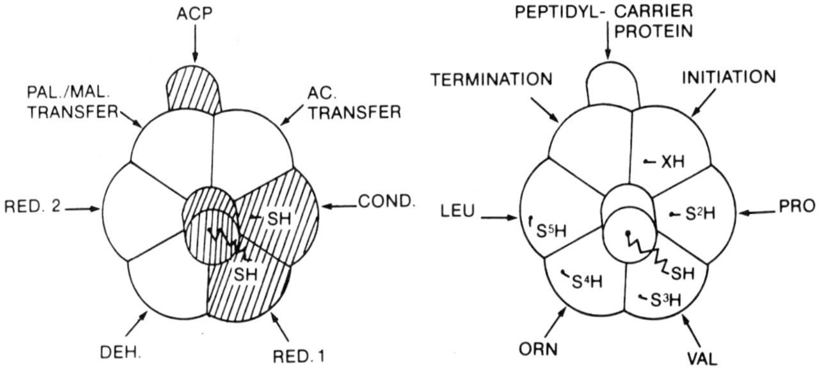

Fig. 8.4 (*a*) Schematic model of fatty acid synthetase, as derived originally from Lynen, with structural details for the yeast enzyme. Hatched areas show the enzymatic activities on the α-chain (ACP: acyl carrier protein, cond: condensation reaction, red 1: 1st reduction). The β-chain contains regions for dehydration (DEH), 2nd reduction (RED 2), transfer of palmitoyl/malonyl-residues (PAL./MAL. Transfer) and acetyl transfer (Ac. Transfer). (*b*) Schematic model of gramicidin S-synthetase indicating functional similarity in sequential reactions with covalent intermediates.

(14, 21) or pig liver fatty acid synthetase (22) show indeed some similarities to the schematic model (Figure 8.5).

8.3 STUDIES ON THE STRUCTURE OF GRAMICIDIN S-SYNTHETASE

It has been pointed out before that the existence of polyfunctional single peptide chains of gramicidin S-synthetase has not yet been established. However, this seems to be a possible interpretation of experimental results available so far. Most extensive studies have attempted to dissociate the multienzyme GS 2, while only preliminary results can be presented of other investigations.

Dissociation Studies

During our attempts to isolate subunits of the multienzyme GS 2 we observed that this enzyme in its fully active form retained a high molecular weight after dodecylsulfate treatment (23, 24). Incubation for 1 h at 95°C in 2% sodium dodecylsulfate and 0.2 *M* dithiothreitol did not reveal any subunit structure (25). To prove the integrity of the functional units, it was shown that the molecular weight of the native multienzyme of 280,000 daltons was not changed upon denaturation. The term integrity is meant to indicate any kind of covalent linkage of peptide chains or a single peptide chain. The molecular weight estimation of the protein-dodecylsulfate complex has been carried out according to Rodbard (26) by polyacrylamide gel electrophoresis. It was shown that (*a*) the electrophoretic mobility at several gel porosities gave a linear Ferguson plot, and (*b*) the extrapolated free mobility of the enzyme and of standard proteins was identical in the gel system used (27). (Figures 8.6 and 8.7). A similar result had been obtained for the 100,000 daltons peptide of GS 1 (28). Other attempts to dissociate GS 2 were also unsuccessful: gel filtration in presence of 6 M urea, sulfonation of thiols in 8 M urea (29) with subsequent dodecylsulfate gel electrophoresis, or modification of carboxymethylated protein with citraconic acid anhydride in 5 M guanidine hydrochloride and gel electrophoresis in 6 M urea (30). Sedimentation analysis of the GS 2-dodecylsulfate complex gave an estimate of 6 S, which indicates a molecular weight of $3 \cdot 10^5$ daltons (25). The covalent linkage of functional units was concluded, since no fragments could be detected in any case, whereas at least partial dissociation has been described for dodecylsulfate "stable" protein associates (for example, glucose oxidase).

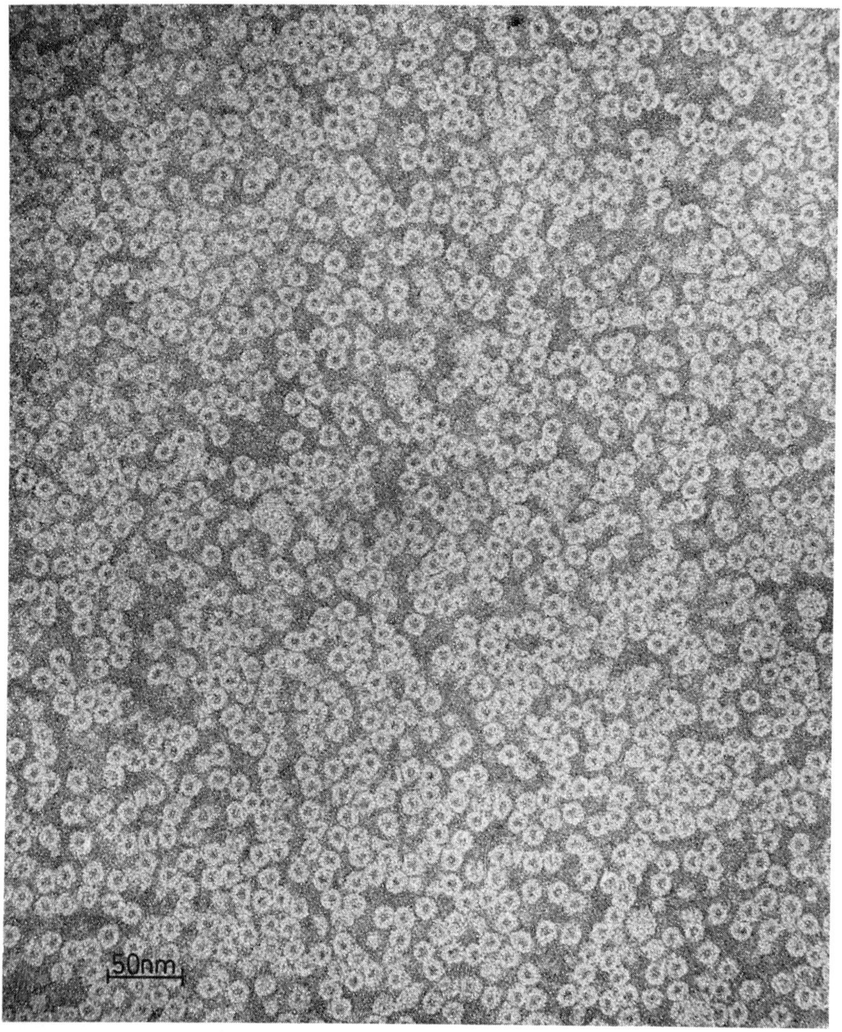

Fig. 8.5 Electron micrograph of purified gramicidin S-synthetase multienzyme GS2. Negative staining, uranylacetate (enlargement 300,000 times). Reproduced from ref. 3 with permission of Springer-Verlag, Heidelberg.

Mutants Producing Defective Multienzymes

Mutants producing no gramicidin S may produce defective synthetase enzymes. Shimura et al. (31) described 11 mutants of GS 2 with apparently unaltered molecular weight (gel filtration of native enzyme). Recently Hori et al. (15) carried out active enzyme sedimentation studies of highly purified multienzymes devoid of either Pro, Val, or Leu activation reaction. Sedimentation rates of wild type enzyme and mutants were identical. Available data are summarized in Table 8.1. All altered GS 2 multienzymes still activate Orn, indicating a central role of this functional unit for the stability of the enzyme structure. Loss of the Pro activation reaction (reaction 4) also leads to loss of the initiation reaction (reaction 6), as could be expected from the Pro dependence of this reaction. A mutant isolated by Kambe et al. (32) apparently produced a fragment of GS 2 of 100,000 daltons that catalyzed Pro activation and dipeptide formation with complementary GS 1.

The loss of initiation reaction (reaction 6) upon loss of Leu activation was not expected but can be correlated with affinity chromatography data (12). Pass et al. described the binding of GS 2 and GS 1 to amino acid derivatives of 4-azaheptamethylene diamine sepharose 4 B with an ATP-dependent association. Binding of GS 1 in the presence of GS 2 on prolyl-, valyl- and ornithyl-columns had been interpreted as a complex formation of both enzymes. A reduced binding of GS 1, when leucine was used as a ligand, indicates the involvement of the functional unit of Leu activation in the

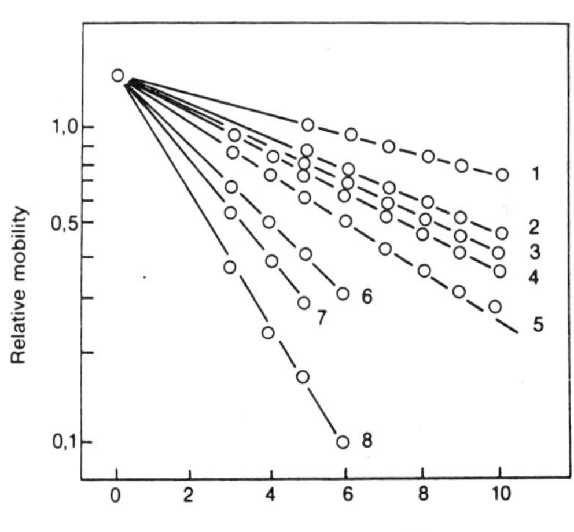

Fig. 8.6 Ferguson plots of sodium dodecyl sulfate protein complexes, showing logarithm of mobility as a function of acrylamide concentration. Proteins used were 1, RNAse; 2 and 3, protein impurities of GS 2 preparations; 4, ovalbumin; 5–7, bovine serum albumin monomer, dimer, trimer; 8, gramicidin S-synthetase multienzyme. Apparent free mobility of bovine serum albumin has been estimated in electrophoresis buffer containing 10% glycerol to 1.35 ± 0.07 relative to bromphenol blue. Plots of all proteins can be extrapolated to approximately this value, indicating a constant charge to mass ratio. Experimental conditions were similar to those of Hayashi et al. [27], who obtained the same result for proteins up to the size of albumin. Reproduced from ref. 25 with permissson of North-Holland Biomedical Press, Amsterdam.

formation of the initiation complex. An enzyme fragment of approximately 260,000 daltons, which did not catalyze leucine activation, also showed a strongly reduced rate of dipeptide formation (32). Thus this functional unit may be directly involved in complex formation between GS 1 and GS 2 in the initiation reaction (reaction 6).

Limited Proteolysis

Limited proteolysis of native enzyme structures may produce catalytically active enzyme fragments. In the case of a multifunctional enzyme information may thus be obtained on the sequence of functional units. The functional units of gramicidin S-synthetase GS 2 are shown in Figure 8.4. The size of the units may be inferred from the dissociation studies of Lee and Lipmann (5) on tyrocidine synthetase multienzymes. A molecular weight of about 70,000 daltons was found for an amino acid activating unit, and 20,000 daltons for the peptidyl-carrier protein.

So the sites of interaction of multienzymes (initiation reaction, transfer of peptides) and sites of termination (cyclization) are thought to be located within the structures of amino acid activating units, contrary to the model shown in Figure 8.4. The sequence of units on a hypothetical single polypeptide chain would be, according to our historical model (Figure 8.1), Pro, Val, Orn, Leu. Two mutant multienzymes of reduced size (hh, n7 of Table 8.1) may originate from preliminary chain termination during translation. If so, the Pro activating unit would be located on the amino terminal end of the chain, while the Leu activating unit would be at its carboxyl terminal.

Very recently we have isolated fragments of GS 2 activating Orn and Leu. Analysis of fragments has not been completed so far. From extracts of

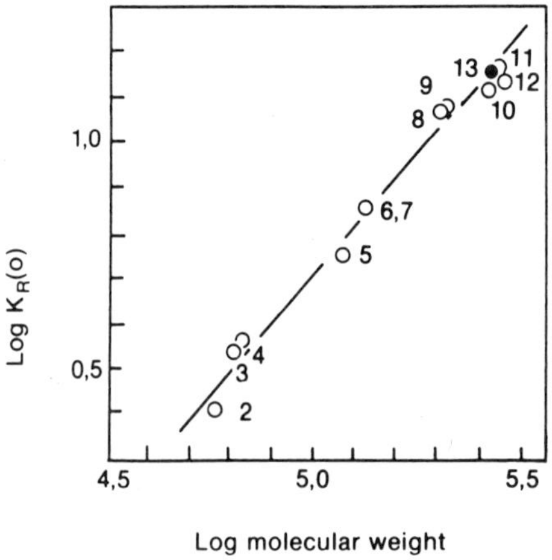

Fig. 8.7 Retardation coefficients obtained from Ferguson plots as a function of molecular weight. Proteins were: 2, catalase; 3, bovine serum albumin (BSA); 4, rabbit serum albumin (RSA); 5, catalase dimer; 6, BSA-dimer; 7, RSA-dimer; 8, BSA-trimer; 9, RSA-trimer; 10, catalase tetramer; 11, BSA-tetramer; 12, hemocyanin; 13, gramicidin S-synthetase multienzyme. K_R-values were obtained from 50 mMphosphate with no assumption on the apparent free mobility. Reproduced from ref. 25 with permission of Elsevier/North-Holland Biomedical Press, Amsterdam.

Table 8.1 Mutants Carrying Defects in Gramicidin S-Synthetase Enzymes (3, 15)

Mutant	Activation of Pro	Val	Orn	Leu	Dipeptide Formation	Peptide Synthesis	$S_{20,w}$	Molecular Weight $\times 10^{-5}$
B II -3	−	+	+	+	−	−	12, 2	2, 8
B I -3	+	−	+	+	+	−	12, 2	2, 8
B I -6	+	−	+	+	+	−		
B II -1	+	−	+	+	+	−		
E -4	+	−	+	+	+	−		
E -5	+	−	+	+	+	−		
B I -9	+	+	+	−	+	−	12, 2	2, 8
B I -4	+	+	+	−	−	−		
C -3	+	+	+	−	−	−		
E -2	+	+	+	−	−	−		
B I -2	+	+	+	+	+	−		
hh	+	+	+	−	(+)	−	11, 3	2, 6
n7	+	−	−	−	+	−	6, 1	1, 0
Wild type	+	+	+	+	+	+	12, 2	2, 8

B. brevis 12 fragments to sizes 35,000 to 220,000 daltons were observed, while 10 fragments of sizes 100,000 to 220,000 daltons were produced by chymotrypsin as analyzed by dodecylsulfate polyacrylamide gel electrophoresis. So far the evaluation of data has not been completed but two preliminary results are shown in Figure 8.8 and Table 8.2.

Chymotrypsin fragmentation does not affect the Orn dependent ATP-PP_i exchange reaction, while other activities were lost (Val) or reduced (Pro, Leu). Chymotrypsin attack may thus be primarily in the Val activation unit. A second type of fragmentation in extracts of *B. brevis* has been observed, where Pro activation is lost almost completely, while Val and Leu activation reactions are moderately reduced. Work is in progress to evaluate maps of fragment structures. The predicted sequence Pro, Val, Orn, Leu can be concluded from the mutants hh and n7, and from the linkage of Orn and Leu activating units in fragments not dissociable by dodecylsulfate (33).

8.4 RELATED MULTIENZYME SYSTEMS

We have recently summarized data on peptide synthesizing multienzyme systems (3). These systems have been characterized by (*a*) activation of

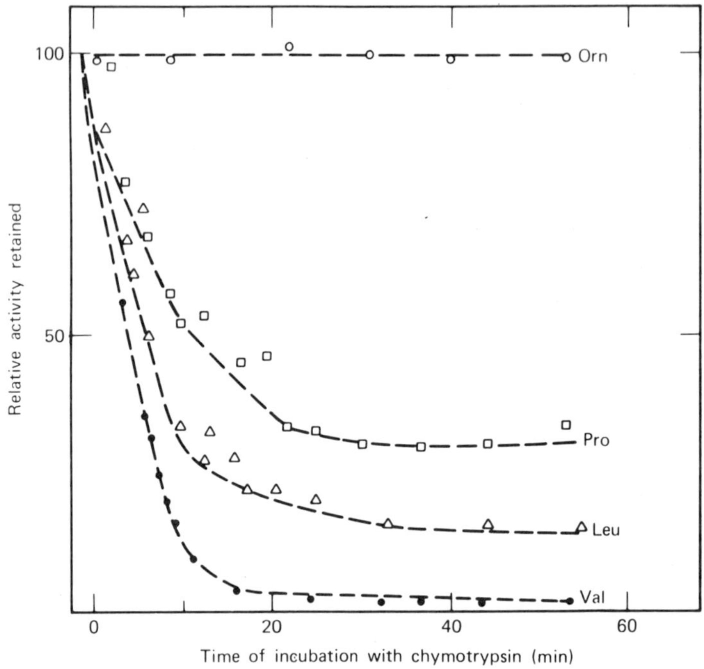

Fig. 8.8 Chymotrypsin-treatment of native GS 2. Effect on the partial reactions of amino acid activation, measured by amino acid dependent ATP-PP_i-exchange. Proteinase had been added to 5% of total GS 2, and incubations have been carried out at 20°C in 50 m*M* TES, pH = 6.5. The reaction was stopped by the addition of PMSF.

amino acids as adenylates and thiolesters, (*b*) no detectable free intermediates, (*c*) covalently bound 4′-phosphopantetheine as coenzyme, which (*d*) mediates a covalent transport of intermediates to the different active centers. There is no requirement yet for covalent subunit assembly or multifunctionality of peptide chains involved. Still the gramicidin S-system and the tyrocidine-system (5) lose their amino acid polymerization activity completely upon enzymatic chain disintegration. So we expect to find more multifunctional single chain enzymes within this group. Comparing peptide synthetases with polyketide synthetases like fatty acid synthetases or 6-methyl-salicylic acid synthetase (6-MSAS) (33) we observe similarities as (*a*) active esters of carboxyl compounds as enzyme bound precursors, (*b*), (*c*), and (*d*) are identical characteristics. Also we observe multifunctional chains, two of 180,000 and 185,000 daltons for yeast FAS, and one of 165,000 daltons for 6-MSAS. Although polyketide synthetases contain a single "condensing enzyme" or synthetase, we find many similarities to polypeptide synthetases. So the synthesis of a repeated peptide sequence would strongly resemble fatty acid biosynthesis. So far no examples are known, but several peptide structures suggest this possibility. A multienzyme synthesizing the depsipeptide enniatin apparently is also a multifunctional protein. Although its pantetheine content has yet to be established, acid activation and covalent transport of intermediates has been demonstrated (35, 36). A molecular weight of 220,000 daltons has been estimated by dodecylsulfate polyacrylamide electrophoresis (37).

8.5 THE ORIGIN OF THE PROTEIN TEMPLATE

Lipmann proposed that the sequence of "activating spots" (Fig. 1) determined the sequence of polymerization. Preliminary results mentioned in this work indeed suggest the "correct" sequence of "functional units." So far we do not know on what level this sequence is determined. Is it the DNA-sequence, is it posttranscriptional processing of mRNA, or do we have a posttranslational fusion of protein units? At present we would favor the idea of posttranscriptional events, since a variety of different peptides of similar composition may be produced, for example by *B. brevis* (3, 34). For reasons of genetic economy similar transcripts may be used in the cell's construction of different polyenzymes.

Table 8.2 Alterations in Amino Acid Activating Activities in Mutant hh and Enzyme Fragments (33)

Purified Enzyme	Amino Acid Dependent ATP-PP_i -Exchange Reaction (Relative Rate as Compared to Orn) Pro	Val	Orn	Leu
GS 1 wild type	1, 1	1, 9	1, 0	1, 7
GS 2 hh[a]	1, 0	2, 1	1, 0	0, 05
GS 2, chymotrypsin fragments[b]	0, 35	0, 0	1, 0	0, 1
GS 2, fragments isolated from *B. brevis* extracts[c]	0, 1	1, 0	1, 0	0, 9

[a] The enzyme has been purified to homogenity. Leu activation is ascribed to Val activation site (33).
[b] Rates are taken from Figure 8.6.
[c] The enzyme preparation corresponds to a fragment mixture that can be obtained by the action of a not yet characterized proteolytic activity during purification of GS 2 (33).

ACKNOWLEDGMENT

The work cited in this article that originated in the authors laboratory was supported by grants Kl 148/13 and Kl 148/16 from the Deutsche Forschungsgemeinschaft and PTB 8013 from the Bundesministerium für Forschung und Technologie.

REFERENCES

1 F. Lipmann (1954), in W. D. McElroy, B. Glass, eds., *The Mechanism of Enzyme Action*, The Johns Hopkins Press, Baltimore, p. 599.

2 F. Lipmann (1971), *Science* **173,** 875–884.

3 H. Kleinkauf, H. Koischwitz (1978), *Prog. Mol. Subcell. Biol.* **6,** 59–112.

4 R. Roskoski, Jr., G. Ryan, H. Kleinkauf, W. Gevers, F. Lipmann (1971), *Arch. Biochem. Biophys.* **143,** 485–492.

5 S. G. Lee, F. Lipmann (1977), *Proc. Natl. Acad. Sci. USA* **74,** 2343–2347.

6 S. G. Lee, F. Lipmann (1974), *Proc. Natl. Acad. Sci. USA* **71,** 607–611.

7 T. E. Barman (1974), *Enzyme Handbook*, Supplement 1, Springer, Berlin.

8 H. Takahashi, E. Sato, K. Kurahashi (1971), *J. Biochem.* **69,** 973–976.

9 H. Kleinkauf, R. Roskoski, Jr., F. Lipmann (1971), *Proc. Natl. Acad. Sci. USA* **68,** 2069–2072.

10 H. Koischwitz, H. Kleinkauf (1976), *Biochim. Biophys. Acta* **429,** 1041–1051.

11 D. I. C. Wang, B. K. Hamilton (1977), *Biotechnol. Bioeng.* **19,** 1225–1232.

12 L. Pass, T.-L. Zimmer, S. G. Laland (1974), *Eur. J. Biochem.* **47,** 607–611.

13 L. Pass, T.-L. Zimmer, S. G. Laland (1973), *Eur. J. Biochem.* **40,** 43–48.

14 H. Koischwitz (1978), Thesis, Berlin.

15 K. Hori, T. Kurotsu, M. Kanda, S. Miura, A. Nozoe, Y. Saito (1978), *J. Biochem.* **84,** 425–434.

16 K. Kurahashi (1974), *Ann. Rev. Biochem.* **43,** 445–459.

17 H. Kleinkauf (1978), *Plant. Med.* **35,** 1–18.

18 F. Lipmann, W. Gevers, H. Kleinkauf, R. Roskoski, Jr. (1971), *Adv. Enzymol.* **35,** 1–34.

19 S. G. Laland, T.-L. Zimmer (1973), *Essays Biochem.* **9,** 31–57.

20 E. Stoll, Ø. Froyshøv, H. Holm, T.-L. Zimmer (1970), *FEBS Lett.* **11,** 348–352.

21 J. Wecke, P. Giesbrecht, H. Koischwitz, H. Kleinkauf, unpublished.

22 W. Folkhard, B. Felser, I. Pilz, O. Kratky, H. Dutler, H. Vogel (1977), *Eur. J. Biochem.* **81,** 173–178.

23 H. Kleinkauf, H. Koischwitz (1974), in D. Richter, ed., *Lipmann-Symposium*, de Gruyter, Berlin, pp. 336–344.

24 C. Christiansen, K. Aarstad, T.-L. Zimmer, S. G. Laland (1977), *FEBS Lett.* **81,** 121–124.

25 H. Koischwitz, H. Kleinkauf (1976), *Biochim. Biophys. Acta* **429,** 1052–1061.

26 D. Rodbard (1976), in N. Catsimpoulas, ed., *Methods of Protein Separation*, Vol. 2, Plenum Press, New York, pp. 145–179 and 181–218.

27 K. Hayashi, E. Matsutera, Y. Ohba (1974), *Biochim. Biophys. Acta* **342,** 185–194.

28 J. Vater, H. Kleinkauf (1976), *Biochim. Biophys. Acta* **429,** 1062–1072.

29 W. W.-C. Chan (1968), *Biochem.* **7,** 4251–4254.

30 F. Wieland (1978), Thesis, München.

31 K. Shimura, M. Iwaki, M. Kanda, K. Hori, E. Kaji, S. Hasegawa, Y. Saito (1974), *Biochim. Biophys. Acta* **338,** 577–587.

32 M. Kambe, Y. Imae, K. Kurahashi (1974), *J. Biochem.* **75,** 481–493.

33 H. Koischwitz, R. Kittelberger, M. Altmann, unpublished results.

34 G. G. Zharikova, A. P. Zarubina, D. M. Kherat, S. P. Myaskovskaya, V. N. Maksimov (1975), in Antibiot. Ikh Produtsenty, Moskau Nauka, pp. 163–186.

35 H. Kleinkauf, H. Koischwitz, J. Vater, R. Zocher, U. Keller, I. Mahmutoglu, K. Bauer, M. Altmann, R. Kittelberger, M. Marahiel, J. Salnikow (1978), *Proc. 12th FEBS Meet.* Vol. 5.

36 R. Zocher, H. Kleinkauf (1978), *Biochem. Biophys. Res. Comm.* **81,** 1162–1167.

37 R. Zocher, unpublished results.

Chapter 9

AMPHIPATHIC PROTEINS

REBECCA P. HUGHEY and NORMAN P. CURTHOYS

Department of Biochemistry
University of Pittsburgh School of Medicine
Pittsburgh, Pennsylvania

9.1. INTRODUCTION

Amphipathic proteins provide one of the clearest examples of a multifunctional protein. By definition an amphipathic molecule is one that contains defined polar and nonpolar regions. In the case of amphipathic proteins these regions are comprised of structurally separate and functionally unique domains. In most cases the nonpolar or hydrophobic domain serves as a membrane anchor, while the polar or hydrophilic domain contains the active site. Such a structure provides a mechanism for the attachment of specific catalytic activities to various membranes or for their entrapment in specific organelles. The subcellular compartmentalization that is achieved frequently contributes to the regulation of cellular metabolism.

The mechanism of cellular localization of the murine β-glucuronidase has been well characterized. Within liver the enzyme is found either as a soluble protein contained within lysosomes or as a particulate protein associated with the membrane of the endoplasmic reticulum (1). The lysosomal enzyme is a tetramer (L) of approximately 75,000 dalton subunits and is more acidic than the corresponding tetramer (X) purified from microsomes (2). The subunits of microsomal enzyme probably contain a small additional sequence of amino acids that is essential for binding the enzyme to the membrane of the endoplasmic reticulum. The X form of the subunits appears to be the primary product of translation (1). Limited proteolysis produces the soluble polar form of the enzyme which is concentrated within the lysosomes. The X form of β-glucuronidase associates with the microsomal membrane. This association is stabilized by its interaction with egasyn (2), an amphipathic glycoprotein, which acts as a specific membrane binding site (3).

The role of egasyn in binding the microsomal β-glucuronidase was established by the characterization that the YBR strain of mice carrying the Eg^{o} mutation lack egasyn and as a result are deficient in microsomal β-glucuronidase (3). In addition the detergent-solubilized microsomal β-glucuronidase has been shown to exist as four larger molecular weight forms (M_1, M_2, M_3, and M_4) which differ due to the binding of 1 to 4 peptides of egasyn, respectively (2). Treatment of this mixture with cold urea produces consecutively the appearance of predominately M_3, M_2, M_1, and X forms of β-glucuronidase.

Egasyn has no known function other than serving as a membrane anchor, although it may serve as a binding site for polar enzymes other than β-glucuronidase (3). The complex of egasyn with a polar enzyme provides a mechanism for membrane association that is analogous to a multifunctional enzyme complex. In contrast, many membrane bound enzymes have been determined to be amphipathic. The condensation of the two domains into a single peptide is by definition a multifunctional protein. In the case of amphipathic proteins this type of multifunctional protein may have resulted from the condensation of two genes, one of which coded for a protein similar to egasyn.

9.2 CYTOCHROME B_5 AND CYTOCHROME B_5 REDUCTASE

The fact that each region of an amphipathic protein is composed of separate domains that are functionally unique has been most firmly established through the detailed study of the electron transport proteins cytochrome b_5 and cytochrome b_5 reductase. These two membrane bound proteins participate in the microsomal system for desaturation of fatty acids. They shut-

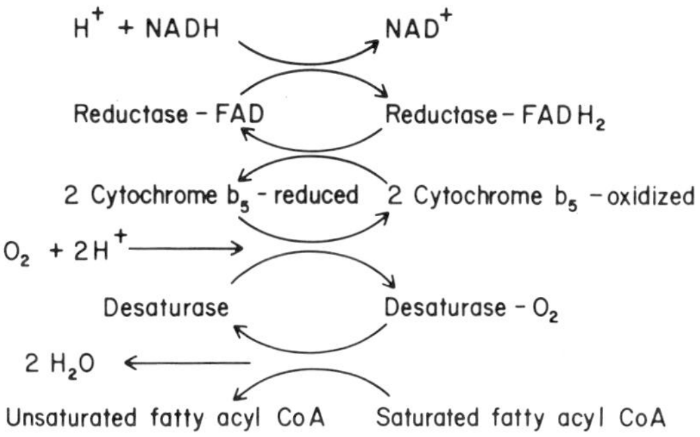

Fig. 9.1 Pathway of fatty acyl CoA desaturation.

tle electrons from NADH to the terminal enzyme, stearyl coenzyme A desaturase as shown in Figure 9.1.

The properties of the cytochrome b_5 and cytochrome b_5 reductase, which led to their classification as amphipathic proteins, also provide one of the simplest models of an intrinsic membrane protein. In contrast to those proteins that interact with membrane lipids by virtue of their overall hydrophobic surfaces, amphipathic proteins use only a minor portion of their primary sequence, the nonpolar domain, for lipid-bilayer interaction. Their polar domain is positioned to function freely in the aqueous environment outside of the membrane. In cytochrome b_5 and cytochrome b_5 reductase the catalytically active polar domains are consistently found on one side of the membrane (4), and even though the desaturase is more deeply immersed in the lipid bilayer its active site is also oriented to the same side of the membrane (5).

Primary Structure and Evidence for Separate Domains

The first purification procedures for cytochrome b_5 (6, 7) and cytochrome b_5 reductase (8, 9) yielded water soluble proteins that retained enzyme activities similar to the membrane bound forms of the two proteins. The cytochrome b_5 was released from liver microsomal membranes either by autolysis (10) due to contaminating lipase disrupted lysosomes or by incubation with trypsin (7). The purified cytochrome b_5 exhibited two major electrophoretic forms (7) that differed only by the presence of the carboxyl-terminus dipeptide, glutamylserine. The first sequencing study of calf liver chrochrome b_5 yielded this partial amino acid sequence (7):

$+NH_3$-Ser-Lys-Ala-(heme core of 79 a.a)-Arg-Ile-Thr-Lys-Pro-Ser-Glu-Ser-COO−

This study confirmed earlier observations that heme binding protected the catalytic domain of the cytochrome b_5 from proteolysis. As a result of this protection a homogeneous preparation of the protein was obtained for further amino acid analysis. The corrected and complete 93 amino acid primary sequence of the bovine protease purified cytochrome b_5 was subsequently established (11). This corresponds to residues 5 to 97 as shown in Figure 9.2.

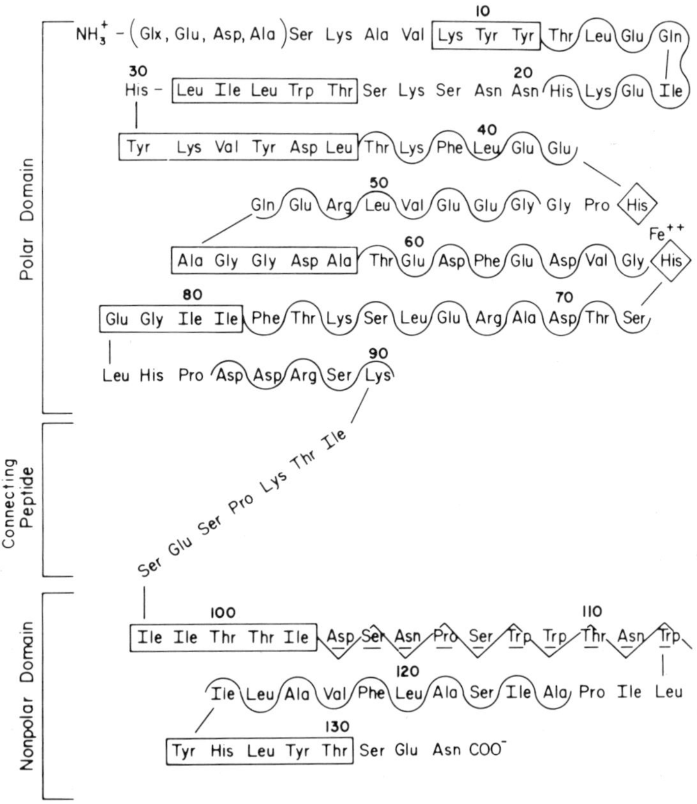

Fig. 9.2 The primary and predicted secondary structure of bovine detergent purified cytochrome b_5. The primary and secondary structure proposed here is a composite of the data reported for the trypsin purified cytochrome b_5 (12), the nonpolar segment (13), and the amino-terminus of the detergent purified cytochrome b_5 (14). In particular the first four residues at the amino terminus are not clearly assigned. the regions of β-pleated sheet are schematically indicated in boxes of grouped residues; the α-helical regions are indicated by a continuous loop above and below the residues; the complex β-turns are indicated by underlining and a continuous zigzag; and the histidines, in diamind boxes, are those coordinating the heme group represented as Fe^{2+}.

About this time researchers (15, 16) began to report on the existence of a larger form of cytochrome b_5. This form of cytochrome b_5 was initially purified from rabbit liver microsomes following its solubilization with detergents (17). Upon removing the detergent the cytochrome b_5 underwent aggregation to form "octamers" of 120,000 daltons. In solutions of 0.4% deoxycholate the protein formed a "dimer" of 35,000 daltons, and in sodium dodecyl sulfate it exhibited a "monomer" molecular weight of 14,500. By amino acid analysis the molecular weight of the monomer was determined to be 16,100 per heme group (17) while that of the protease-purified cytochrome b_5 was reported as 11,000 (11, 17). The comparison of the amino acid analysis of the two forms attributed the difference in size to the proteolytic release of 44 amino acids, 60% of which were hydrophobic. Treatment of the detergent purified cytochrome b_5 with trypsin produced a smaller form of cytochrome b_5, which by amino acid analysis was apparently identical to the protease purified cytochrome b_5. A single peptide containing 40 of the 44 extra amino acids was also recovered. It was determined to have an amino acid content nearly identical to that predicted by comparing the large and small forms of the cytochrome b_5 (Table 9.1). The structural difference between the two forms of cytochrome b_5 was also consistent with the finding that the visible spectrum of the two forms were identical while the UV spectrum changed noticeably (17). This was because trypsin treatment removed 3 tryptophans and 2 tyrosines. The high hydrophobic amino acid content of this single peptide (molecular weight ~5000) led to the initial proposal (17) that this domain of the cytochrome b_5 was responsible for anchoring the protein to its native membrane, while the remainder of the peptide contained all the structural features necessary for electron transport.

Early studies had supported localization of the nonpolar domain of cytochrome b_5 at the amino terminus of the polar domain (17). However, the comparison of the primary amino acid sequence of detergent purified human cytochrome b_5 and the protease purified rabbit cytochrome b_5 clearly established its localization at the carboxyl terminus (18). As shown in Figure 9.2, the primary sequence of bovine cytochrome b_5 can be divided into three regions: the amino terminal polar domain, which contains the heme binding site; the carboxy terminal nonpolar domain; and the connecting peptide between the two domains.

Later sequencing studies of detergent purified cytochrome b_5 from different mammals (human, equine, and bovine) revealed a common carboxyl terminal dipeptide of glutamylaspartate (14). This highly charged peptide at the carboxyl terminus of the nonpolar domain was the first evidence to indicate that both ends of the nonpolar peptide (and of the native cytochrome b_5) may be external to the membrane. However, more recent reports showed

Table 9.1 Amino Acid Composition of Different Forms of Rabbit Cytochrome b_5

Amino Acid	(A) Detergent Purified	(B) Protease Purified	(A-B)	Trypsin-Isolated Nonpolar Peptide
Lys	11	10	1	1
His	7	7	0	0
Arg	4	3	1	1
Asp	16	10	6	4
Thr	10	7	3	3
Ser	10	7	3	3
Glu	15	14	1	1
Pro	5	3	2	2
Gly	7	6	1	1
Ala	10	5	5	4
Cys	0	0	0	0
Val	7	4	3	3
Met	3	1	2	2
Ile	8	4	4	4
Leu	15	9	6	5
Tyr	5	3	2	2
Phe	4	3	1	1
Trp	4	1	3	3
Total	141	97	44	40
Molecular Weight	16,072	11,079	4993	4579

that porcine (19) and bovine (13) cytochrome b_5 have a slightly less negatively charged terminus due to substitution of asparagine for the aspartate. To date the complete amino acid sequences of the nonpolar domains of bovine (13), equine (20), and porcine (19) cytochrome b_5 have been published. As shown in Table 9.2 they exhibit remarkable homology. In addition, the carboxyl terminus of human detergent purified cytochrome b_5 has been reported as Leu-Tyr-Met-Ala-Glu-Asp-COO– (14). The amino acid content of the human carboxyl terminus 43-residue peptide (18) also shows remarkable similarity to the amino acid content of the corresponding peptide already reported from other sources.

Assuming that membrane attachment by a nonpolar domain of the same or similar amino acid sequence could be a common mechanism of protein-lipid interaction, the structure of cytochrome b_5 reductase was also investigated. Reductase, purified after solubilization with lysosomal cathepsins, was a hydrophilic globular protein containing one FAD molecule per peptide of 33,000 daltons (21). Purification of the reductase after solubilization with Triton X-100 (22) produced a form of the enzyme that aggregated (600,000 daltons) in the absence of detergents while the "monomer" exhibited a molecular weight of 43,000. The difference between the protease and detergent purified forms was due to the removal of 99 amino acids, 65% of which were found to be hydrophobic. Treatment of the detergent purified reductase with chymotrypsin produced a form of the enzyme that was identical in amino acid content to the protease purified reductase (Table 9.3). Upon sodium dodecyl sulfate polyacrylamide gel electrophoresis, the chymotrypsin treated reductase exhibited an additional band that corresponded to a 10,000 dalton peptide. Protease treatment also altered the UV spectrum of the reductase (22). These changes were due to the loss of 4 tryptophans and 5 tyrosines in the hydrophobic peptide. More recent work (23) localized this peptide to the carboxyl terminus. The remarkable similarity between the structures on the cytochrome b_5 reductase and

Table 9.2 Amino Acid Sequence of Nonpolar Peptide Domain and Connecting Peptide of Cytochrome b_5

		91									100
Equine		Ile	Ala	Lys	Pro	Val	Glu	Thr	Leu	Ile	Thr
Bovine		Ile	Thr	Lys	Pro	Ser	Glu	Ser	Ile	Ile	Thr
Porcine		Ile	Ala	Lys	Pro	Ser	Glu	Thr	Leu	Ile	Thr
Conserved:		X		X	X		X			X	X
									110		
Thr	Val	Asp	Ser	Asn	Ser	Ser	Trp	Trp	Thr	Asn	Trp
Thr	Ile	Asp	Ser	Asn	Pro	Ser	Trp	Trp	Thr	Asn	Trp
Thr	Val	Glu	Ser	Asn	Ser	Ser	Trp	Trp	Thr	Asn	Trp
X			X	X		X	X	X	X	X	X
							120				
Val	Ile	Pro	Ala	Ile	Ser	Ala	Val	Val	Val	Ala	Leu
Leu	Ile	Pro	Ala	Ile	Ser	Ala	Leu	Phe	Val	Ala	Leu
Val	Ile	Pro	Ala	Ile	Ser	Ala	Leu	Val	Val	Ser	Leu
	X	X	X	X	X	X			X		X
					130						
Met	Tyr	Arg	Ile	Tyr	Thr	Ala	Glu	Asp —COO^-			
Ile	Tyr	His	Leu	Tyr	Thr	Ser	Glu	Asn			
Met	Tyr	His	Phe	Tyr	Thr	Ser	Glu	Asn			
	X			X	X		X				

Table 9.3 Amino Acid Composition of Different Forms of Bovine Cytochrome b_5 Reductase

Amino Acid	(A) Detergent Purified	(B) Protease Purified	(A-B)	Chymotrypsin Treated Detergent Purified
Lys	25	20	5	20
His	12	9	3	9
Arg	20	16	4	16
Asp	34	28	6	28
Thr	16	14	2	11
Ser	19	13	6	12
Glu	38	29	9	26
Pro	33	29	4	28
Gly	27	20	7	19
Ala	21	15	6	14
Cys	7	6	1	
Val	28	16	12	19
Met	10	8	2	8
Ile	24	19	5	19
Leu	40	27	13	29
Tyr	13	8	5	9
Phe	18	13	5	13
Trp	6	2	4	2
Total	391	292	99	
Molecular weight	43,320	33,000	10,140	

cytochrome b_5 led to the classification of the reductase as an amphipathic membrane protein.

The fact that the nonpolar domains of cytochrome b_5 and the cytochrome b_5 reductase serve as membrane anchors has been firmly established by the reconstitution experiments of Strittmatter. Early experiments demonstrated that only the detergent purified form of cytochrome b_5 could be reconstituted into the electron transport system of liver microsomes (24). Reconstitution of electron transport was later achieved utilizing preformed vesicles prepared from microsomal lipids (25), egg lecithin (26), or dimyristoyl lecithin (27). In all cases the reconstitution was absolutely dependent upon the presence of the nonpolar domain of both the cytochrome b_5 and cytochrome b_5 reductase.

Secondary and Tertiary Structures

Because of its small molecular size and its availability, the secondary and tertiary structure of protease purified cytochrome b_5 was established by x-ray crystallography (12). The primary interest at the time was a comparison of this cytochrome to the structures established previously for cytochrome c and myoglobin. More recently this study has provided an invaluable description for comparison of the polar and nonpolar domains of a multifunctional amphipathic protein.

Protease purified cytochrome b_5 from calf liver was shown to be an ellipsoid of dimensions 25 × 25 × 32 Å. Approximately 75% of the 93 total amino acids are involved in six short helicies or five pleated sheet structures that are found in the peptide (see Figure 9.2). The heme iron is coordinated by residues His 43 and His 67 and is held in a hydrophobic crevasse of two sets of parallel helicies (residues 37–42 and 46–53; 59–66 and 68–78) with a floor of pleated sheets. The side chains of the polar amino acids are either oriented toward the outer surface of the protein or are involved in the helicies of the heme crevasse. In this study the carboxyl terminus appeared to be unstructured suggesting a "flexible link" to the membrane anchor. However, later analysis of the CD spectra of the detergent purified cytochrome b_5 indicated that the connecting peptide between the polar and nonpolar domains was highly ordered (28). This structure positions the polar domain of the cytochrome b_5 away from the membrane surface.

The first detailed study of the biochemistry of the nonpolar domain was carried out by Tanford (29). He performed direct detergent binding studies in order to investigate the previous observations that aggregates of the detergent purified cytochrome b_5 could be dispersed by addition of deoxycholate or sodium dodecyl sulfate (29). Detergent binding was found to be a cooperative process that occurred near the critical micelle concentrations of deoxycholate, Triton X-100 or sodium dodecyl sulfate. The isolated nonpolar peptide exhibited an identical ability to act as a nucleus for detergent micelle formation. This study indicated that the nonpolar domain of cytochrome b_5 was a distinct functional segment that maintained its structural integrity in the absence of the catalytically active polar domain. A separate study of its CD spectra (28) indicated that this nonpolar peptide, with its generally diffuse hydrophobic surface, contained some helical structure. This is consistent with the theoretical considerations of Guidotti which predict that an α-helix should be a stable structure within a lipid bilayer (30).

More detailed studies of the bovine cytochrome b_5 have investigated the molecular structure of the hydrophobic peptide in order to determine how it coordinates so tightly with the membrane lipids. It was discovered that the detergent purified cytochrome b_5 exhibited an increase in fluorescence at 335 nm when bound to preformed phospholipid vesicles (31). The increased fluorescence was consistent with the spectral characteristics of a tryptophan residue in a nonpolar environment. Treatment of vesicle bound cytochrome b_5 with trypsin released the polar domain containing the heme but did not affect the fluorescence, indicating that the tryptophan residue of interest was one of the three in the nonpolar peptide. The 43 amino acid isolated nonpolar peptide produced the same increase in fluorescence when bound to phospholipid vesicles. This again suggested that the nonpolar peptide retained its native conformation even in the absence of the polar domain.

Treatment of the vesicle bound nonpolar peptide with *N*-bromosuccinimide (NBrS) reduced the fluorescence nearly 100% when 1 *M* equivalent of tryptophan had reacted. The sequential degradation of a preparation of isolated nonpolar peptide containing 0.7 *M* equivalents of NBrS-reacted tryptophan with carboxypeptidase Y released 1 equivalent of unreacted tryptophan (Trp 112) before releasing NBrS-reacted tryptophan (see Figure 9.2). Therefore, the fluorescent and chemically reactive tryptophan residue was most likely to be Trp 109. The participation of Trp 108 could not be unequivocally ruled out because of the potential kinetic effects of a modified pentultimate amino acid.

Treatment of the vesicle bound nonpolar peptide with trinitrobenzene sulfonate showed rapid reaction of the amino terminus and the side chain of Lys 93 (31). This information along with localization of glutamate at position 96 and the discovery that lysosomal solubilized cytochrome b_5 often contains residues 91–97 (11), confirmed the orientation of the nonpolar peptide in the membrane with its amino terminus 7 amino acids exposed to the outside and apparently not essential for membrane binding.

The secondary structure of the nonpolar peptide of cytochrome b_5 has been predicted by using the algorithms of Chou and Fasman (32–34) to analyze the amino acid sequence. The proposed structure (13, 35) consisted of an 11 residue α-helix formed from residues Pro 115 through Ile 125; two 5 residue β sheets formed from Ile 98 to Ile 102 and Try 126 to Thr 130; and a 10 residue structure of three β turns formed from Asp 103 to Trp 112 (see Figure 9.2). These predictions were consistent with the analysis of the CD spectra of the nonpolar peptide. By treating the detergent purified cytochrome b_5 with controlled carboxypeptidase digestion, derivatives shortened by 6, 18, and 27 amino acids were formed. They were studied in order to define a limited size needed for membrane binding (35). Only the two largest cytochrome b_5 derivatives were still capable of binding to lecithin vesicles and reacting in electron transport with membrane bound stearyl CoA desaturase. Thus removal of the carboxyl terminal β sheet and α helix regions does not effect membrane binding even though 33 of the 43 nonpolar peptide amino acids are naturally found buried in the membrane. This predicted structure also allowed the negative charges of the carboxyl terminus dipeptide (glutamylaspartate or glutamylasparagine) to surface on the same side of the membrane as the polar domain of the cytochrome b_5.

All of the available information firmly supports the classification of cytochrome b_5 as an amphipathic protein. The hydrophilic globular polar head containing the active prosthetic group appears to function as a normal soluble protein while the hydrophobic nonpolar domain is a stable structured peptide that behaves as an integral membrane protein. Only the protease sensitive connecting peptide relates these two functionally different proteins. The similarity of the general physical characteristics of the cytochrome b_5 reductase would lead us to predict that its primary, secondary, and tertiary structures are quite similar to the established structure of the cytochrome b_5.

Interactions between Polar and Nonpolar Domains

During the course of the purification of the two forms of cytochrome b_5 and the cytochrome b_5 reductase, an additional area of investigation was initiated. The catalytic properties of the detergent purified and protease purified reductase with NADH as electron donor and ferricyanide as electron acceptor were only slightly altered from that observed with the membrane bound form. In contrast, the kinetics of the reaction of the detergent purified reductase with the protease purified cytochrome b_5 as electron acceptor, were significantly changed (22). Addition of deoxycholate at a very low concentration improved the altered kinetics but did not return them to normal. The decreased interaction of aggregated reductase with soluble cytochrome b_5 was not unexpected, but it did introduce a model system for future investigation of the mechanism of interaction of the two catalytically active membrane proteins. These findings suggested that the nonpolar domains were not only essential in anchoring the polar domains to the membrane but also in orienting the active sites for optimal interaction.

In order to characterize the lipid-protein interactions, various reconstitutions of a functional electron transport system in membranes of defined lipid content were developed. The first such approach was Strittmatter's discovery (36) that the detergent purified cytochrome b_5 could be bound in a 10- to 20-fold excess to its native microsomal membrane. At saturation the cytochrome b_5 became 20% of the total membrane bound protein, and exogenously added cytochrome b_5 could not be distinguished from the endogenous cytochrome b_5. This suggested that the cytochrome b_5 binding was due to an interaction with the membrane lipids rather than being limited by the availability of a membrane protein binding site.

After saturation of the membrane with cytochrome b_5 the catalytic activity (as measured by reduction of cytochrome b_5 by cytochrome b_5 reductase with electrons from NADH) was increased 10^4 over the rate observed with the aggregated and unbound detergent purified cytochrome b_5 (36). This large difference in the reduction rate provided an easy assay for the binding of cytochrome b_5 to microsomal membranes. It also supported the idea that membrane binding by the nonpolar domain did orient the proteins in the optimal arrangement for interaction.

Although the rate of reduction of soluble exogenous cytochrome c by reduced cytochrome b_5 increased linearly with the amount of cytochrome b_5 incorporated into microsomes, the kinetics of cytochrome b_5 reduction by cytochrome b_5 reductase did not follow simple first order kinetics (37). The deuterium rate effect found with NAD^2H-ferricyanide cytochrome b_5 reductase activity was not evident when endogenous cytochrome b_5 was the electron acceptor (37). Together, these results suggested that the rate limiting step in electron transfer was the translational diffusion time of the cytochrome b_5 and the reductase.

Study of the rate limiting step was also carried out following reconstitution of detergent purified cytochrome b_5 and reductase with extracted microsomal lipids (25). This membrane was constructed to simulate the same cytochrome b_5 to reductase ratio (13 to 1) as seen in native membranes. It also exhibited identical rates of cytochrome b_5 and cytochrome c reduction as determined in microsomes. Experiments with an NADH analog of reduced electron transfer activity and with NAD^2H confirmed the proposal that the diffusion of the nonpolar domains of the cytochrome b_5 and the reductase in the lipid bilayer was the rate limiting step in electron transfer between the polar domains. Such a proposal assumed that the two different proteins were randomly distributed in the membrane and that electron transfer occurs only from the reductase to the cytochrome b_5 and not between two molecules of cytochrome b_5.

This latter assumption was established by studying independently the electron transfer to endogenous and to exogenously added cytochrome b_5 (37). Addition of a ten-fold excess of cytochrome b_5 to microsomes increased four-fold the overall reduction rate of cytochrome b_5 by cytochrome b_5 reductase. In contrast, addition of a tenfold excess of deuteroheme cytochrome b_5 (an analog with a reduction rate 70% of that seen with cytochrome b_5) resulted in an increased reduction rate of the excess analog (read as ΔA 409) but no change in the reduction rate of the endogenous cytochrome b_5 (read as ΔA 423). Thus the increased rate of cytochrome b_5 reduction was due to an increased frequency of collision with and electron transfer from the reductase and was not due to electron transfer between cytochrome b_5 molecules. Since the less reactive analog of cytochrome b_5 did not effect the reactivity of the endogenous cytochrome b_5, the data also suggested that the cytochrome b_5 and the reductase formed only a transient complex during electron transfer.

Further confirmation that electron transfer between the cytochrome b_5 and the reductase is dependent upon translational diffusion was found by studying the effect of lipid fluidity and protein concentrations on electron transfer in vesicles composed of pure dimyristoyl lecithin (28). Vesicles were prepared with a constant amount of reductase but with an excess of cytochrome b_5 to yield ratios of cytochrome b_5 to reductase of 2.0, 5.3, and 45. While the energy of activation for the NADH-ferricyanide reductase activity was independent of temperatures between 5 and 30°C, the energy of activation for electron transfer from NADH to soluble cytochrome c was altered at temperatures corresponding to the transition temperatures of the lecithin (23.7–23.8°C and 12.5–13.5°C). The vesicle preparations containing the higher ratios of cytochrome b_5 to reductase showed smaller changes in the energy of activation in response to the changes in lipid fluidity as well as a smaller decrease in the maximal rate of the NADH-cytochrome c reductase activity. Retention of activity even at the lowest temperatures may have reflected some freedom of rotation of the protein polar domain due to the flexibility of the connecting peptide. In addition, at the highest ratio of cytochrome b_5 to reductase used the isotope rate effect of the NAD^2H-cytochrome c reductase activity approached the maximum value seen with NAD^2H-ferricyanide reductase activity, implying that the reductase was saturated with the cytochrome b_5 and that the rate limiting step had once again become reductase reduction by NAD^2H (37).

In summary, the nonpolar domains of the cytochrome b_5 and the reductase function by both anchoring the two proteins to the same membrane and by orienting them for optimal electron transfer. This fact was also established by observing that the reactivity of the vesicle bound reductase with the protease purified cytochrome b_5 is only 13% of that found with both proteins bound to the same membrane (38). In addition, unbound detergent purified cytochrome b_5 is only 4.5% as reactive while cytochrome b_5 bound to a different membrane has only 0.1% of the maximal activity. On the other hand, the activity of the two protease purified proteins was similar to the activity found when both were membrane bound.

The need for the membrane anchor becomes even more apparent when one reconsiders the fact that these two proteins are the electron suppliers for the microsomal stearylCoA desaturation system. The protein catalyzing the desaturase reaction was found to be a hydrophobic integral membrane protein (62% nonpolar amino acids) containing one atom of nonheme iron/53,000 dalton peptide (39). The desaturase could be purified only after solubilization with detergents, but it could not be incorporated into preformed lecithin vesicles. However, removal of deoxycholate from solutions containing enzyme and lecithin resulted in the formation of vesicles in which the desaturase was oriented with its reactive iron on the surface of the membrane (40). Complete activity of the desaturase with its hydrophobic substrate was dependent upon electron transfer from NADH by either soluble or membrane bound cytochrome b_5 reductase and membrane bound cytochrome b_5. The cytochrome b_5 could react with the desaturase only when bound to the same membrane surface. In addition to their localization in the endoplasmic reticulum, the cytochrome b_5 and the reductase were also found in low amounts in outer mitochondrial membrane, golgi, and nuclear membranes (24). It was also reported that, at least *in vitro*, cytochrome b_5 can transfer between membrane surfaces (38). Therefore, some unknown mechanism related to their reactivity in the desaturase pathway or a particular membrane lipid composition may be responsible for concentration of the cytochrome b_5 and reductase in the microsomal membrane. This is of particular importance when assessing the function of the nonpolar domain of these two amphipathic proteins. The primary function may be in orientation of the polar domains for optimal electron transfer, but lipid structuring around the hydrophobic peptides may have a secondary effect on the stability and reactivity of the three proteins in the desaturase pathway. In particular, the study of the interaction of the cytochrome b_5 and the desaturase may yield an even better understanding of the mechanism of interaction of integral membrane proteins.

9.3 BRUSH BORDER MEMBRANE HYDROLASES

The brush border membranes of the enterocytes of the small intestine and of the proximal tubule cells of the kidney are rich in various peptidase, disaccharidase, and phosphatase activities. These hydrolases are responsible for the final steps in the digestion of nutrients in the small intestine or the degradation of small peptides contained in the glomerular filtrate. Their activities are localized on the outside surface of their respective cells, exposed to the lumen of either the small intestine or the renal nephron. As summarized in Table 9.4, at least five of these hydrolases have been shown to be amphipathic enzymes. They each contain two distinguishable domains: a large heavily glycosylated hydrophilic domain that contains the active site,

Table 9.4 Comparison of the Protease and Detergent Purified Forms of Amphipatic Brush Border Membrane Hydrolases

Hydrolase	Subunit Structure (mol. wt.): Protease Purified	Subunit Structure (mol. wt.): Detergent Purified		Unique Characteristics of Detergent Purified Form	
1. Aminopeptidase (hog)	α_2-(140,000)	α_2-(150,000)	(41)[a]	Aggregates minus detergent	(41)[a]
		α_2-(162,000)	(46)	Reconstitution into lipid vesicles	(42)
				Different N-terminus	(43)
				Additional antigenic determinants	(44)
				Isolated hydrophobic peptide	(43)
				Photo labeling	(45)
				Charge-shift immunoelectrophoresis	(46)
(rat)	α_2-(135,000)	α_2-(150,000)		Aggregates minus detergent	(47)
				Reconstitutes into lipid vesicles	
2. γ-Glutamyltrans-peptidase (rat)	$\alpha\beta$-(27,000, 41,000)	$\alpha\beta$(27,000, 60,000)		Aggregates minus detergent	(48)
				Binds micelles of detergent	
				Reconstitutes into lipid vesicles	(49)
3. Sucrase-Isomal-tase (rabbit)	α_2-(112,000)	uncharacterized		Additional antigenic determinant	(50)
				Aggregates minus detergent	
				Reconstitutes into lipid vesicles	(51)
4. Dipeptidyl Pepti-dase (hog)	α_2-(130,000)	α_2-(137,000)		Aggregates minus detergent	(52)
				Different N-terminus	
				Labeling with lactoperoxidase	(53)
				Charge shift immunoelectrophoresis	(54)
5. Alkaline phospha-tase (hog)	α_2-(60,000)	α_2-(64,000)		Aggregates minus detergent	(55)

[a] Numbers refer to appropriate references.

and a relatively small (M_r = 4000–20,000) hydrophobic domain that interacts with the lipid bilayer.

None of the brush border membrane hydrolases have been characterized as well as the microsomal cytochrome b_5. Very little information regarding the primary sequence, the secondary or tertiary structure of the separate domains, or the details of the spacing between the domains is available for the brush border membrane proteins. But a discussion of the recurrent features of their characterization should illustrate the generality of the utilization of an amphipathic structure as a mechanism to attach hydrophilic enzymes to a hydrophobic membrane.

Each of the hydrolases can be quantitatively solubilized from the membrane either by detergent extraction or by limited proteolysis. Detergent solubilization yields the native amphipathic enzyme, whereas protease solubilization releases only the catalytic domain. The two solubilized forms of the enzyme can usually be purified by similar procedures, except that the purification of the detergent solubilized form requires the continued presence of detergent. Removal of the detergent causes the detergent solubilized enzyme to undergo aggregation. The detergent and protease purified forms exhibit different mobilities on sodium dodecyl sulfate polyacrylamide gel electrophoresis. In addition, protease treatment of the detergent purified hydrolase produces a form of the enzyme that no longer aggregates in the absence of detergent and that is usually indistinguishable from the protease purified enzyme during sodium dodecyl sulfate polyacrylamide gel electrophoresis.

γ-Glutamyltranspeptidase

Rat renal γ-glutamyltranspeptidase is a broad specificity transferase that catalyzes the transfer of γ-glutamyl groups from a variety of γ-glutamyl peptides, including glutathione, to a wide range of amino acid and peptide acceptors (56). This lack of specificity and its localization in membranes principally involved in absorption or secretion have led to the proposal that this enzyme participates in amino acid transport via the γ-glutamyl cycle (57–59). This proposal assumes that γ-glutamyltranspeptidase is a transmembranous protein and that it is capable of reacting with intracellular glutathione and extracellular amino acids to yield intracellular γ-glutamyl amino acids.

Recent experiments have suggested that γ-glutamyltranspeptidase activity is primarily extracellular (60–63) and that at physiological concentrations of amino acids, hydrolysis of glutathione is the preferred reaction (64). In addition, a patient who lacks γ-glutamyltranspeptidase activity, has been reported to exhibit glutathionemia and glutathionuria, but not amino aciduria (65). Based on these observations, it has been proposed that the physiological function of γ-glutamyltranspeptidase is the degradation of extracellular glutathione (61). A fundamental difference in the proposed functions of γ-glutamyltranspeptidase is its mechanism of attachment and its orientation within membranes.

γ-Glutamyltranspeptidase from rat kidney has been purified following solubilization with either proteolytic enzymes or detergents (48, 66). The papain purified transpeptidase is soluble in aqueous buffers and has a molecular weight of 68,000. In contrast, the Triton purified transpeptidase is soluble only in the presence of detergents. This form of the enzyme undergoes aggregation just below the critical micelle concentration of Triton X-100, and it binds 1.16 mg of Triton X-100/mg of protein (48). The molecular weight of the Triton X-100 protein complex was calculated to be 169,000, but the glycoprotein portion of the complex is 52% of the total mass (87,000). The mass of Triton X-100 (82,000) associated with the complex suggests that the transpeptidase binds to micelles of detergent. Treatment of the Triton purified transpeptidase with papain or bromelain results in a form of the enzyme identical in all respects to the papain purified enzyme.

Both the Triton and papain purified transpeptidase exhibit two protein bands on sodium dodecyl sulfate polyacrylamide gel electrophoresis. The smaller subunits of the two forms appear identical (M_r = 27,000) while the larger subunits of the Triton and papain purified enzymes have apparent molecular weights of 54,000 and 51,000, respectively. Therefore, only the larger of the two subunits of the transpeptidase appears to be amphipathic. An accurate estimation of the difference in molecular weights of the two large subunits was complicated by the presence of the micelle binding site that probably bound a large amount of charged detergent and therefore caused the large subunit of the Triton purified transpeptidase to migrate faster than in proportion to its molecular weight. Thus a more accurate analysis of the subunit molecular weight of an amphipathic protein was obtainable only by a method such as Ferguson plot analysis of its mobility during sodium dodecyl sulfate polyacrylamide gel electrophoresis. Using this approach, Horiuchi et al. (67) recently confirmed that the molecular weight of the smaller subunits of the two forms of the transpeptidase were identical. The large subunit of the Triton purified enzyme was estimated to be 21,000 daltons greater than the corresponding subunit of the papain purified enzyme. This determination is in close agreement with the difference of 19,000, obtained by comparison of the calculated molecular weights of intact Triton and papain purified enzymes (48).

γ-Glutamyltranspeptidase is rapidly inactivated by the γ-glutamyl analog, 6-diazo-5-oxo-L-norleucine, which binds covalently to the small subunit of the transpeptidase (68, 69). Inactivation is competitively inhibited by glutathione but not by acceptor substrates. These results suggest that the γ-glutamyl site is localized on the small subunit of the transpeptidase. At least one function of the large subunit is to provide an anchor for the attachment of the transpeptidase to the membrane. The hydrophilic domain of this subunit may provide the binding site for the attachment of the catalytic subunit or, alternatively, it may also contain the amino acid acceptor portion of the active site.

The interaction of γ-glutamyltranspeptidase with membranes was studied further by characterizing the association and orientation of this enzyme within egg lecithin vesicles (49). The association of the Triton purified transpeptidase with single layered [^{14}C]lecithin vesicles was established by comparing enzyme activity and radioactivity profiles during Sepharose 4B chromatography and isopycnic sucrose gradient centrifugation. The enzyme-vesicle complex exhibits a density corresponding to that of a single enzyme molecule bound to a single vesicle. γ-Glutamyltranspeptidase purified following solubilization with papain does not bind to vesicles. In addition, papain treatment of vesicles containing the Triton purified transpeptidase results in the release of 95% of the transpeptidase activity without release of internally trapped [^{3}H]sucrose. The released transpeptidase is chromatographically identical to the papain purified transpeptidase. γ-Glutamyltranspeptidase activity associated with both native membranes and with lecithin vesicles exhibits a temperature induced transition in its energy of activation. In contrast, the proteolytic and detergent solubilized forms of the enzyme exhibit a single energy of activation over the entire temperature range. These results suggest that the enzyme-vesicle complex closely approximates the interaction and orientation of γ-glutamyltranspeptidase with brush border membranes and provides further evidence that γ-glutamyltranspeptidase binding to membranes is due to a papain sensitive sequence of amino acids.

Aminopeptidase

The initial classification of the aminopeptidase from pig intestine as an amphipathic protein resulted from the extensive characterization of its association and orientation within the membrane. Greater than 95% of the aminopeptidase activity was rapidly solubilized by papain treatment of closed, right-side-out brush border membrane vesicles (70). This treatment resulted in solubilization of approximately 50% of the membrane protein and carbohydrate, while none of the membrane phospholipid or cholesterol was released. In addition, a comparison of their behavior during isopycnic sucrose-and Ficoll-gradient centrifugation indicated that both the treated and the concentrated vesicles were impermeable to macromolecules. Therefore, it was concluded that the catalytic domain of the aminopeptidase is localized external to the lipid bilayer.

These results were confirmed and extended by using various immunological techniques (44). F_{ab} monovalent antibodies were shown to inactivate the aminopeptidase activity in sealed brush border membrane vesicles to nearly the same extent as that demonstrated for the soluble, papain purified aminopeptidase. Recent freeze-fracture electron microscopic data indicated that the slight difference in the extent of inactivation by antibody resulted from the fact that 5–10% of the isolated vesicles were inverted (71). The concept that membrane attachment results from the presence of an additional domain was indicated by the use of peroxidase (44) or I^{125} labeled (72) antibodies to quantitate the number of determinants per aminopeptidase molecule. The Triton purified enzyme was shown to exhibit two to four additional determinants compared to the membrane bound aminopeptidase.

By use of a macromolecular photolabeling probe, it was demonstrated that the hydrophobic domain of the aminopeptidase spans the membrane (45). The photoactivatable reagent consisted of 4-fluoro-3-nitrophenyl azide covalently bound to the F_{ab} fragment of a human myeloma (IgG) protein. The myeloma protein was either prelabeled with I^{125} or, following photolysis, it was reacted with peroxidase conjugated sheep F_{ab} anti-human IgG. Following solubilization of the membranes with Triton X-100, the extent of photolabeling was quantitated by determining the amount of radioactivity or peroxidase activity precipitable with anti-papain purified aminopeptidase antibodies.

When sealed brush border membrane vesicles were treated from the outside, 7% of the total labeled membrane proteins solubilized with Triton X-100 were precipitable with the anti-aminopeptidase antibody. In additon, papain treatment of the Triton solubilized complex had no effect on the percent of specific label precipitated. Therefore all of the labeling observed with closed vesicles was directed at the external hydrophilic domain of the aminopeptidase. In separate experiments the probe was trapped inside closed vesicles before photolysis. Under these conditions pretreatment with papain resulted in a 65% decrease in the immunoprecipitable label, indicating that a large proportion of the label was associated with the hydrophobic domain of the aminopeptidase.

Sodium dodecyl sulfate polyacrylamide gel analysis of the initial preparation of trypsin purified aminopeptidase obtained from the pig intestine indicated that the enzyme was composed of three subunits with molecular weights of 130,000, 97,000 and 49,000 (73). Subsequent characterization of the protease (autolysis) purified pig renal aminopeptidase indicated that it was composed only of subunits of 140,000 molecular weight, while the subunits of the detergent purified enzyme appeared 10,000 daltons larger (41). In addition, trypsin treatment of the protease purified renal enzyme generated the lower molecular weight peptides of 140,000, 95,000, and 48,000.

The recent experiments of Sjöström et al. (46) have established that *in vivo* intestinal proteolysis is responsible for the conversion of the large peptide of the intestinal aminopeptidase into the smaller ones found in the purified enzymes. They developed an immunoadsorbent technique for the rapid purification of the peptidase. The detergent purified intestinal enzyme, isolated from a pig in which the pancreatic duct had been ligated for three days before sacrifice, migrated in sodium dodecyl sulfate polyacrylamide gel electrophoresis as a single polypeptide chain of 162,000 mol wt. In contrast, the detergent purified enzyme isolated from a normal pig exhibited three polypeptides of molecular weights 162,000, 123,000, and 61,000.

The hydrophobic sequence of the aminopeptidase has been isolated by trypsin treatment of the detergent purified enzyme (43). In order to facilitate subsequent detection of the small peptide the aminopeptidase was initially labeled with I^{125}. Following trypsin treatment the peptide was purified by chromatography on Sephadex G-50 and DEAE-cellulose. The isolated peptide had a high content of hydrophobic amino acids, and, as determined by sodium dodecyl sulfate polyacrylamide gel electrophoresis, it had an apparent molecular weight of 10,000. Amino terminal analysis of the isolated peptide and of the detergent and papain purified enzymes suggested that the hydrophobic domain of the aminopeptidase is associated with the amino terminus of the native protein.

Charge-shift immunoelectrophoresis was recently developed as a method to screen for amphipathic proteins (74). According to this procedure, a protein is initially electrophoresed in one dimension in the presence of either a cationic, an anionic or a nonionic detergent. It is then electrophoresed in the presence of a nonionic detergent in a second dimension where it reacts with antiserum. Amphipathic proteins bind enough detergent that their mobility is significantly different in each of the three determinations. In contrast, a nonhydrophobic protein exhibits the same mobility with all three detergents. By charge-shift immunoelectrophoresis both the *in vivo* modified and the unmodified intestinal aminopeptidase exhibited amphipathic properties immediately after isolation (46). However, during storage the enzyme was converted to a hydrophilic protein by a process that was inhibited by $HgCl_2$.

A similar aminopeptidase has been recently isolated from rat renal brush border membranes (75). It was initially purified and characterized as the cysteinylglycine dipeptidase involved in processing precursors of mercapturic acids (N-acetyl S-derivatized cysteines). Further study of the substrate specificity of the enzyme revealed its similarities to the porcine aminopeptidases. Purification and characterization of the rat renal aminopeptidase after solubilization with Triton X-100 or papain clearly showed it to be an amphipathic protein. Only the Triton purified enzyme could be incorporated into lecithin vesicles (47). By sodium dodecyl sulfate polyacrylamide gel electrophoresis this form of the enzyme had an apparent subunit molecular weight of 150,000. Upon storage the enzyme was degraded to a form with an apparent subunit molecular weight of 139,000, which could no longer be incorporated into lecithin vesicles. This form of the enzyme was still larger than the form purified after solubilization from native membranes with papain (apparent subunit molecular weight of 135,000).

The molecular weight of the intact papain purified aminopeptidase was calculated from its Stoke's radius and its sedimentation coefficient to be 203,000, whereas sedimentation equilibrium analysis indicated a molecular weight of 190,000. The disparity between these two determinations and the apparent subunit molecular weights illustrates the fact that glycoproteins yield anomalously high subunit molecular weights during sodium dodecyl sulfate polyacrylamide gel analysis (76). Therefore, the comparison of the subunits of the various forms of the renal aminopeptidase indicated only that they differed in size. Characterization of the size of the membrane binding peptide will require its careful analysis after isolation.

Sucrase-Isomaltase

The Triton purified rabbit intestinal sucrase-isomaltase can be incorporated into lecithin vesicles (51). NMR and ESR analysis indicated that the incorporation of the small hydrophobic portion of the protein in the lipid bilayer is accompanied by a minimal perturbation of the packing of the lipids in the bilayer. The incorporation of the protein was assymetric. Both the ability of papain to release the catalytic activity and electron micrographic analysis indicated that the protein was orientated toward the external aqueous phase. The data also suggested that the hydrophilic domain was neither in contact with the lipid bilayer nor with the lipid surface but sat on a stalk so that a gap of greater then 10 Å was produced between the lipid surface and the bulk of the protein.

For each of the three hydrolases that have been reconstituted, enzyme-vesicle complexes could be formed only by the procedure involving cholate removal (42, 49, 51). In no case has it been possible to demonstrate the efficient reassociation of a brush border membrane hydrolase with preformed vesicles. This could result if, as demonstrated with the aminopeptidase (45), the hydrophobic domain of the hydrolases spans the membrane and terminates with a hydrophilic sequence that is exposed to the internal medium. Movement of such a hydrophilic sequence through a preformed lipid bilayer could constitute an energy barrier sufficient to prevent reconstitution. In contrast the suggestion that, in the case of cytochrome b_5, the polar carboxyl terminus of the hydrophobic domain is situated on the outside of the membrane along with the amino terminus hydrophilic domain could explain the ease with which this protein reassociates with preformed vesicles.

It has recently been established that the sucrase-isomaltase is anchored in the intestinal brush border membrane by a hydrophobic segment which is located near the amino terminus of the isomaltase subunit (77). Papain solubilization of the enzyme results from cleavage of only the polypeptide chain of the isomaltase subunit. In addition, citraconylation of intact brush border membranes results in the preferential solubilization of the sucrase subunit. The "anchoring peptide" of the isomaltase subunit was isolated following papain solubilization of the enzyme reconstituted in lecithin vesicles. Its amino acid sequence was partially determined starting from the amino terminus and it was found to include a long, extremely hydrophobic sequence (78).

9.4 PERSPECTUS

From the few examples of amphipathic enzymes described here, it is obvious that not only multifunctional proteins but also many more complex proteins of multiple peptide composition utilize a structure consisting of two domains (polar-nonpolar) as a mechanism for membrane association. With the increasing number of reports of amphipathic proteins and enzymes in the literature, enough examples should soon be available to permit the study of the homology of membrane anchors. Such a comparison should include the nonpolar domains of cytochrome b_5 and cytochrome b_5 reductase as well as the corresponding sequences of more complex enzymes such as γ-glutamyltranspeptidase, with its two functionally distinct subunits, and aminopeptidase, with its two subunits both of which apparently contain a hydrophobic domain. In addition these sequences should be compared to that of egasyn, the possible "genetic precursor" of the hydrophobic domains in amphipathic enzymes.

Now that the hydrophobic domain of cytochrome b_5 has been shown to have a definite secondary structure, it should be predicted that similar hydrophobic sequences will also exhibit similar secondary structure. This would be consistent with earlier predictions that suggested that a thermodynamically stable peptide structure in a hydrophobic environment would require concealment of the polar peptide bonds and a generally hydrophobic surface composed of the nonpolar portions of the amino acids (30).

Future studies should also yield further information on the interaction of polar and nonpolar domains of amphipathic proteins. Potential interactions include primary effects such as positioning of the functional sites near "activating" lipids or near other proteins involved in the same metabolic pathway. While some hydrophobic domains are likely to be stabilized in specific membranes as a result of the interactions of their polar domains on both sides of the membrane, we still do not understand how proteins with more superficial nonpolar anchors are retained in the correct membrane. Stabilization of this latter type of membrane protein may require interaction with proteins such as egasyn in order to achieve the appropriate subcellular compartmentalization.

Finally, the future study of amphipathic proteins and enzymes may uncover additional and more complex functions for the nonpolar domains. One possibility is the transmembrane transport of the products of their respective catalytic reactions. Degradative enzymes of the epithelial brush border membranes should be prime targets for such an analysis.

REFERENCES

1 R. T. Swank and K. Paigen (1973), *J. Mol. Biol.* **77,** 371–389.

2 S. Tomino and K. Paigen (1975), *J. Biol. Chem.* **250,** 1146–1148.

3 A. J. Lusis, S. Tomino, and K. Paigen (1976), *J. Biol. Chem.* **251,** 7753–7760.

4 P. Strittmatter, M. J. Rogers, and L. Spatz (1972), *J. Biol. Chem.* **247,** 7188–7194.

5 H. G. Enoch, A. Catalá, and P. Strittmatter (1976), *J. Biol. Chem.* **251,** 5095–5103.

5 P. Strittmatter (1960), *J. Biol. Chem.* **235,** 2492–2497.

7 P. Strittmatter and J. Ozols (1966), *J. Biol. Chem.* **241,** 4787–4792.

8 P. Strittmatter (1961), *J. Biol. Chem.* **236,** 2329–2335.

9 P. Strittmatter (1971), *J. Biol. Chem.* **246,** 1017–1024.

10 P. Strittmatter (1967), in S. P. Colowick and N. O. Kaplan, eds., *Methods in Enz.* Academic Press, New York, pp. 553–556.

11 J. Ozols and P. Strittmatter (1969), *J. Biol. Chem.* **244,** 6617–6618.

12 F. S. Matthews, P. Argos, and M. Levine (1971), *Cold Spring Harbor Symp. Quant. Biol.* **36,** 387–395.

13 P. J. Fleming, H. A. Dailey, D. Corcoran, and P. Strittmatter (1978), *J. Biol. Chem.* **253,** 5369–5372.

14 J. Ozols (1974), *Biochemistry* **13,** 426–434.

15 A. Ito and R. Sato (1968), *J. Biol. Chem.* **243,** 4922–4923.

16 P. Strittmatter (1963), in P. D. Boyer, H. Lardy, and K. Myrback, eds., *The Enzymes,* Vol. 8, 2nd ed., Academic Press, New York, p. 113–145.

17 L. Spatz and P. Strittmatter (1971), *Proc. Natl. Acad. Sci.* **68,** 1042–1046.

18 J. Ozols (1972), *J. Biol. Chem.* **247,** 2242–2245.

19 J. Ozols and C. Gerard (1977), *Proc. Natl. Acad. Sci.* **74,** 3725–3729.

20 J. Ozols and C. Gerard (1977), *J. Biol. Chem.* **252,** 8549–8553.

21 P. Strittmatter (1971), *J. Biol. Chem.* **246,** 1017–1024.

22 L. Spatz and P. Strittmatter (1973), *J. Biol. Chem.* **248,** 793–799.

23 K. Mihara, R. Sato, R. Sakakibara, and H. Wada (1978), *Biochemistry* **17,** 2829–2834.

24 P. Strittmatter, M. J. Rogers, and L. Spatz (1972), *J. Biol. Chem.* **247,** 7188–7194.

25 M. J. Rogers and P. Strittmatter (1973), *J. Biol. Chem.* **248,** 800–806.

26 M. J. Rogers and P. Strittmatter (1975), *J. Biol. Chem.* **250,** 5713–5718.

27 P. Strittmatter and M. J. Rogers (1975), *Proc. Natl. Acad. Sci.* **72,** 2658–2661.

28 L. Visser, N. C. Robinson, and C. Tanford (1975), *Biochemistry* **14,** 1194–1199.

29 N. C. Robinson and C. Tanford (1975), *Biochemistry* **14,** 369–378.

30 G. Guidotti (1977), *J. Supraml. Struct.* **7,** 489–497.

31 P. J. Fleming and P. Strittmatter (1978), *J. Biol. Chem.* **253,** 8198–8202.

32 P. Y. Chou and G. D. Fasman (1977), *Trends Biochem. Sci.* **2,** 128–131.

33 P. Y. Chou and G. D. Fasman (1977), *J. Mol. Biol.* **115,** 135–175.

34 P. Y. Chou and G. D. Fasman (1974), *Biochemistry* **13,** 222–245.

35 H. A. Dailey and P. Strittmatter (1978), *J. Biol. Chem.* **253,** 8203–8209.

36 P. Strittmatter, M. J. Rogers, and L. Spatz (1972), *J. Biol. Chem.* **247,** 7188–7194.

37 M. J. Rogers and P. Strittmatter (1974), *J. Biol. Chem.* **249,** 895–900.

38 H. G. Enoch, P. J. Fleming, and P. Strittmatter (1977), **252,** 5656–5660.

39 P. Strittmatter, L. Spatz, D. Corcoran, M. J. Rogers, B. Setlow, and R. Redline (1974), *Proc. Natl. Acad. Sci.* **71,** 4565–4569.

40 H. G. Enoch, A. Catalá, and P. Strittmatter (1976), *J. Biol. Chem.* **251,** 5095–5103.

41 H. Wacker, P. Lehky, F. Vanderhaeger, and E. A. Stein (1976), *Biochim. Biophys. Acta* **429,** 546–554.

42 H. Wacker, F. Muller, and G. Semenza (1976), *FEBS Lett.* **68,** 145–152.

43 S. Maroux, and D. Louvard (1976), *Biochim. Biophys. Acta* **419,** 189–195.

44 D. Louvard, S. Maroux, and P. Desnuselle (1975), *Biochim. Biophys. Acta* **389,** 389–400.

45 D. Louvard, M. Semeriva, and S. Maroux (1976), *J. Mol. Biol.* **106,** 1023–1035.

46 H. Sjöström, O. Noren, L. Jeppesen, M. Staun, B. Svensson, and L. Christiansen (1978), *Eur. J. Biochem.* **88,** 503–511.

47 B. B. Rankin and N. P. Curthoys, manuscript in preparation.

48 R. P. Hughey and N. P. Curthoys (1976), *J. Biol. Chem.* **251,** 7863–7870.

49 R. P. Hughey and N. P. Curthoys (1979), *J. Biol. Chem.* **254,** 1124–1128.

50 Y. Takesue (1969), *J. Biochem.* (*Tokyo*) **65,** 545–552.

51 J. Brunner, H. Hauser, and G. Semenza (1978), *J. Biol. Chem.* **253,** 7538–7546.

52 A. G. Booth and A. J. Kenny (1976), *Biochem. J.* **159,** 395–407.

53 A. J. Kenny, A. G. Booth, and R. D. C. Macnair (1978), in W. G. Guder and U. Schmidt, eds., *Biochemical Nephrology*, Vol. 8, Hans Huber Publishers, Bern, pp. 46–58.

54 B. Svensson, M. Danielsen, M. Staun, L. Jeppesen, O. Noren, and H. Sjöström (1978), *Eur. J. Biochem.* **90,** 489–498.

55 A. Colbeau and S. Maroux (1978), *Biochim. Biophys. Acta* **511,** 39–51.

56 S. S. Tate and A. Meister (1974), *J. Biol. Chem.* **249,** 7593–7602.

57 M. Orlowski and A. Meister (1970), *Proc. Natl. Acad. Sci.* **67,** 1248–1255.

58 A. Meister (1973), *Science* **180,** 33–39.

59 A. Meister and S. S. Tate (1976), *Ann. Rev. Biochem.* **45,** 559–604.

60 T. Kuhlenschmidt and N. P. Curthoys (1975), *Arch. Biochem. Biophys.* **167,** 519–524.

61 R. Hahn, A. Wendel, and L. Flohe (1978), *Biochim. Biophys. Acta* **539,** 324–337.

62 A. Novodgrosky, S. S. Tate, and A. Meister (1976), *Proc. Natl. Acad. Sci.* **73,** 2414–2418.

63 M. Inoue, S. Horiuchi, and Y. Morino (1977), *Eur. J. Biochem.* **78,** 609–615.

64 T. M. McIntyre and N. P. Curthoys (1979), *J. Biol. Chem.* **254,** 6499–6504.

65 J. D. Schulman, S. I. Goodman, J. W. Mace, A. D. Patrick, F. Tietze, and E. J. Butler (1975), *Biochem. Biophys. Res. Commun.* **65,** 68–74.

66 S. S. Tate and A. Meister (1975), *J. Biol. Chem.* **250,** 4619–4627.

67 S. Horiuchi, M. Inoue, and Y. Morino (1978), *Eur. J. Biochem.* **87,** 429–437.

68 M. Inoue, S. Horiuchi, and Y. Morino (1977), *Eur. J. Biochem.* **73,** 335–342.

69 S. S. Tate and A. Meister (1977), *Proc. Natl. Acad. Sci.* **74,** 931–935.

70 D. Louvard, S. Maroux, C. Vannier, and P. Desnuelle (1975), *Biochim. Biophys. Acta* **375,** 236–248.

71 W. Haase, A. Schafer, H. Murer, and R. Kinne (1978), *Biochem. J.* **172,** 57–62.

72 C. Vannier, D. Louvard, S. Maroux, and P. Desnuelle (1976), *Biochim. Biophys. Acta* **455,** 185–199.

73 S. Maroux, D. Louvard, and J. Baratti (1973), *Biochim. Biophys. Acta* **321,** 282–295.

74 S. Bhakdi, B. Bhakdi-Lehnen, and O. J. Bjerrum (1977), *Biochim. Biophys. Acta* **470,** 35–44.

75 R. P. Hughey, B. B. Rankin, J. S. Elce, and N. P. Curthoys (1978), *Arch. Biochem. Biophys.* **186,** 211–217.

76 J. P. Segrest, R. L. Jackson, E. P. Andrews, and V. T. Marchesi (1971), *Biochim. Biophys. Res. Commun.* **44,** 390–395.

77. J. Brunner, H. Hauser, H. Braun, K. H. Wilson, H. Wacker, B. O'Neill, and G. Semenza (1979), *J. Biol. Chem.* **254,** 1821–1828.

78. G. Frank, J. Brunner, H. Hauser, H. Wacker, G. Semenza, and H. Zuber (1978), *FEBS Lett.* **96,** 183–188.

Chapter 10

ADP-RIBOSYLATING EXOTOXINS

R. JOHN COLLIER

Department of Microbiology
College of Letters and Science
and The Molecular Biology Institute
University of California, Los Angeles

JOHN J. MEKALANOS

Department of Microbiology
College of Letters and Science
University of California, Los Angeles

10.1 INTRODUCTION

A major benefit of the wealth of fundamental biochemical knowledge gathered in recent years is that it provides a basis for understanding mechanisms of pathogenesis at the molecular level. In bacterial diseases one of the most clearly understood aspects of pathogenesis pertains to the actions of protein exotoxins. In certain cases it has been possible to define in detail the molecular actions of toxins known to be responsible for the major symptoms of the diseases in which they are produced.

Although bacterial exotoxins have long appeared on the basis of gross characteristics to be a diverse group of unrelated toxic proteins, recent discoveries have altered this view. It is now clear that at least several important exotoxins have striking similarities in biochemical modes of action. Diphtheria toxin, exotoxin A from *Pseudomonas aeruginosa*, cholera toxin, and probably also *Escherichia coli* heat labile toxin and a similar enterotoxin from *Salmonella typhimurium*, all act by mechanisms that involve transfer of the ADP-ribosyl moiety of nicotinamide adenine dinucleotide (NAD^+) into covalent linkage with specific target proteins (TP) in susceptible animal cells. These toxins therefore define an emerging class of *ADP-ribosylating exotoxins* that may well be ultimately shown to be evolutionarily related.

$$\overset{\displaystyle N}{\overset{|}{R}}—P—P—\overset{\displaystyle A}{\overset{|}{R}} + TP \rightleftharpoons TP—R—P—P—\overset{\displaystyle A}{\overset{|}{R}} + \text{nicotinamide} + H^+$$

Inclusion of these ADP-ribosylating toxins in this volume on multifunctional enzymes rests on the fact that part or all of the toxin molecule must cross the plasma membrane of a sensitive cell before it can act. This follows from the fact that the substrates of the toxins are either cytoplasmic components, or, in at least one case, membrane bound, but accessible from the cytosolic face. Although the mechanism of membrane penetration is not known, those toxins that have been studied in some detail (diphtheria, *Pseudomonas*, and cholera toxins) all contain a domain, fragment, or subunit distinct from that carrying the catalytic center that is somehow required for transfer of the catalytic center across the membrane. The toxins are therefore multifunctional in the sense that they contain separate moieties devoted to catalysis and transmembrane transfer. [Some of the subject matter covered here has been treated in other reviews (diphtheria toxin, refs. 1–6; diphtheria and pseudomonas toxins, refs. 7, 8; cholera toxin, refs. 9–11).]

10.2 TOXINS ACTING ON PROTEIN SYNTHESIS

Diphtheria Toxin

Diphtheria Diphtheria is a potentially fatal disease in which the causative bacterium, *Corynebacterium diphtheriae*, usually resides in the upper respiratory tract. There the bacteria proliferate on mucosal surfaces, commonly in the throat, forming in moderate or severe cases a leathery *pseudomembrane*, which is the most characteristic diagnostic feature of the disease. Less commonly the diphtheria bacillus may cause infections at other sites in the body including the skin, other mucous membranes, and surface wounds.

The existence of a soluble toxin produced by *C. diphtheriae* was first postulated by Loeffler in 1884 on the basis of the observation that severe tissue damage occurred at sites distant from the localized area of the primary infection (12). Thus visible or microscopic lesions are frequently observed in the heart, adrenals, kidneys, lungs, and other organs, despite the fact that the causative bacteria are rarely found within these organs or in the blood stream. Subsequently it was shown that the diphtheria bacilli did indeed produce a highly toxic substance in culture, which produced death in experimental animals with a pattern of lesions similar to that observed in clinical diphtheria. Furthermore, immunization against culture filtrates containing the toxin protected against severe forms of the disease.

The toxin was obtained in sufficiently pure form by the late 1930s to determine that it was protein in nature and has since been purified, crystallized, and studied in many laboratories (13). Estimates of the molecular weight are about 60,000 or slightly higher (1). No carbohydrate or other nonprotein moieties have been found in the purest preparations.

Production of diphtheria toxin by *C. diphtheriae* shows many interesting features. It is known that release of the toxin does not require lysis of the cells and that the concentration of iron in the medium is crucial, with high concentrations causing a suppression of toxin secretion. Perhaps most intriguing is the relationship of toxin production to infection of *C. diphtheriae* by bacteriophage. It has long been known that only strains of *C. diphtheriae* that are lysogenic for, or vegetatively infected with, certain bacteriophage are capable of secreting diphtheria toxin, and it has recently become clear that such phage actually carry the structural gene for diphtheria toxin (14–17). Thus the toxin is synthesized in appropriate cell-free systems from *E. coli* to which DNA from corynephage β has been added (16), and certain mutant bacteriophage have been shown to encode altered forms of the toxin (17).

Diphtheria toxin shows toxicity for various animals such as guinea pigs, rabbits, and monkeys, about equivalent to that for man (per unit body weight), but rats or mice show comparable symptoms only at toxin concentrations some three to four orders of magnitude higher. Effects of the toxin in animals are manifested after a delay of hours to days, depending on the dose, and death is never observed in animals until at least 10 h after injection, even with massive doses of toxin. The gross morphological and physiological changes after injection of lethal doses of toxin indicate that the toxin has little tissue specificity, but the changes in animals are so complex that it has been impossible to infer from such studies which aspect of metabolism that is primarily affected.

Inhibition of Protein Synthesis The use of cultured mammalian cells has proven extremely valuable in analyzing the toxin's action (18). It was reported in 1957 that the toxin was lethal for cultured cells from sensitive animals (19, 20), and it was later found that cells from rats and mice showed resistance corresponding to that of the parent animals (21).

In 1959 Strauss and Hendee reported that diphtheria toxin blocked protein synthesis in HeLa cells (22). This effect, which began one to several hours after toxin addition depending on the concentration, was the earliest metabolic effect observed among several processes studied, including glycolysis, respiration, and nucleic acid synthesis; and no earlier effect has been observed in more recent studies (1).

Work in cell-free systems confirmed the hypothesis that the toxin blocked protein synthesis directly (22, 23). Incorporation of radioactive amino acids in cell-free systems from HeLa cells was inhbited rapidly and almost completely by the toxin, and treatment of cells with toxin before extraction produced a similar effect on cell-free incorporation.

Subsequent work in such systems revealed an interesting requirement for NAD^+ for the inhibition of protein synthesis (24). If NAD^+ was removed from the cell-free system by dialysis, gel filtration, and other fractionation methods, or by treatment with NADase (for example, streptococcal NAD^+ glycohydrolase), then addition of toxin even to high concentrations produced no inhibition of amino acid incorporation. Addition of NAD^+ to the systems once again restored toxin sensitivity.

By fractionating cell-free systems from HeLa cells and rabbit reticulocytes it was possible to localize the site of the toxin's action to elongation factor 2 (EF-2), a 100,000 dalton protein that is required for the translocation step of protein synthesis (25, 26). In this step peptidyl-tRNA is shifted from the A to the P site on the ribosome, and messenger RNA is translocated by one triplet codon in a reaction involving hydrolysis of GTP to GDP and inorganic phosphate. After incubation with the toxin and

NAD^+, the altered EF-2 retains its capacity to bind GTP, but is inactive in translocation.

ADP-Ribosyltransferase Activity In 1968 Honjo and co-workers reported that when partially purified EF-2 was incubated with diphtheria toxin plus radioactive NAD^+ labeled within the adenosinediphosphate-ribose moiety of the molecule (ADP-ribose), label was incorporated into trichloroacetic acid-insoluble material (27–29). No radioactivity was incorporated from nicotinamide labeled NAD^+. The protein bound label from the ADP-ribose moiety could be shown to be covalently attached to EF-2, and the inactivation of the elongation factor paralleled incorporation of label. On the basis of these results and others it was proposed that the toxin acted as an [NAD^+:EF-2]-ADP-ribosyltransferase that inactivated EF-2 by the following reaction:

$$NAD^+ + \text{EF-2} \rightleftharpoons \text{ADP-ribosyl-EF-2} + \text{nicotinamide} + H^+$$

There is now firm evidence that this reaction occurs in cells treated with diphtheria toxin and that it is in fact responsible for the observed inhibition of protein synthesis. Thus a decline in the activity of EF-2 in toxin treated cultured cells has been reported, and a similar effect has been observed in certain organs of animals injected with toxin (30). More convincing data have come from studies on toxin resistant variants of Chinese hamster ovary cells (31). Certain of these have been shown to contain altered EF-2 that is incapable of acting as substrate in the ADP-ribosylation reaction. Addition of diphtheria toxin to very high concentrations causes no inhibition of protein synthesis or any other manifestation of toxicity in these cells.

EF-2 contains a single attachment site for ADP-ribose, and Maxwell and co-workers have isolated a 15 residue tryptic peptide containing the site from rat liver EF-2 (32, 33). The residue to which ADP-ribose is attached, termed X, does not correspond to any of the commonly occurring amino acids in proteins. Inasmuch as X is present in EF-2 that has not been treated with diphtheria toxin, it is presumably a product of posttranslational modification of the protein, quite independent of the ADP-ribosylation reaction. Whatever the nature of X, it seems certain that the ADP-ribosyl group is attached by way of the NMN ribose moiety, inasmuch as ribose 5′-phosphate remains attached to EF-2 after removal of the adenosine 5′-phosphate portion of ADP-ribose with snake venom phosphodiesterase.

EF-2 from all eucaryotic organisms tested, including that from animal, plant, and fungal sources, acts as an efficient acceptor of ADP-ribose, whereas other proteins tested, including EF-G, the functional analog of EF-

2 in bacteria, do not. The fact that EF-2 from rats and mice is fully active as a substrate of the toxin implies that the toxin resistance of these animals results from inhibition of the intoxication process at a step preceeding ADP-ribosylation (34, 35). Residue X has been found in EF-2 from yeast as well as that from rat liver (36), and may be a major determinant of enzyme substrate recognition. It has been reported that other proteins, including bovine serum albumin, RNA polymerase core enzyme, and diphtheria toxin itself, may act as weak acceptors of ADP-ribose in the presence of very high concentrations of toxin and NAD^+, but it is possible that this apparent activity is an artifact resulting from the formation of Schiff bases between free ADP-ribose and the proteins (37). Diphtheria toxin and certain peptides thereof have NAD^+-glycohydrolase activity, which cleaves NAD^+ to ADP-ribose, nicotinamide, and H^+.

The specificity with respect to the pyridine nucleotide substrate is also high and suggests that the oxidized form of NAD^+ is the only donor of ADP-ribose *in vivo* (29, 38, 39). Only certain unnatural analogs of NAD^+ with minor structural alterations have any ADP-ribosyl donor activity *in vitro*. NADH and oxidized and reduced NADP are inactive as substrates. NMN may have slight activity, but his remains uncertain.

The equilibrium of the ADP-ribosylation of EF-2 is far to the right under physiological conditions; a standard free energy change has been calculated of −5.2 kcal/mole at pH 7 and 25°C (29). From this value and the known free energy of hydrolysis of the nicotinamide-ribose linkage of NAD^+ (−9.2 kcal/mole) the value for the ADP-ribosyl-EF-2 linkage may be calculated to be about −4.0 kcal/mole under the conditions specified. Inasmuch as nicotinamide is a product of the reaction it would be expected to promote the reverse reaction and hence shift the equilibrium to the left. Reversal of the reaction may in fact be demonstrated in the presence of high concentrations of nicotinamide (for example, 30 mM) and is facilitated by reducing the pH. This explains an earlier observation that EF-2 inactivated in the presence of toxin and NAD^+ could be reactivated by incubation with high concentrations of nicotinamide (in the presence of toxin) (40).

Structure-Activity Relationships

AN OVERVIEW Soon after the ADP-ribosylation reaction was elucidated, crude or partially purified toxin was found to contain a peptide less than half as large as the whole toxin that was highly active in catalyzing this reaction (41). Further studies revealed that the peptide was a fragment of the toxin (Fragment A) released after mild proteolytic digestion and reduction (42–45). Moreover, it could be shown that fragmentation of the toxin was in fact necessary for expression of ADP-ribosyltransferase activity.

The secreted form of diphtheria toxin is a single polypeptide chain (60,000–63,000 mol wt) containing two disulfide bridges and no free sulfhydryls. Although this intact form is toxic for animals and cultured animal cells, it appears to be devoid of ADP-ribosyltransferase activity. This enzymic activity is expressed *in vitro* after treatment with trypsin or other proteases of similar specificity, followed by reduction of the molecule (Figure 10.1). Trypsin preferentially cleaves the intact toxin at any of three closely spaced arginine residues located within one of the two disulfide loops, and thereby generates two large fragments, A (21,145 mol wt) and B (about 39,000 mol wt) linked by a dissulfide bridge. This nicked form of toxin resembles the intact form in its toxicity and lack of ADP-ribosyltransferase activity. Following reduction of the disulfide bridge linking Fragments A and B, the two may become dissociated, and the ADP-ribosyltransferase activity of Fragment A is expressed. Fragment B has no known enzymic activity.

We assume that the free Fragment A in impure toxin preparations results from cleavage by bacterial proteases followed by reduction or disulfide interchange promoted by traces of thiols in the medium (the thiols perhaps arising from cell lysis). Toxin isolated by routine methods contains both intact and nicked forms, and the ratio of nicked to intact increases on storage, presumably as a result of the action of trace amounts of proteases.

Inasmuch as the intact toxin has not been found to contain ADP-ribosyltransferase activity in cell-free systems under any conditions tested, it is generally assumed that scission of the intact toxin chain must occur either before or after attachment to susceptible cells. We know that trypsin nicked toxin inhibits protein synthesis in HeLa cells slightly more rapidly than a control preparation containing 80% intact toxin, thus suggesting that Fragment A can act as the enzymically active moiety within cells (43). However, it must be borne in mind that there is at present no direct evidence regard-

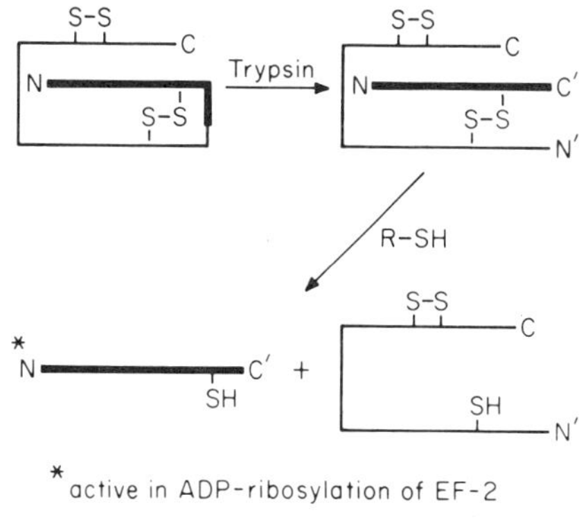

Fig. 10.1 Production of Fragment A (*; heavy line) and Fragment B (thin line) from intact diphtheria toxin by trypsinization and reduction.

ing the molecular nature of the enzymically active moiety generated intracellularly when cells are treated with intact toxin.

Despite its activity in catalyzing ADP-ribosyl transfer, Fragment A is essentially devoid of toxicity for whole animals or cultured cells (that is, it has toxicity on the order of 10^{-6} that of whole toxin). Similarly, there is evidence that the B moiety alone is not toxic. Thus both A and B moieties of the toxin are required, and apparently must be physically linked, in order to block protein synthesis in animal cells. As described below, it is believed that the B moiety somehow facilitates penetration of A into cells.

FRAGMENT A AND ADP-RIBOSYLTRANSFERASE ACTIVITY A prominent feature of Fragment A, and one which greatly facilitates handling it, is its stability (1). The fragment shows minimal loss of activity after brief exposure to a temperature of 100°C at neutral pH, or to pH 2 or 12 at room temperature. Also the fragment is markedly resistant to trypsin or chymotrypsin, and this resistance is enhanced in the presence of saturating concentrations of NAD^+ (39). This resistance to proteolysis may be a factor in the high toxicity of the molecule, in that it may facilitate longevity of the fragment in the intracellular milieu.

The amino acid sequence of Fragment A is presented in Figure 10.2, and the amino acid composition in Table 10.1 (46). The overall composition is distinctive in that there is only a single residue each of histidine (residue 21) and cysteine (residue 186), two of tryptophan (residues 50 and 153), and a large number of aromatic residues (19 out of 193). The amino-terminal sequence is identical with that of intact toxin. The carboxy-terminal sequence is interesting because of its heterogeneity. One may isolate from CNBr digests of the fragment, three carboxy-terminal peptides, with partial sequences, ---Asn-Arg-COOH, ---Asn-Arg-Val-Arg-COOH, and ---Asn-Arg-Val-Arg-Arg-COOH. These three peptides are apparently derived from three forms of Fragment A containing the corresponding C-terminal sequences, and indeed three major, evenly spaced bands are observed when Fragment A is subjected to isoelectric focusing (pI's 4.56, 4.72, and 4.85) or electrophoresis on polyacrylamide gels under nondenaturing conditions (47). All three forms are similar in specific activity. The carboxy-terminal heterogeneity implies that trypsin attacks intact toxin with similar frequencies at each of the three arginine residues (residues 190, 192, and 193).

The series of arginines at which trypsin attacks is located within a 14 residue disulfide loop formed by the half-cystine in Fragment A (residue 186) and another near the amino-terminus of Fragment B (2). The sulfhydryl group in Fragment A formed upon reduction of this disulfide may be blocked by any of a number of sulfhydryl specific reagents without significantly affecting ADP-ribosyltransferase activity.

10
H_2N-Gly-Ala-Asp-Asp-Val-Val-Asp-Ser-Ser-Lys-Ser-Phe-Val-Met-Glu-

20 30
Asn-Phe-Ser-Ser-Tyr-His-Gly-Thr-Lys-Pro-Gly-Tyr-Val-Asp-Ser-Ile-

40
Gln-Lys-Gly-Ile-Gln-Lys-Pro-Lys-Ser-Gly-Thr-Gln-Gly-Asn-Tyr-Asp-

50 60
Asp-Asp-Trp-Lys-Gly-Phe-Tyr-Ser-Thr-Asp-Asn-Lys-Tyr-Asp-Ala-Ala-

70
Gly-Tyr-Ser-Val-Asp-Asn-Glu-Asn-Pro-Leu-Ser-Gly-Lys-Ala-Gly-Gly-

80 90
Val-Val-Lys-Val-Thr-Tyr-Pro-Gly-Leu-Thr-Lys-Val-Leu-Ala-Leu-Lys-

100 110
Val-Asp-Asn-Ala-Glu-Thr-Ile-Lys-Lys-Glu-Leu-Gly-Leu-Ser-Leu-Thr-

120
Glu-Pro-Leu-Met-Glu-Gln-Val-Gly-Thr-Glu-Glu-Phe-Ile-Lys-Arg-Phe-

130 140
Gly-Asp-Gly-Ala-Ser-Arg-Val-Val-Leu-Ser-Leu-Pro-Phe-Ala-Glu-Gly-

150
Ser-Ser-Val-Glu-Ser-Tyr-Ile-Asn-Asn-Trp-Glu-Gln-Ala-Lys-Ala-Leu-

160 170
Ser-Val-Glu-Leu-Glu-Ile-Asn-Phe-Glu-Thr-Arg-Gly-Lys-Arg-Gly-Gln-

180 190
Asp-Ala-Met-Tyr-Glu-Tyr-Met-Ala-Gln-Ala-Cys-Ala-Gly-Asn-Arg-

193
Val-Arg-Arg-COOH

Fig. 10.2 Amino acid sequence of Fragment A from diphtheria toxin.

Much of our knowledge about the enzymic properties of Fragment A comes from studies of its interaction with NAD^+. Binding of NAD^+ has been studied by several methods. By flow dialysis (dynamic dialysis) it has been shown that the fragment contains a single NAD^+ binding site with K_d of 8.3 μM (47, 48). Occupancy of the site produces significant changes in the optical properties of the molecule that may be used to monitor NAD^+ binding. The intrinsic tryptophan fluorescence is quenched by up to 80%, and a weak, broad absorbance band ($\lambda_{max} \sim 360$ nm; $\epsilon \sim 500$) is generated. Both the nicotinamide and adenine moieties contribute to the affinity of NAD^+ for Fragment A (K_d's 200 μM and 30 μM, respectively), but the ribose and phosphate moieties contribute little, as judged by the facts that (*a*) NMN binds with a lower affinity than nicotinamide, and (*b*) decreasing affinity is observed as one progresses in the series, adenine, adenosine, AMP, ADP, ATP.

Table 10.1 Amino Acid Composition of Fragment A from Diphtheria Toxin

	Number/Molecule	
Residue	Analysis[a]	Sequence
Lys	15.9 (16)	16
His	0.93 (1)	1
Arg	6.03 (6)[b]	7[b]
Asx[c]	22.6 (23)	
Asp		13
Asn		10
Thr	8.25 (8)	9
Ser	17.0 (17)	17
Glx[c]	22.8 (23)	
Glu		15
Gln		7
Pro	6.16 (6)	6
Gly	20.0 (20)	20
Ala	14.2 (14)	14
Val	16.2 (16)	16
Met	4.24 (4)	4
Ile	5.78 (6)	6
Leu	12.4 (12)	12
Tyr	9.8 (10)	10
Phe	6.96 (7)	7
Cys	1.08 (1)[d]	1
Trp	2.1 (2)[e]	2
Total	192	193

[a] Average values (extrapolated values for serine and threonine) from two 24 h and two 72 h hydrolysates in 6 *M* HCl at 110°.

[b] Seven arginyl residues are present in the 193 residue Fragment A, but the 190 and 192 residue molecules contain only five and six residues of arginine, respectively.

[c] Asx includes aspartic acid and aspararine; Glx includes glutamic acid and glutamine.

[d] Determined as S-carboxymethylcysteine.

[e] Determined spectrophotometrically.

Both the quenching of fluorescence and the appearance of the new absorbance band result from interaction of the protein with the nicotinamide moiety of NAD^+ (47). Thus at concentrations that compete for the binding of NAD, *N*-methyl nicotinamide and NMN produce both optical effects, whereas adenine does not. These effects may result from direct interaction of the positively charged nicotinamide ring with the indole side chain of one or both tryptophans and may involve the formation of a charge-transfer complex. Nicotinamide itself, being uncharged, produces neither of the optical effects.

Fragment A may be shown to have low NAD^+ glycohydrolase activity in the absence of EF-2 (47). Correlation of the profiles of this activity with ADP-ribosylation activity eluted from polyacrylamide gels run on native Fragment A indicates that the reaction is due to the fragment itself and not to traces of other contaminating NAD^+ glycohydrolases. The existence of this reaction apparently reflects the fact that the ADP-ribosyl moicty of NAD^+ bound to Fragment A is transferred to water with a low efficiency. It seems highly unlikely that this reaction contributes significantly to toxicity, but it is a convenient enzymological marker.

Although EF-2 has a K_M much lower than that of NAD^+ (0.15 μM, versus 1.4 μM), no interaction of EF-2 with Fragment A has been detected in the absence of NAD^+ (47, 49). From this result and others, including data from studies of the kinetics of ADP-ribosyl transfer to EF-2, Chung and Collier have proposed an ordered sequential mechanism for the reaction, with the sequence of binding of substrates being NAD^+, followed by EF-2 (49). The reaction apparently proceeds through a central ternary complex containing Fragment A and both substrates. Attempts to isolate a covalent intermediate containing ADP-ribose linked to Fragment A have been unsuccessful, as expected from the proposed mechanism (47).

FRAGMENT B AND THE ENTRY PROCESS It is clear that in order to inactivate EF-2 within an animal cell, the catalytic center on Fragment A must somehow become exposed to the cytosol, since both EF-2 and NAD^+ are cytosolic constituents. At the very least this must involve projection of the active site of Fragment A through the internal surface of the plasma membrane, and it is generally assumed, without hard evidence, that Fragment A may be released free in the cytosol. Study of the mechanism of this penetration has proven difficult, but facts are slowly yielding to several approaches.

Although the precise function of Fragment B in toxicity is obscure, its major role is almost certainly to facilitate attachment and entry of the A fragment. Evidence for this came initially from studies of CRM 197, a nontoxic, immunologically cross-reactive form of toxin (50, 51). CRM 197, which is produced by *C. diphtheriae* lysogenized with a mutant form of β phage, contains an enzymically inactive A moiety, but fully functional B. Although CRM 197 does not block the enzymic activity of Fragment A, it has the interesting property of inhibiting the action of diphtheria toxin on whole cells. This effect can be most easily explained by postulating that CRM 197 binds competitively to specific cells surface receptors used by the toxin. A K_d of circa 10^{-8} *M* was calculated for the interaction. More recently Fragment B has been shown to inhibit toxicity on whole cells in a manner similar to CRM 197 (52).

Direct evidence for the toxin receptors has proven difficult to obtain. The major problem has been the inability to demonstrate competitive binding between ^{125}I-labeled toxin and excess unlabeled toxin. This apparently reflects nonspecific binding by toxin (or perhaps only iodinated toxin) to a large array of sites on cell surfaces.

Although this problem has been experienced with several different cell lines and in various laboratories, Middlebrook and co-workers have recently reported competetive binding to various monkey kidney cell lines (including Vero cells) (53). These cells were reported to contain a high concentration of toxin receptors, which correlates with their increased sensitivity to toxin over most other cultured cells.

The number and nature of toxin receptors remains uncertain. Boquet and Pappenheimer estimated that HeLa cells contain about 4000 diphtheria toxin receptors, based on experiments in which cells were incubated with iodinated toxin and then washed with antitoxin to remove nonspecifically bound toxin (54). Unfortunately this estimate involves untested assumptions. Draper et al. have recently suggested that the toxin receptor may contain a complex oligosaccharide as the essential element recognized by the toxin (55). They found that Conconavalin A (Con A), succinylated Con A, and wheat germ agglutinin antagonized the action of the toxin on Chinese hamster V79 cells, as did a glycopeptide from ovalbumin and cell wall polysaccharide from *Salmonella cholera suis*. From comparison of the structures of inhibitory and noninhibitory polysaccharides it was concluded that mannose and terminal *N*-acetylglucosamine residues may be important in the recognition by toxin. Unfortunately no data are available to demonstrate direct binding of the toxin to inhibitory oligo- or polysaccharides.

Although the mechanism by which Fragment B functions in the entry of A is obscure, some data are available that are possibly relevant. First, CRM 45, a mutant form of toxin that lacks the carboxy terminal 17,000 daltons of the molecule, does not inhibit the action of toxin on whole cells (56). This suggests that the receptor binding site is located within the carboxy-terminal region of Fragment B. Second, Boquet et al. have shown that Fragment B, CRM 45, and that portion of B contained within CRM 45 (termed B 45), all bind Triton X-100, a nonionic detergent known to form complexes with many membrane proteins but few nonmembrane proteins. Whole toxin formed similar complexes after treatment with sodium dodecyl sulfate, but Fragment A did not under any conditions tested.

Extrapolating from the Triton binding properties of Fragment B, Boquet et al. have proposed a model of B function in which it is postulated that B forms a channel for direct traversal of the membrane by the A moiety (57). According to the model the initial event is binding to the toxin receptor by way of the specific site on the carboxy-terminal half of B. The hydrophobic domain on the amino-terminal half of B then inserts into the phospholipid bilayer where it forms a channel, either by itself or in association with that part of the receptor molecule already lying within the membrane. The A domain may be conceived to be drawn through the channel as it is formed until the A-B disulfide bridge and the polypeptide loop spanned by the disulfide reach the inner surface of the membrane. It is assumed that nicking and reduction take place at the inner surface and that Fragment A renatures as it enters the cytoplasm. A conceptually similar model has been proposed for the entry of cholera toxin (see below).

At present it is not possible to discriminate between direct penetration models such as that described above and others involving some form of endocytosis as an obligatory step in entry. Nicolson has proposed on the basis of electron microscopic evidence obtained with ferritin-ricin conjugates that ricin, a toxic plant lectin, enters by receptor mediated endocytosis and is released into the cytosol after disintegration of the vesicles (58). The ricin is then accessible to target ribosomes. Presumably a similar model might be applicable to diphtheria toxin. Although studies with various metabolic inhibitors might be expected to yield relevant data, those performed to date have not permitted firm conclusions to be drawn.

Exotoxin A from *Pseudomonas aeruginosa*

Pseudomonas aeruginosa is strikingly different from *C. diphtheriae* both in gross properties and characteristics as an infectious agent. Thus, for example, in contrast to *C. diphtheriae*, *P. aeruginosa* is gram-negative and is capable of growing in water containing very low levels of nutrients. *P. aeruginosa* is virtually harmless for healthy individuals and most commonly infects immune compromised patients, such as those with extensive burns or under treatment with immunosuppressants. This property, together with the organism's resistance to most antibiotics and its capacity to multiply in moist, inorganic environments, makes *P. aeruginosa* a major causative agent of infections acquired in hospitals.

Localized infections may occur at various sites in the body including the respiratory and urinary tracts, the intestine, the ear, and the eye, and generalized infections may ensue in patients with impaired resistance. Mortality approaches 100% in cases of generalized infection.

Clinical isolates of *Pseudomonas aeruginosa* commonly produce several extracellular products that can cause damage to humans or animals; these include proteases, hemolytic factors, a leukocidic factor, and a lethal toxin. The last of these, Exotoxin A (60), is a protein of molecular weight about 66,000. Although the importance of the toxin in human infections remains uncertain, most clinical isolates (about 90%) produce the toxin *in vitro* (61), and survivors of Pseudomonas infections generally have significant levels of circulating antibodies against the toxin (62).

Following an earlier report that Exotoxin A inhibited protein synthesis in cultured cells (63) Iglewski and Kabat showed in 1975 that the toxin blocked protein synthesis in cell-free systems from eucaryotic cells by an NAD^+-dependent mechanism identical to that of diphtheria toxin (64). That ADP-ribosylation occurs at the same site on EF-2 and with identical stereospecificity is indicated by the fact that reversal of the ADP-ribosylation of EF-2 may be catalyzed with either toxin, regardless of the specific toxin used for the forward reaction (65, 66). Furthermore, studies with inhibitors and measurements of kinetics imply close similarities in the mechanism of reaction and the configurations of the active sites on the two toxins (65; Stephen Lory, unpublished data).

Despite the apparent identity in their intracellular reactions, the two toxins differ in other respects. They show no immunologic cross-reactivity, and there is no cross-neutralization of the two (64, 66). Also their spectra of action on various animals and cell lines in culture are different (67). Thus, for example, mice and mouse cells in culture are markedly sensitive to Exotoxin A, in contrast to diphtheria toxin. This finding may indicate that the two toxins use different cell surface receptors for attachment.

Structure-activity relationships in the two toxins show both similarities and differences (Figure 10.3). Exotoxin A is secreted as a single chain (mol wt 66,000; pI 5.1) containing four disulfide bridges and no free sulfhydryl groups (68). Like diphtheria toxin, the unmodified form of Exotoxin A is a proenzyme with little or no activity in catalyzing the ADP-ribosylation of EF-2 (68, 69). This activity and the NAD^+-glycohydrolase activity may be elicited by reduction of at least two of the four disulfide bridges in the presence of a chaotropic agent, such as urea (68, 69; Stephen Lory, unpublished data). This apparently permits the molecule to assume a configuration in which the catalytic center is exposed. Such treatment destroys

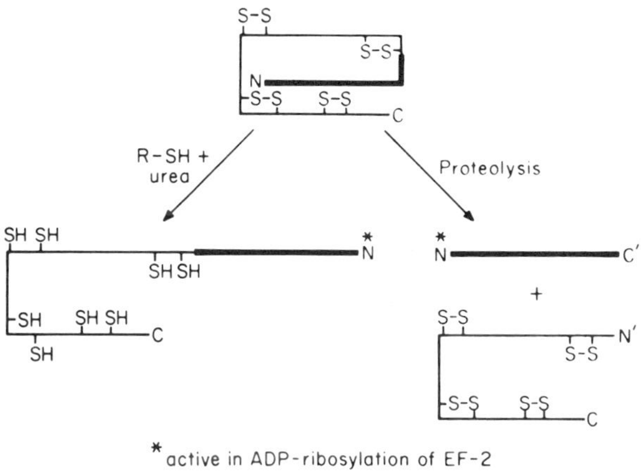

Fig. 10.3 Pathways of activation of Exotoxin A from *Pseudomonas aeruginosa* by reduction or proteolysis by unidentified bacterial proteases. An enzymically active 26,000 dalton fragment similar or identical to that generated by bacterial proteases may be produced by treating reduced toxin with chymotrypsin in the presence of NAD^+.

the toxicity of the molecule, however, presumably because the integrity of the receptor binding site is destroyed.

Treatment of the reduced or partially reduced molecule with chymotrypsin in the presence of NAD^+ yields an enzymically active peptide, mol wt 26,000, that strongly resembles Fragment A from diphtheria toxin in enzymic properties (70). NAD^+ apparently serves to protect the domain containing the NAD^+ binding site from proteolysis, but the remainder of the molecule is degraded to small peptides. There have been no reports of isolation of the complementary portion of the toxin analagous to Fragment B from diphtheria toxin.

An enzymically active peptide similar or identical to that generated by chymotrypsin treatment has been isolated from culture fluids of *P. aeruginosa* and has also been found in purified toxin preparations after storage (65). Production of the fragment is apparently mediated by bacterial proteases, but the pathway of its formation is unclear.

Neither the mechanism of attachment of Exotoxin A to cell surfaces nor the events that transpire after attachment have been studied in detail, and one knows even less about these aspects of Exotoxin A action than of diphtheria toxin action. It has been reported that the action of Exotoxin A in the burned mouse model system may be enhanced by the effects of pseudomonas protease injected simultaneously, or in live infections, produced by the infectious agent (71). However, no details as to the mechanism of enhancement are available.

10.3 TOXINS ACTING ON ADENYLATE CYCLASE

Cholera Toxin

Cholera and Cyclic AMP The major symptoms of Asiatic cholera are so different from those of diphtheria or Pseudomonas infections that it is surprising that there is any relationship whatsoever to these latter diseases. Cholera results from an infection of the gastrointestinal tract of man by *Vibrio cholerae*, a gram-negative, comma shaped organism [the pathological aspects of this disease are reviewed by Pierce (72), Carpenter (73), and Finkelstein (9, 10)]. The bacteria are noninvasive and remain localized, as in diphtheria; large numbers are found absorbed to the mucosal surfaces as well as in the luminal fluid. It is now recognized that the mortality from cholera results directly from the massive diarrhea that occurs. Loss of fluid (as high as 20 liters per day) and dissolved electrolytes causes severe dehydration which may result in death within a few hours of the first signs of illness. Intravenous replacement of fluid and electrolytes reduces the mortality rate to less than 1%.

Although the causative agent of cholera was identified before the turn of the century, recognition that a potent exotoxin plays a key role in pathogenesis is a relatively recent development [for reviews focusing on cholera toxin see Bennett and Cuatrecasas (74) and Gill (75)]. In 1959 De (76) reported the existence of a heat labile factor present in cell-free filtrates of *Vibrio cholerae* cultures that could elicit choleralike symptoms when injected into the ligated intestines of rabbits. In the intervening years the toxin has been purified and studied intensively, and now rivals diphtheria toxin in the level of understanding of its mode of action.

Roughly a decade after the discovery of cholera toxin studies by Field and co-workers (77) demonstrated the role of cyclic AMP as a regulator of ion flux and fluid secretion across the intestinal mucosa, and soon thereafter it was found that cholera toxin mimicked the effects of cyclic AMP (78). Subsequently it was shown that cholera toxin induced elevated levels of cyclic AMP in virtually all types of mammalian cells (9, 72–75), and it is now clear that this effect is due to stimulation of the synthesis of cyclic AMP rather than inhibition of its degradation.

Structure and Secretion of Cholera Toxin Cholera toxin is a multimeric protein (M_r 84,000) (79). In the form commonly isolated it is composed of two protomers, A and B, containing a total of seven polypeptide chains (Figure 10.4). The A protomer (mol wt 29,000) is composed of two polypeptide chains, A_1 (mol wt 23,000) and A_2 (mol wt 6000) covalently linked by a single disulfide bond (80, 81). The B protomer (M_r 54,000)

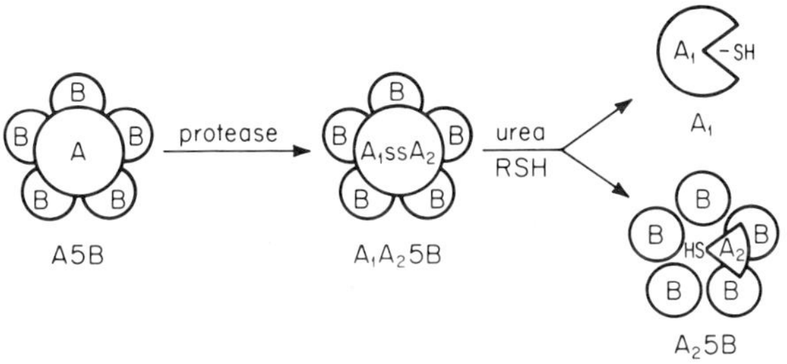

Fig. 10.4 Schematic representation of cholera toxin showing pathway for production of enzymically active A_1 chain and ganglioside-binding A_25B complex.

consists of five, noncovalently associated B chains (M_r 11,500) arranged in a highly stable ringlike structure. Thus, the toxin may be designated A_1A_25B with respect to its peptide composition (80).

The A protomer is associated with the B by noncovalent forces arising mainly from the interaction of the A_2 peptide with the B ring (80). This interaction, although disrupted by sodium dodecyl sulfate, is strong, as evidenced by the stability of the A_25B complex in the presence of 6 *M* urea and 20 m*M* dithiothreitol (at neutral pH) (82). The disulfide bond linking A_1 and A_2 is easily cleaved in the presence of reducing agents, but noncovalent forces between A_1 and A_2 tend to retard dissociation of the structure (80, 83). Subsequent removal of the reducing agent by dialysis results in reformation of the A_1-A_2 disulfide bond and efficient reconstitution of the native toxin.

The B protomer of the toxin is remarkably stable and withstands dissociation during exposure for several hours to either sodium dodecyl sulfate or 6 *M* urea at room temperature and neutral pH (80, 83). Each B peptide of the toxin contains a single intrachain disulfide bridge, which resists reduction while the B protomer is intact (80, 81, 83). Choleragenoid, a nontoxic, immunologically cross-reactive protein that occurs naturally in cell-free culture fluid of *V. cholerae*, is probably identical to the B protomer of the toxin (80).

Proteolytic processing apparently plays an important role in the structural organization of the toxin. The existence of a precursor to the A protomer has been demonstrated by growing *V. cholerae* in the presence of protease inhibitors (84) or by employing protease deficient mutants (85). The A protomer of toxin produced under these conditions has the same apparent molecular weight as the A_1A_2 form commonly isolated (29,000), but it can not be dissociated into lower molecular weight chains in the presence of sodium dodecyl sulfate and reducing agents (85). Limited digestion of "unnicked" toxin with bacterial proteases or trypsin rapidly converts the intact A chain into a form that appears identical to the A_1-A_2 protomer commonly isolated (85). Presumably, therefore, the toxin is normally secreted in a form containing intact A chain and is rapidly converted into the nicked form commonly isolated by attack of bacterial proteases (84, 85). The importance of this proteolytic processing to the enzymic activity of the toxin is discussed below.

The complete amino acid sequence of the B chain of cholera toxin has been reported (86, 87), but only partial sequence data are available for the A_1 and A_2 fragments (88, 89). Since the N-terminal amino acid residue of the A_2 peptide is methionine, this peptide probably represents the N-terminal fragment of the intact A chain described above (75).

Little is known about the mechanism of secretion or assembly of cholera toxin. Hyperproduction of the toxin can be induced with subinhibitory concentrations of certain antibiotics like lincomycin (90). Since these antibiotics also cause hyperproduction of known periplasmic enzymes (e.g. ribonuclease and streptomycin adenylate transferase) it is likely that cholera toxin is actively secreted rather than released during cell lysis. Bacterial mutants altered in the production of cholera toxin, both hypertoxinogenic (91) and hypotoxinogenic (91–95), have been isolated, but little is known about the organization of the toxin's structural genes. The role of bacteriophage has been investigated, but no correlation between lysogeny and toxinogenicity established (96). The existence of toxinogenic strains of *Vibrio cholerae* that contain no plasmids suggests that the toxin structural genes may reside on the bacterial chromosome, at least in some strains (83). One regulatory toxin locus reportedly shows weak genetic linkage to genes coding for histidine biosynthesis (97). The genetic origin of cholera toxin and its function in nonpathogenic cells both remain to be determined.

Biological Activities of Cholera Toxin and Its Subunits Original observations by van Heyningen and co-workers (98) indicated that cholera toxin had a strong affinity for gangliosides and led to the implication of ganglioside GM_1 as the attachment site for the toxin on cell surfaces. Preincubation of the toxin with GM_1 inactivates the toxin and abolishes its cell surface binding properties (98–100). Furthermore exogenously added GM_1 may be shown to be incorporated into cell membranes and to increase the sensitivity of cells to the toxin (100). This effect is most dramatic in certain transformed fibroblast lines that normally contain no detectable GM_1 and are therefore markedly insensitive to the toxin (101, 102). In addition, treatment of cells with neuraminidase produces a shift in the ganglioside content from disialo- and trisialogangliosides to GM_1, and a concomitant increase in the cell's sensitivity to cholera toxin is observed (103).

The role of the B subunits in the interaction of the toxin with GM_1 was first demonstrated by showing that choleragenoid blocks the action of the toxin on whole cells (104). Pure B protomer has no effect on adenylate cyclase (9, 74, 75). Other studies have confirmed the fact that the B protomer recognizes and binds to the oligosaccharide moiety of the ganglioside.

Dissection of the mechanism of activation of adenylate cyclase by cholera toxin has relied chiefly on studies with broken cells. Gill originally reported that adenylate cyclase activity present in avian erythrocyte membranes could be elevated by cholera toxin only in the presence of NAD^+ (107). The requirement for NAD^+ as a cofactor in the activation of adenylate cyclase by cholera toxin was subsequently established in a variety of broken cell preparations responsive to cholera toxin (75). Requirements for other constituents of broken cell systems have also been reported including ATP, GTP, and a soluble protein (75, 107–110). The role of the protein remains obscure but that of the nucleoside triphosphates is becoming clear, as described below.

Gill (108) demonstrated that either the toxin or the A_1 peptide catalytically activated adenylate cyclase in the presence of NAD^+, ATP, and erythrocyte cytosol. The activation proceeded immediately upon addition of thiol-treated cholera toxin or the A_1 peptide to the disrupted cells. This contrasted with whole cells in which a lag of 30–60 minutes preceeded the toxin's effects on the cyclase. The reaction in broken cells was uninhibited by choleragenoid or GM_1 and was therefore considered to be analogous to an intracellular event. It was independent of receptor binding and possible membrane penetration functions contributed by the B subunits of the toxin. The lag in toxin effects with intact cells is presumably due to the time required for penetration of the catalytic A_1 peptide of the toxin through the membrane.

Subsequent studies by Moss and co-workers (111, 112) demonstrated that cholera toxin and its A protomer exhibited NAD^+ glycohydrolase activity in the presence of reducing agents. This activity was stimulated by arginine and other compounds containing a guanidinium group, and in fact it could be shown that ADP-ribosyl-arginine was formed in the presence of the toxin and NAD^+. Thus, cholera toxin exhibited ADP-ribosyl transferase activity. By analogy with the activities of diphtheria toxin and Exotoxin A of *P. aeruginosa*, it was suggested that cholera toxin activated adenylate cyclase by ADP-ribosylation of some protein component of the cyclase system.

Mechanism of Activation of Adenylate Cyclase by Cholera Toxin Rodbell and co-workers first recognized the importance of GTP in restoring hormone sensitivity to adenylate cyclase in isolated membranes (113). Cuatrecasas and Bennett found that activation of adenylate cyclase present in toad erythrocyte plasma membranes was almost entirely dependent on the presence of GTP (114). The requirement for GTP could be eliminated by high concentrations of ATP, and this fact may explain why the GTP requirement has not been recognized in all broken cell systems sensitive to cholera toxin. It is now clear that GTP is a requirement for toxin activation of adenylate cyclase; the apparent substitution by high concentrations of ATP results from production of GTP from endogenous GMP and GDP present in membranes by way of nucleotide phosphoryl transferase reactions (115).

The role of GTP in the toxin activation process has been established in part by the use of nonhydrolyzable GTP analogs, such as guanyl-5′ylimidodiphosphate (GMP-P(NH)P). Levinson and Blume (115) observed that cholera toxin activation of adenylate cyclase present in neuroblastoma cell membranes was dependent on GTP, as well as NAD^+. The magnitude of activation in the presence of cholera toxin, NAD^+, and GTP was quantitatively similar to the activation obtained with GMP-P(NH)P alone. GTP alone had no effect on membranes not treated with toxin.

Other studies indicated that activation of adenylate cyclase by hormones involved both the interaction of the hormone with its receptor and the subsequent interaction of a regulatory component of the cyclase with GTP (113). Removal of the hormone from the receptor resulted in deactivation of the cyclase in the presence of GTP but not analogs of GTP resistant to hydrolysis. Thus deactivation of the cyclase apparently involved hydrolysis of GTP at a regulatory site, and persistent activation by GMP-P(NH)P was due to the resistance of this analog to hydrolysis (116). Since cholera toxin treated membranes responded to GTP in much the same way that untreated membranes responded to GMP-P(NH)P, it was suggested that cholera toxin interfered with the hydrolysis of GTP at the regulatory site (115).

Direct evidence for this model of cholera toxin's action was obtained by Cassel and Selinger (117) utilizing turkey erythrocyte membranes. A catecholamine stimulated GTPase activity was observed in these membranes, and the properties of this activity suggested that it was associated with the guanyl nucleotide regulatory site of adenylate cyclase. Treatment of these membranes with cholera toxin, NAD^+, ATP, and cytosol resulted in a decrease in hormone stimulated GTPase activity, which paralleled the elevation of adenylate cyclase activity in these membranes. Thus toxin activation of adenylate cyclase was associated with inhibition of the GTPase "turn off" reaction. The guanyl nucleotide regulatory site of adenylate cyclase was presumably the site of both GTP dependent activation of the cyclase and the GTPase turn off reaction.

Pfeuffer (118) reported the purification of a protein mediating guanyl nucleotide regulation of adenylate cyclase activity in pigeon erythrocyte membranes. This protein (mol wt 42,000) had a high affinity for GTP affinity columns and remained active after purification from detergent extracts of membranes in reconstituting hormone-GTP or GMP-P(NH)P responsive adenylate cyclase activity.

Cassel and Pfeuffer (119) subsequently showed that cholera toxin in the presence of ^{32}P-labeled NAD^+ catalyzed incorporation of radioactive label into this 42,000 dalton protein and two other higher mol wt proteins (mol wt 86,000 and 200,000) present in pigeon erythrocyte membranes. The 42,000 and 86,000 dalton proteins could be extracted from membranes treated with toxin and ^{32}P-NAD^+; they were retained by GTP affinity columns and could be eluted with GTP. The radioactivity associated with these two GTP binding proteins was identified as ADP-ribose, and reversal of this modification could be accomplished with high concentrations of nicotinamide and cholera toxin. These data strongly suggest that cholera toxin catalyzed the ADP-ribosylation of these proteins.

Cross-reconstitution experiments showed that addition of the purified ADP-ribosylated, GTP binding proteins to untreated membranes, previously depleted of their intrinsic GTP binding proteins, resulted in activation of the adenylate cyclase. Since the GTP-binding proteins purified from untreated membranes had little effect on the activity of the same membrane preparation, these studies confirmed the role of the ADP-ribosylated proteins in mediating the toxin's effect on adenylate cyclase activity. Presumably the 42,000 dalton protein comprises the guanyl nucleotide regulatory site, and ADP-ribosylation of this protein by cholera toxin results in activation of the cyclase through the inhibition of GTP hydrolysis at this regulatory site. It follows that this protein is probably a GTPase and that ADP-ribosylation of the protein inhibits its GTPase activity without impairing its ability to bind GTP and interact with the cyclase. Thus cholera toxin catalyzes persistent activation of the cyclase through the modification by ADP-ribosylation of the guanyl nucleotide regulatory protein in a manner that freezes it in its GTP bound, activated regulatory mode.

Similar results have been reported by Gill and Meren (120) utilizing the pigeon erythrocyte system. These workers found it necessary to decrease background incorporation of ^{32}P-ribose into proteins and poly ADP-ribose by enzymes present in the broken erythrocyte preparation before toxin specific incorporation could be observed. Conditions were developed whereby ADP-ribosylation of a 42,000 mol wt protein was observed only in the presence of the A_1 peptide, NAD^+, ATP, and cytosol. GTP also stimulated ADP-ribosylation of this protein. Activation of adenylate cyclase occurred in parallel with the modification of the protein, while reversal of the reaction with nicotinamide decreased the activity of the cyclase.

Other evidence reported by Johnson et al. (121) supports this model of cholera toxin's action. A genetic variant of S 49 mouse lymphoma cells was shown to lack an adenylate cyclase system sensitive to activation by hormones, guanyl nucleotides, fluoride ion, or cholera toxin. Adenylate cyclase activity responsive to these agents could be reconstituted by treatment of the mutant membranes with extracts of wild type membranes. Chemical inactivation studies indicated that a key component distinct from the catalytic unit of the cyclase was present in these extracts, and extracts derived from cholera toxin treated wild type membranes activated the mutant adenylate cyclase in the presence of GTP. These and other studies indicated that the mutant cell line lacked the guanyl nucleotide regulatory protein. Together these data and those previously discussed leave little doubt that the GTP binding protein is the substrate of cholera toxin and that this protein is intimately associated with the regulation of adenylate cyclase activity by a variety of biological agents.

Entry and Activation of the Toxin Although the guanyl nucleotide regulatory protein is an intrinsic membrane protein, its interaction with GTP implies that at least part of the molecule must be exposed to the cytosol. A similar argument arises from the fact that the activation of adenylate cyclase by cholera toxin requires NAD^+ and possibly other cytosolic factors. Thus penetration of the A_1 peptide, or at least its catalytic center, through the lipid bilayer must precede activation of adenylate cyclase. As in diphtheria toxin and Exotoxin A the transmembrane dynamics of this entry process are still the least understood aspect of the action of cholera toxin on cells.

Activation of adenylate cyclase by cholera toxin in whole cells is temperature dependent and displays critical breaks in activity below certain temperatures (122). These temperatures, which vary with the type of cell tested, are thought to correspond to transition temperatures associated with the freezing of lipid components in the cell membrane. Since both activation of the cyclase in broken cells (108) and binding to GM_1 (104) do not show this same pattern of temperature sensitivity, lipid fluidity must be important in the entry process. Aggregation and capping of GM_1 receptors by the multivalent cholera toxin molecules have been observed using fluorescent toxin derivatives (123). This process would certainly be dependent on membrane fluidity, but the importance of aggregation to the entry of the toxin's A protomer is unknown.

The participation of phagocytosis or pinocytosis in the toxin entry mechanism is equally uncertain. The fact that certain cell types, like erythrocytes, which are markedly reduced in these internalization processes, still respond strongly to cholera toxin, suggests that a phagocytic mechanism is not involved. However, active transport of the toxin or its active subunits through the membrane by endocytic cellular processes, which for example are triggered by the aggregation of a receptor (124), cannot be ruled out.

The possibility exists that the B protomer of the toxin mediates direct penetration of the lipid bilayer by the A chain. Gill (80) has suggested that the ringlike B protomer might form a hydrophilic channel or pore in the lipid bilayer through which the A subunit could pass into the cell. The observations of Moss et al. (75) on the interaction of cholera toxin with liposomes are of interest with respect to this model. Cholera toxin or its B protomer (choleragenoid) promoted leakage of ^{14}C-labeled glucose from sealed multilamellar liposomes. This process occurred only when GM_1 was incorporated into the vesicles; other gangliosides could not substitute. These studies indicate that the B subunits of the toxin upon binding GM_1 inserted in a membrane disrupt the lamellar structure in such a way as to alter permeability to small molecules at least.

Cuatrecasas (76) has suggested that a translocation of the toxin within the plane of the membrane might allow for contact of the A subunit with the cyclase by way of lateral diffusion of the toxin-ganglioside complex within the membrane. The A protomer has been reported to possess hydrophobic characteristics that may explain its interaction with liposomes, cell membranes, and the ceramide portion of gangliosides. The A_1 peptide of the toxin might therefore remain associated with the membrane while exposing its catalytic surface to the cytoplasmic milieu, and thus to NAD^+, GTP, and other factors required for activation of the cyclase. Regardless of whether the A_1 chain remains bound to the inner surface of the plasma membrane or is released into the cytosol, the major question remains how to explain penetration of the bilayer by the active site.

Evidence from this and other laboratories indicates that both limited proteolytic cleavage and reduction of the toxin's A protomer are required for expression of ADP-ribosyltransferase activity. Reduction of the A_1-A_2 disulfide bond is necessary for manifestation of both NAD^+-glycohydrolase and ADP-ribosyltransferase activities (83, 111, 112). This may conccivably occur intracellularly upon contact of the A protomer with reduced glutathione in the cytosol. Recently the enzymic activity of unnicked cholera toxin, containing an intact A protomer, has been examined in our laboratory. This toxin and its isolated A chain are enzymically inactive even in the presence of reducing agents, but such activity is expressed (in the presence of thiols) after limited digestion with trypsin or elastase. The activation by protease correlates with the production of a nicked form of the A protomer composed of the disulfide-linked A_1 and A_2 fragments.

The unnicked and nicked forms of cholera toxin exhibit identical specific toxicities, however. This result is similar to that reported for the unnicked and nicked forms of diphtheria toxin in which the unnicked form is also enzymically inactive *in vitro* prior to trypsin nicking but is identical in specific toxicity to nicked toxin in cell culture assays. These results suggest that the unnicked forms of these toxins are proteolytically processed by the target cell to yield the enzymically active fragments. With regard to the possibility that this process might occur at the cell membrane it is interesting that Matuo et al. (127) have reported peptidase activity associated with sarcoma cell membranes that is capable of producing enzymically active fragments of cholera toxin.

Escherichia coli and *Salmonella typhimurium* Heat-Labile Enterotoxins

Certain strains of *Escherichia coli* associated with enteropathogenic disease in man and other animals are known to produce a diarrhea inducing, exoenterotoxin similar in many respects to cholera toxin (9, 72–75). This toxin is heat-labile and cross-reacts antigenically with cholera toxin (75, 128). The *E. coli* toxin activates adenylate cyclase both in intact cells and in broken cell systems (129). In cell-free systems the activation process requires NAD^+, ATP, and cytoplasmic factors (129). Furthermore, the toxin exhibits NAD^+-glycohydrolase activity and ADP-ribosyltransferase activity with L-arginine as acceptor (130). Structural data on the purified toxin are not consistent, and the size of the molecule remains uncertain. The structural gene for production of this toxin is located on a variety of plasmids (131) and even bacteriophage (132) harbored by enteropathogenic strains of *E. coli*. This gene has recently been cloned and its products analyzed in *E. coli* "minicell" systems (133). Two proteins that appear to comigrate on SDS polyacrylamide gels with the A protomer (A_1A_2) and the B chain of cholera toxin are produced by the *E. coli* toxin gene. Thus it appears that this toxin may well be closely related to cholera toxin both structurally and biochemically. It is not clear whether the subtle differences noted thus far between the *E. coli* and cholera toxins are due to fundamental differences at the genetic level or simply due to differences in post-translational processing, secretion, and assembly of the toxin in the different bacterial species.

An apparently similar heat-labile enterotoxin has been described in certain strains of *Salmonella typhimurium* (134–136). Besides evoking fluid secretion in the intestine, the toxin induced elongation of Chinese hamster ovary cells, an effect dependent on cyclic AMP, and was neutralized by antisera against cholera toxin or choleragenoid (136). Heterogeneity in size was found similar to that observed with the *E. coli* toxin (135).

10.4 CONCLUDING REMARKS

The existence of several distinct toxins with ADP-ribosyltransferase activity raises many questions and thereby provides fertile areas for future investigation.

Are the various toxins evolutionarily related? It seems certain that cholera toxin is closely related to the *E. coli* heat-labile toxin (and probably to the *Salmonella typhimurium* toxin as well, although data are sparse on this toxin at present). These toxins appear to have similar modes of action; they are immunologically cross-reactive; and they are produced by related, gram-negative organisms occupying similar ecological niches. Furthermore, the *E. coli* toxin has been found on transmissible genetic elements, which facilitates our thinking about dissemination of the gene for this toxin. It is less evident that there must be homology between diphtheria and pseudomonas toxins, or between these and the toxins affecting adenylate cyclase, but the similarities in enzymological properties among the toxins would strongly favor this idea. As noted in an earlier review on diphtheria and pseudomonas toxins (7), it seems highly unlikely that two unrelated proteins would undergo convergent evolution to form nonhomologous toxins that transferred the same group (ADP-ribosyl) of the same substrate (NAD^+) onto an identical sidechain (residue X) of the same protein (EF-2) of eucaryotic cells. Thus, although diphtheria and pseudomonas toxins do not cross-react immunologically, have dissimilar amino acid compositions, and are produced by unrelated organisms, it seems likely that they evolved from a common evolutionary precursor. The same reasoning may be applied to the possible homology between the toxins ADP-ribosylating EF-2 and those acting on adenylate cyclase, but with less weight because of the differences in protein substrate specificity.

The argument for homology between the EF-2 and adenylate cyclase specific toxins is supported by other evidence as well. Recent studies by Mekalanos, Collier, and Romig (85) have shown that the mechanism of the cholera toxin catalyzed reaction appears to be the same as that of diphtheria and pseudomonas toxins, that is, an ordered sequential mechanism. This conclusion is based on kinctic studies involving the use of an artifical low molecular weight ADP-ribosyl acceptor, iodinated *N*-guanyl-tyramine. In addition one may cite similarities in the protein substrates of the two classes of toxins. Both EF-2 and the 42,000 dalton GTPase are GTP binding proteins, and the attachment site for the ADP-ribosyl group appears to be a basic side chain in both cases. Inasmuch as Fragment A from diphtheria toxin does not ADP-ribosylate arginine or iodinated *N*-guanyl tyramine, as does cholera toxin, there are clearly significant differences in configuration of the active site on the toxins. Nevertheless, it will be interesting to know if there are amino acid sequence similarities between EF-2 and the adenylate cyclase GTPase.

In relation to the evolution of the ADP-ribosylating toxins one may also pose questions of the existence of nontoxic ADP-ribosyltransferases in bacteria and of possible functions of the toxic ADP-ribosyltransferases in bacterial structure or physiology. Two [NAD^+:protein] ADP-ribosyltransferases, presumably nontoxic, have been described in *E. coli* infected with bacteriophage T_4, but there is no definite indication at present of any relationship to the toxins described (137–139). One of the enzymes (alteration enzyme) is associated with the T_4 virion, whereas the other (modification enzyme) is not. Both catalyze attachment of the ADP-ribosyl group to an arginine side chain of the α subunit of RNA polymerase, and poly(L-arginine) can also act as substrate for both. Like diphtheria toxin, these enzymes appear to be dispensible for the phage replication cycle. An enzyme apparently similar to the T_4 alteration enzyme has been described in phage N4 (140). ADP-Ribosyltransferase activity has also been detected in uninfected *E. coli* with egg-white lysozyme as acceptor, but the enzyme responsible has not been characterized (138).

Finally, the question may be posed as to potential applications of the knowledge of the ADP-ribosylating toxins. Diphtheria, of course, represents the prime example of the efficacy of immunization against a toxin; mass immunization against formaldehyde treated diphtheria toxin (that is, diphtheria toxoid) has led to the virtual eradication of clinical diphtheria in countries with advanced medical care. Although field trials with chemically inactivated cholera toxin have not been encouraging, work is continuing to develop an appropriate vaccine and to determine the optimal route of administering it. The facts that the major antigenic determinants of cholera toxin are present on the B protomer and that antibodies against B neutralize the biological action of the toxin on whole cells suggest that the ideal immunogen might be a mutant cholera vibrio that produced the B protomer, but not the A. Such mutants, which might be used as a live oral vaccine, are presently being sought. Because of the demonstrated cross-reactivity between cholera and *E. coli* heat-labile toxins a vaccine against one might well be protective against either. Although widespread vaccination against Exotoxin A would not seem advisable, a vaccine against this toxin might well be useful in protecting certain, high risk groups. Leppla and co-workers have recently reported encouraging results in developing such a vaccine (141).

A potential and less evident application of the cytocidal ADP-ribosylating exotoxins is in the selective killing of cells. Although both diphtheria toxin and Exotoxin A have little specificity, it is quite conceivable that the actions of these toxins, or of enzymically active moieties therefrom, might be precisely targeted by forming conjugates with other proteins that bind specifically to particular cell types. Encouraging work along these lines has recently emerged from three laboratories. Thorpe et al. synthesized a conjugate of diphtheria toxin with angilymphocytic globulin and demonstrated an apparent enhancement in cytotoxicity for lymphoblastoid cells of 1000-fold over the toxin alone (142). In this laboratory we have prepared a disulfide linked conjugate of diphtheria toxin Fragment A with Conconavalin A, and demonstrated a marked enhancement in toxicity compared with controls (143). Similar results have been obtained in the laboratory of Uchida with a Fragment A conjugate of a different lectin (144). Although such studies are in their infancy, it appears that this approach holds promise for selective targeting of toxicity in laboratory applications and perhaps eventually in certain types of medical therapy.

REFERENCES

1 R. J. Collier (1975), *Bacteriol. Rev.* **39,** 54.

2 A. M. Pappenheimer, Jr. (1977), *Ann. Rev. Biochem.* **46,** 69.

3 J. R. Murphy (1976), "Structure Activity Relationships in Diphtheria Toxin," in A. W. Bernheimer, ed., *Mechanisms in Bacterial Toxinology*, John Wiley, New York, p. 31.

4 A. M. Pappenheimer Jr. and D. M. Gill (1973), *Science* **182,** 353.

5 D. M. Gill, A. M. Pappenheimer, Jr., and T. Uchida (1973), *Fed. Proc.* **32,** 1508.

6 J. Dirkx, *Arch. Int. Physiol. Biochim.* **82,** 157 (1974).

7 R. J. Collier (1976), "Inhibition of Protein Synthesis by Exotoxins from *Corynebacterium diphtheriae* and *Pseudomonas aeruginosa*," in P. Cuatrecasas, ed., *The Specificity and Action of Animal, Bacterial and Plant Toxin*, Chapman and Hall, London, p. 69.

8 R. J. Collier, "Diphtheria Toxin and and Exotoxin A from *Pseudomonas aeruginosa*," in F. E. Hahn, ed., *Antibiotics V*, Springer-Verlag, Berlin, in press.

9 R. A. Finkelstein (1973), *CRC Crit. Rev. Microbiol.* **2,** 553.

10 R. A. Finkelstein (1976), "Progress in the Study of Cholera and Related Enterotoxins," in A. W. Bernheimer, ed., *Mechanisms in Bacterial Toxicology*, John Wiley, New York, p. 53.

11 D. M. Gill (1977), *Adv. Cyclic Nucl. Res.* **8,** 85.

12 F. Loeffler (1884), *Mitt. Klin. Gesund. Berlin* **2,** 421.

13 A. M. Pappenheimer, Jr. (1937), *J. Biol. Chem.* **120,** 543.

14 D. M. Gill, T. Uchida, and R. A. Singer (1972), *Virology* **50,** 664.

15 R. K. Holmes and L. Barksdale (1969), *J. Virol.* **3,** 586.

16 J. R. Murphy, A. M. Pappenheimer, Jr., and S. T. Deborms (1974), *Proc. Nat. Acad. Sci USA* **71,** 11.

17 T. Uchida, D. M. Gill, and A. M. Pappenheimer, Jr. (1971), *Nat. N. Biol.* **233,** 8.

18 M. Solotorovsky and W. Johnson (1970), "Tissue Culture and Bacterial Protein Toxins," in S. J. Ajl, S. Kadis, and T. C. Montie, eds., *Microbial Toxins*, Academic Press, New York, p. 277.

19 E. S. Lennox and A. S. Kaplan (1957), *Proc. Soc. Exp. Biol. Med.* **95,** 700.

20 C. Placido-Sousa and D. G. Evans (1957), *Br. J. Exp. Pathol* **38,** 644.

21 J. Gabliks and M. Solotorovsky (1962), *J. Immunol.* **88,** 505.

22 N. Strauss and E. D. Hendee (1959), *J. Exp. Med.* **109,** 144.

23 R. J. Collier and A. M. Pappenheimer, Jr. (1964), *J. Exp. Med.* **120,** 1007.

24 R. J. Collier and A. M. Pappenheimer, Jr. (1964), *J. Exp. Med.* **120,** 1019.

25 R. J. Collier (1967), *J. Mol. Biol.* **25,** 83.

26 R. S. Goor and A. M. Pappenheimer, Jr. (1967), *J. Exp. Med.* **126,** 899.

27 T. Honjo, Y. Nishizuka, O. Hayaishi, and I. Kato (1968), *J. Biol. Chem.* **243,** 3553.

28 T. Honjo, Y. Nishizuka, and O. Hayaishi (1969), *Cold Spring Harbor Symp. Quant. Biol.*, **34,** 603.

29 T. Honjo, Y. Nishizuka, I. Kato, and O. Hayaishi (1971), *J. Biol. Chem.* **246,** 4251.

30 D. M. Gill, A. M. Pappenheimer, Jr., R. Brown, and J. J. Kurnick (1969), *J. Exp. Med.* **129,** 1.

31 T. J. Moehring and J. M. Moehring (1977), *Cell* **11,** 447.

32 E. A. Robinson, O. Henriksen, and E. S. Maxwell (1974), *J. Biol. Chem.* **249,** 5088.

33 E. A. Robinson, O. Henriksen, and E. S. Maxwell (1974), "The Site of ADP-Ribosylation of Elongation Factor 2," in *Poly (ADP-Ribose): An International Symposium*, U.S. Govt. Printing Office, Washington, D.C., p. 315.

34 N. Goto, I. Kato, and H. Sato (1968), *Jap. J. Exp. Med.* **38,** 185.

35 J. M. Moehring and T. J. Moehring (1968), *J. Exp. Med.* **127,** 541.

36 B. G. Van Ness, J. B. Howard, and J. W. Bodley (1978), *Fed. Proc.* **37,** 1496.

37 E. Kun, A. C. T. Chang, M. L. Sharma, A. M. Ferro, and D. Nitecki (1976), *Proc. Natl. Acad. Sci USA* **73,** 3131.

38 R. S. Goor and A. M. Pappenheimer, Jr. (1967), *J. Exp. Med.* **126,** 913.

39 J. Kandel, R. J. Collier, and D. W. Chung (1974), *J. Biol. Chem.* **249,** 2008.

40 R. S. Goor, A. M. Pappenheimer, Jr., and E. Ames (1967), *J. Exp. Med.* **126,** 923.

41 R. J. Collier and H. A. Cole (1969), *Science* **164,** 1179.

42 R. Drazin, J. Kandel, and R. J. Collier (1971), *J. Biol. Chem.* **246,** 1504.

43 R. J. Collier and J. Kandel (1971), *J. Biol. Chem.* **246,** 1496.

44 D. M. Gill and L. L. Dinius (1971), *J. Biol. Chem.* **246,** 1485.

45 D. M. Gill and A. M. Pappenheimer, Jr. (1971), *J. Biol. Chem.* **246,** 1492.

46 T. J. Moehring and J. M. Moehring (1977), *Cell* **11,** 447.

47 R. A. Finkelstein (1973), *CRC Crit. Rev. Microbiol.* **2,** 553.

48 A. Michel and J. Dirkx (1974), *Biochim. Biophys. Acta* **365,** 15.

49 D. W. Chung and R. J. Collier (1977), *Biochim. Biophys. Acta* **483,** 248.

50 T. Uchida, A. M. Pappenheimer, Jr., and A. A. Harper (1972), *Science* **175,** 901.

51 T. R. Ittelson and D. M. Gill (1973), *Nature* **242,** 330.

52 T. Zanen, G. Muyldermans, and N. Bengnier (1976), *FEBS Letters* **66,** 261.

53 J. L. Middlebrook, R. B. Dorland, S. H. Leppla, and R. A. Altenbern (1978), *Abstr. Ann. Meet. Amer. Soc. Microbiol.*, p. 25.

54 P. Boquet and A. M. Pappenheimer, Jr. (1976), *J. Biol. Chem.* **251,** 5770.

55 R. K. Draper, D. Chin, and M. I. Simon (1978), *Proc. Nat. Acad. Sci. USA* **75,** 261.

56 T. Uchida, A. M. Pappenheimer, Jr., and R. Gregory (1973), *J. Biol. Chem.* **248,** 3838.

57 P. Boquet, M. S. Silverman, A. M. Pappenheimer, Jr., and W. B. Vernon (1976), *Proc. Nat. Acad. Sci. USA* **73,** 4449.

58 G. L. Nicolson (1974), *Nature* **251,** 628.

59 P. H. Clarke and M. H. Richmond, eds. (1975), *Genetics and Biochemistry of Pseudomonas*, John Wiley, London.

60 P. V. Liu (1966), *J. Infect. Dis.* **116,** 112.

61 M. J. Bjorn, M. L. Vasil, J. C. Sadoff, and B. H. Iglewski (1977), *Infect. Immun.* **16,** 362.

62 M. Pollack, L. T. Callahan III, and N. S. Taylor (1976), *Infect. Immun.* **14,** 942.

63 O. R. Pavlovskis and F. B. Gordon (1972), *J. Infect. Dis.* **125,** 631.

64 B. H. Iglewski, and D. Kabat (1975), *Proc. Natl. Acad. Sci. USA.* **72,** 2284.

65 D. W. Chung and R. J. Collier (1977), *Infect. Immun.* **16,** 832.

66 B. H. Iglewski, L. P. Elwell, P. V. Liu, and D. Kabat (1975), in S. Shaltiel, ed., *Proc. Fourth Int. Symp. on Interconversion of Enzymes*, Springer-Verlag, Berlin and New York.

67 J. L. Middlebrook and R. B. Dorland (1977), *Can. J. Microbiol.* **23,** 183.

68 S. H. Leppla (1976), *Infect. Immun.* **14,** 1077.

69 M. L. Vasil, D. Kabat, and B. H. Iglewski (1977), *Infect. Immun.* **16,** 353.

70 S. Lory and R. J. Collier (1978), *J. Supramolec. Struct.*, **52,** 139.

71 K. Snell, I. A. Holder, S. H. Leppla, and C. B. Saelinger (1978), *Infect. Immun.* **19,** 839.

72 N. F. Pierce, W. B. Greenough, and C. C. J. Carpenter, Jr. (1971), *Bacteriol. Rev.* **35,** 1.

73 C. C. J. Carpenter (1972), *J. Inf. Dis.* **126,** 551.

74 V. Bennett and P. Cuatrecasas (1977), "Cholera Toxin: Membrane Gangliosides and Activation of Adenylate Cyclase," in P. Cuatrecasas, ed., *The Specificity of Action of Animal, Bacterial and Plant Toxins* (*Receptors and Recognition*), Chapman and Hall, London, p. 1.

75 D. M. Gill (1977), "Mechanism of Action of Cholera Toxin," in P. Greengard and G. A. Robinson, eds., *Advances in Cyclic Nucleotide Research*, Raven Press, New York, p. 85.

76 S. N. De (1959), *Nature* **183,** 1533.

77 M. Field, G. Plotkin, and W. Silen (1968), *Nature* **217,** 469.

78 M. Field, D. Fromm, C. K. Wallace, and W. B. Greenough, III (1969), *J. Clin. Invest.* **48,** 24a.

79 J. J. Lospallute and R. A. Finkelstein (1972), *Biochim. Biophys. Acta* **257,** 158.

80 D. M. Gill (1976), *Biochemistry* **15,** 1242.

81 C. Y. Lai, E. Mendez, and D. Chang (1976), *J. Infect. Dis.* **S133,** 523.

82 J. J. Mekalanos, R. J. Collier, and W. R. Romig (1978), *Infect. Immun.* **20,** 552.

83 J. J. Mekalanos, R. J. Collier, and W. R. Romig, unpublished.

84 D. M. Gill and R. S. Rappaport (1977) in H. Fukumi and M. Ohashi, eds., 12th Joint Conference, U.S.-Japan Cooperative Medical Program Cholera Panel, National Institute of Health, Tokyo.

85 J. J. Mekalanos, R. J. Collier, and W. R. Romig, *J. Biol. Chem.*, **254,** 5849–5861.

86 A. Kurosky, D. E. Markel, and J. W. Peterson (1977), *J. Biol. Chem.* **252,** 7257.

87 C. Y. Cai (1977), *J. Biol. Chem.* **252,** 7249.

88 E. Mendez, C. Y. Lai, and A. Wodnar-Filipowicz (1975), *Biochem. Biophys. Res. Commun.* **67,** 1435.

89 J. W. Jacobs, H. D. Niall, and G. W. G. Sharp (1974), *Biochem. Biophyis. Res. Commun.* **61,** 391.

90 M. Levner, F. P. Wiener, and B. A. Rubin (1977), *Infect. Immun.* **15,** 132.

91 J. J. Mekalanos, R. J. Collier, and W. R. Romig (1978), *Proc. Natl. Acad. Sci. USA* **75,** 941.

92 B. D. Howard (1971), *Nature* **230,** 97.

93 R. A. Finkelstein, M. L. Vasil, and R. K. Holmes (1974), *J. Infect. Dis.* **129,** 117.

94 R. K. Holmes, M. L. Vasil, and R. A. Finkelstein (1975), *J. Clin. Invest.* **55,** 551.

95 F. E. Ruch, J. R. Murphy, L. Graf, and M. Field (1978), *J. Infect. Dis.* **137,** 747.

96 J. C. Gerdes, and W. R. Romig (1974), *Infect. Immun.* **11,** 445.

97 M. L. Vasil, R. K. Holmes, and R. A. Finkelstein (1975), *Science* **187,** 849.

98 W. E. Van Heyningen, C. C. J. Carpenter, N. F. Pierce, and W. B. Greenough, III (1971), *J. Infect. Dis.* **124,** 415.

99 P. Cuatrecasas (1973), *Biochemistry* **12,** 3547.

100 P. Cuatrecasas (1973), *Biochemistry* **12,** 3558.

101 M. D. Hollenberg, P. H. Fishman, V. Bennett, and P. Cuatrecasas (1974), *Proc. Natl. Acad. Sci. USA* **71,** 4224.

102 P. H. Fishman, J. Moss, and M. Vaughn (1976), *J. Biol. Chem.* **251,** 7610.

103 A. Haksar, D. V. Maudsley, and F. G. Peron (1974), *Nature* **251,** 514.

104 P. Cuatrecasas (1973), *Biochemistry* **12,** 3577.

105 J. Holmgren, I. Lonnroth, and L. Svennerholm (1973), *Infect. Immun.* **8,** 208.

106 J. Staerk, H. J. Ronneberger, H. Wiegandt, and W. Ziegler (1974), *Eur. J. Biochem.* **48,** 103.

107 D. M. Gill (1975), *Proc. Natl. Acad. Sci. USA* **72,** 2064.

108 D. M. Gill (1978), *J. Infect. Dis.* **S133,** 555.

109 G. L. Johnson and H. R. Bourne (1977), *Biochem. Biophys. Res. Commun.* **78,** 792.

110 J. Moss and M. Vaughan (1977), *Proc. Natl. Acad. Sci. USA* **74,** 4396.

111 J. Moss, V. C. Manganiello, and M. Vaughan (1976), *Proc. Natl. Acad. Sci. USA* **73,** 4424.

112 J. Moss and M. Vaughan (1977), *J. Biol. Chem.* **252,** 2455.

113 M. Rodbell, L. Birnbaumer, S. L. Pohl, and H. M. J. Krans (1971), *J. Biol. Chem.* **246,** 1877.

114 V. Bennett and P. Cuatrecasas (1975), *J. Membrane Biol.* **22,** 1.

115 S. L. Levinson and A. J. Blume (1977), *J. Biol. Chem.* **252,** 3766.

116 M. Schramm and M. Rodbell (1975), *J. Biol. Chem.* **250,** 2232.

117 D. Cassel and Z. Selinger (1977), *Proc. Natl. Acad. Sci. USA* **74,** 3307.

118 T. Pfeuffer (1977), *J. Biol. Chem.* **252,** 7224.

119 D. Cassel and T. Pfeuffer (1978), *Proc. Natl. Acad. Sci USA* **75,** 2669.

120 D. M. Gill and R. Meren (1978), *Proc. Natl. Acad. Sci. USA* **75,** 3050.

121 G. L. Johnson, H. R. Kaslow, and H. R. Bourne (1978), *Proc. Natl. Acad. Sci. USA* **75,** 3113.

122 V. Bennett and P. Cuatrecasas (1975), *J. Membrane Biol.* **22,** 29.

123 S. Craig and P. Cuatrecasas (1975), *Proc. Natl. Acad. Sci. USA* **72,** 3844.

124 J. F. Ash, D. Lourard, and S. J. Singer (1977), *Proc. Natl. Acad. Sci. USA* **74,** 5584.

125 J. Moss, P. J. Fishman, R. L. Richards, C. R. Alving, M. Vaughan, and R. O. Brady (1976), *Proc. Natl. Acad. Sci. USA* **73,** 3480.

126 N. Sahyoun, and P. Cuatrecasas (1975), *Proc. Natl. Acad. Sci. USA* **72,** 3438.

127 Y. Matuo, M. A. Wheeler, and M. W. Bitensky (1976), *Proc. Natl. Acad. Sci. USA* **73,** 2653.

128 C. L. Gyles (1974), *Infect. Immun.* **9,** 564.

129 D. M. Gill, D. J. Evans, Jr., and D. G. Evans (1976), *J. Infect. Dis.* **S133,** 5103.

130 J. Moss and S. Richardson, *J. Clin. Invest.*, submitted for publication.

131 C. Gyles, M. So, and S. Falkow (1974), *J. Infect. Dis.* **130,** 40.

132 Y. Takeda and J. R. Murphy (1978), *J. Bacteriol.* **133,** 172.

133 W. S. Dallas and S. Falkow, 13th Joint Conference, U.S.-Japan Cooperative Medical Program, Microbiology and Infectious Disease Program, National Institute of Allergy and Infectious Diseases, National Institute of Health.

134 L. R. Koupal and R. H. Deibel (1975), *Infect. Immun.* **11,** 14.

135 D. M. Sedlock, L. R. Koupal, and R. H. Deibel (1978), *Infect. Immun.* **20,** 375.

136 P. D. Sandefur and J. W. Peterson (1977), *Infect. Immun.* **15,** 988.

137 C. G. Goff (1974), *J. Biol. Chem.* **249,** 6181.

138 R. Skorko, W. Zillig, H. Rohrer, H. Jufiki, and R. Mailhammer (1977), *Eur. J. Biochem.* **79,** 55.

139 H. Rohrer, W. Zillig, and R. Mailhammer (1975), *Eur. J. Biochem.* **60,** 227.

140 A. Pesce, C. Casoli, and G. Schito (1976), *Nature* **262,** 412.

141 S. H. Leppla, O. C. Martin, and O. R. Pavlovskis (1978), *Abstr. Ann. Meet. Amer. Soc. Microbiol.*, 29.

142 P. E. Thorpe, W. C. J. Ross, A. J. Cumber, C. A. Hinson, D. C. Edwards, and A. J. S. Davies (1978), *Nature* **271,** 752.

143 D. G. Gilliland, R. J. Collier, J. M. Moehring, and T. J. Moehring, *Proc. Natl. Acad. Sci. USA*, in press.

144 T. Uchida, personal communication.

Chapter 11

MULTIFUNCTIONAL PROPERTIES OF ANTIBODIES

DIETMAR G. BRAUN and WOLFGANG SCHALCH

Basel Institute for Immunology
Basel, Switzerland

11.1 INTRODUCTION

Antibodies have two basic functions: (*a*) they bind antigens, and (*b*) they mediate effector functions as a consequence of antigen binding. Interaction of antibodies with antigens depends on recognition and binding of immunodeterminant groups (the ligands) on molecules either in solution or

Wolfgang Schalch is recipient of an EMBO-Long-Term Fellowship

on particulate matter such as parasites, microbes, viruses, and cells. Antierythrocyte antibodies, for example, which determine blood group specificities, may be of questionable value to the individual; antibodies specific for microbes or viruses, however, are unquestionably of protective value guaranteeing survival of the individual. The protective value of antibodies not only depends on specific binding to the antigen, for example on the surface of pathogenic bacteria, but also, it involves at least one additional function of the antibody molecule (referred to as the effector function) that is linked to a part of the molecule distinct from the binding region and has to do with the clearing of bacteria. Such assignment of different functions to structurally distinct parts of the antibody molecule makes it a multifunctional protein. Antibodies belong to a group of proteins, the immunoglobulins, which themselves must be subdivided on the basis of structural and functional properties, and the above statement concerning binding and effector functions should be further differentiated. That is the binding specificities to ligands, because of their multiplicity, must embrace a diverse array of properties relating both to the seemingly infinite antigen specificities (diversity of ligands) on one hand and the subtle effector specificities on the other hand. Both are displayed in terms of protein structure differences in class and subclass properties of the individual molecules.

In this chapter we review the basic structural elements of antibody molecules first and then turn to a discussion of the central topic "antigen-antibody interactions" and what is known about the diverse mechanisms that occur in a second step reaction subsequent to this interaction, such as complement fixation and interaction with cell surfaces (cytotropic reactions). Further, antibodies function as receptor molecules on cells (lymphocytes), they have the ability to interact with lectins and with bacterial proteins (protein A) in a nonspecific fashion; they are able to mediate cytophilic reactions. Antigen-antibody reactions are not always strictly specific, they often show a pattern of cross-reactivity. This finding describes the fact that the antibody ligand fit need not be perfect for interaction but may merely rely on certain basic structural elements shared by several antigens. Such a recognition pattern also finds its expression in the observation of multispecificity. Nevertheless, the binding of a ligand to the antibody generally leads to functional consequences.

It is not the topic of this chapter to review and evaluate in great detail the variety of physiochemical methods employed to deduce by indirect criteria the possible conformational changes the molecule might undergo when interacting with antigen. In short, we wish to describe the antibody as a functional unit. As such it has at least two functions linked to two different anatomical sites, the variable domain and the constant domain. It must be stated at this point that functional consequences (effector functions) depend on the structural properties on the class of antibody molecules involved.

11.2 THE STRUCTURE OF ANTIBODIES

A discussion of the antibody structure in functional terms relies on three basic phenomena: (*a*) antibody polypeptide chains consist of variable and of constant regions (1,2): (*b*) various homology regions are identified in immunoglobulin polypeptide chains (2—7); (*c*) these are recognized as globular structural units and supply the basis of the domain hypothesis (8).

The basic structure of antibody molecules (9) is described as a protein composed of two identical light (22,000—23,000 mol wt) and heavy (50,000—70,000 mol wt) polypeptide chains covalently held together by disulfide bonds formed by cysteine residues (Figure 11.1). Since these sulfhydryl bonds are exposed in aqueous solutions they interact with thiol reducing agents, a procedure that allows one to separate light from heavy chains (10). In contrast, intrachain disulfide bonds are not accessible to free sulfhydryl groups in aqueous solution unless urea (8 *M*) or guanidine hydrochloride (6 *M*) are added to disrupt the noncovalent forces folding the polypeptide chains into globular units, the chain domains.

There are several kinds of light chain types and heavy chain classes (11) that are serologically typable by means of specific heteroantisera (Table 11.1). In structural terms both the light chain types (κ and λ) and the heavy chain classes (γ, μ, α, δ, and ϵ) relate to unique amino acid sequences of the constant regions (11, 12). Two immunoglobulin molecules, the IgM and the IgA, occur naturally as polymers with subunits each consisting of two light and two heavy chains. The association of an additional polypeptide, called J (for joining) chain, with polymeric IgM and IgA was interpreted to influence the polymerization (13). These constant regions, part of which were described as the fragment crystallizing (Fc) in Porter's original paper on rabbit IgG (14), are the sites that mediate the effector functions of antibodies. Their number is small; in man and in most species there are two light chain types that may be subdivided into allotypes (κ) and isotypes (λ), and five heavy chain classes (Table 11.1). The γ class of heavy chains accounts for 75 percent of the immunoglobulin protein, while all other classes are of considerably lower concentration. The γ class consists of four subclasses (γ_1, γ_2, γ_3, and γ_4 molecules); the α class consists of the α_1 and α_2 subclass (11).

The regions conferring ligand specificity for antigens are located in the variable parts of antibodies. Both the variable regions of the light and of the

Table 11.1 Human Immunoglobulins: Physical, Chemical, and Biological Properties[a]

Properties	IgG	IgA	IgM	IgD	IgE
Sedimentation coefficient	7S	7S, monomer 11S, dimer[b]	19S	7S	8S
Molecular weight	160,000	170,000, monomer 340,000 dimer	890,000		185,000
Serum level (mg/ml)	12.4 ± 2.7[c]	2.8 ± 0.35	1.2 ± 0.35	0.003–0.3	0.000016–0.0018
Carbohydrate content	2.9	7.5	11.8	13	12
Valence	2	2, monomer	10	?	2
Light chain	κ, λ	κ, λ	κ, λ	κ, λ	κ, λ
Heavy chain sublcasses	$\gamma_1, \gamma_2, \gamma_3, \gamma_4$	α_1, α_2	μ	δ	ϵ
J chain	–	+, dimers	+	–	–
Molecular formulas	$\gamma_2\kappa_2, \gamma_2\lambda_2$	$\alpha_2\kappa_2, \alpha_2\lambda_2$ $(\alpha_2\kappa_2)_2J, (\alpha_2\lambda_2)_2J$ $(\alpha_2\kappa_2)_2SJ, (\alpha_2\lambda_2)_2SJ$[b]	$(\mu_2\kappa_2)_5J$ $(\mu_2\lambda_2)_5J$	$\delta_2\kappa_2, \delta_2\lambda_2$	$\epsilon_2\kappa_2, \epsilon_2\lambda_2$

Half-life in serum (days)	23	5.8	5.1	2.8	2.5^{d}
Occurrence	Serum amniotic fluid	Serum secretions colostrum saliva tears	Serum	Serum	Serum
Complement fixation	γ_1, $(\gamma_2)^e$, γ_3	–	+	?	–
Heterologous skin sensitizing	γ_1, γ_3, γ_4	–	–	–	–

[a] Adopted in part from E. A. Kabat (1976), *Structural Concepts in Immunology and Immunochemistry*, 2nd ed., Holt, Rinehart, and Winston, New York. Reprinted with permission from M. J. Hobart and I. McConnall (1975), *The Immune System*, Blackwell Scientific Publications.

[b] In secretion dimers are associated with an extra 4.2S (58,000 mol wt) piece, called the "secretory piece."

[c] Standard deviation.

[d] IgE fixed to basophils and mast cells; in tissues the half-life is much longer.

[e] Complement fixation ±.

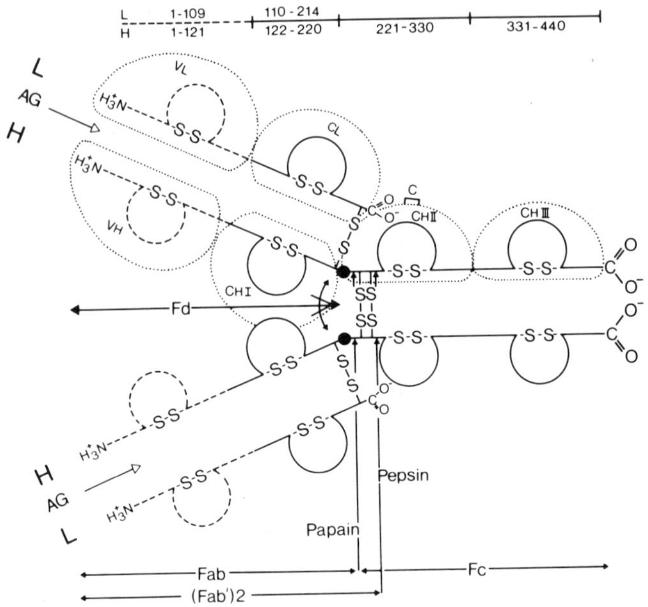

Fig. 11.1 Schematic diagram of an IgG molecule involving the four chain structure with two identical light (residue positions 1–214) and two identical heavy chains (residue positions 1–440). The approximate length in amino acid residues is given at the top of the figure, also the subdivision in domains formed by intrachain disulfide bonded loops that bring together distant residue positions in the linear sequence. L indicates light and H heavy chain; V_L and C_L indicate variable and constant light chain regions; V_H indicates variable, and C_HI, C_HII, C_HIII constant heavy chain regions. The papain-cleaved Fab and Fc fragments are indicated as is the pepsin-derived $(Fab')_2$ fragment. AG indicates the site where antigen binds, C the site where complement binds. The Fd, drawn as a pseudo-axis, is the designation for the stretch V_H and C_HI.

heavy chains form a grove or cavity shaped antibody combining site and are part of the Fab (antigen binding) fragment (15). The variable regions of immunoglobulin polypeptide chains have attracted the greatest attention by immunologists and molecular biologists because of their seemingly infinite structural variability in terms of amino acid sequence (16, 17). Nonetheless, this variability is encoded in well recognizable structural patterns grouping several structures into subgroups and further into subsubgroups owing to linked amino acid substitutions shared by several proteins (18). The recognition of the so-called framework associated amino acid residues determining the common structural features for each κ, λ and the heavy chain variable regions, and the recognition of three hypervariable or antigen complementarity determining regions in the variable region furthermore implies a genetically determined organization of these structures. Further, the demonstration of identical variable regions of polypeptide chains derived from antibodies of two littermate rabbits elicited by and specific for the same antigen indicates that variable regions are confidently encoded in the germ-line DNA of the species (19); this finding is strong evidence against a random pattern of variation. For these and other considerations antigen specificities are not accumulated from a small set of inherited variable region structures by somatic mutation and selection (20–22) but are contained as fully programmed complementary structures in the germ-line DNA, which have been selected during evolutionary times (16, 18). Hence, the antibody problem is basically similar to the diversity of enzyme specificities.

It is distinct from this by the fact that a large number of variable, antigen complementarity conferring regions of the heavy chain can be linked to certain heavy chain molecules with different constant regions of the γ, μ, α, δ, or ϵ class. These composite molecules in turn have the possibility to combine with λ or κ light chains. Thus a variety of different constant parts becomes available to the antibody molecule so that a certain antigen specificity may be associated with different antibody functions (2).

This very brief description of the antibody structure may be sufficient. Our next concern relates to the folding of the antibody molecule, which allows the expression of its functional properties. The description of the antibody folding as it comes out from both amino acid sequence data and x-ray crystallography is a proof of the domain hypothesis (24). The reader may be reminded that light chains, both κ and λ, have two domains: a variable and a constant one. γ Chains have four domains: the variable (V_H) and three constant ones ($C_\gamma1$, $C_\gamma2$, and $C_\gamma3$). μ Chains have five domains (V_H, $C_\mu1$, $C_\mu2$, $C_\mu3$, and $C_\mu4$); so have ϵ chains (V_H, $C_\epsilon1$. $C_\epsilon2$, $C_\epsilon3$, and $C_\epsilon4$), whereas the α chain has four homology regions (V_H, $C_\alpha1$, $C_\alpha2$, and $C_\alpha3$). Before we can briefly describe the principal folding pattern of antibody polypeptide chains within the whole molecule a region, called the hinge, needs mentioning (17). Some of the structural properties of the hinge regions of human immunoglobulins are given in Table 11.2. This table also contains the sequences found in the hinge region. This stretch of polypeptide links together the $C_\gamma1$ and $C_\gamma2$ domains conferring flexibility to the Fab fragments, and it therefore is interpositioned between the antigen binding part of the molecule and the part conferring effector functions. Structurally rich in proline residues, it is otherwise very diverse, for example, in its length and in the number of interhinge disulfide bonds. The fact that in contrast to the α and γ chains the μ and ϵ chains have no hinge region bears no apparent relationship to the effector functions of the molecules. The α chain hinge is peculiar though in that it contains O-linked carbohydrate (25).

Table 11.2 Amino Acid Sequences of Hinge Regions of Human Ig Heavy Chains in One Letter Code[a,d]

		Hinge	
		No. Residues	No. Prolines
	——— Cγ1 ——→←——— *hinge* ———→←——— Cγ2 ———→ 217 233		
IgGl[b]	... TKVDKKVEPKSC [ ] DKTHTCPPCPAPELLGGPSVFLF ... (217 at V; 233 at A)	13	4
IgG2	... ——— T —— R — CC [ ] VE — PP — P —(PPGAV)SF	10	5
IgG3	... ——↓—— [(47 residues[c])] — TPPP — PP — P — P ——— ...	13 (65)[c]	7 (22)[c]
IgG4	... ——— R —— S — YG [ ] PP — PP — P — S — F ——— ...	10	5
IgA1	——— Cα1 ——→ ←—— Cα2 —— ... BPSZBVTV — CPVPSTPPTPSPSTPPTPSPSCCH — RLS — H ... (repeats 1, 2)	17	10
IgA2	... NSSQDVTV — CRVPPPPP [ ] CCH — RLS — H ...	5	5
IgM	——— Cμ1 ——→ ←—— Cμ2 —— ... KEK — VPLPVIA [ ] Z — PPKV —— VP ...	0	0
IgE	——— Cε1 ——→ ←—— Cε2 —— ... SSTBNVKTFSV — [ ] SRBFTP — T — KIL ...	0	0

[a] In part adopted from Beale and Feinstein (1976), *Q. Rev. Biophys.* **9,** 2, pp. 135–180. With permission of Cambridge University Press. $C_{\gamma}1$, $C_{\gamma}2$, $C_{\alpha}1$, $C_{\alpha}2$, $C_{\mu}1$, $C_{\mu}2$, $C_{\epsilon}1$, and $C_{\epsilon}2$ designate amino acid sequence stretches adjacent to the hinge region; the residue positions 217 and 233 conform to the numbering of IgG1 NIE. A continuous line indicates sequence identity with IgG1 NIE except for the proline residues; because of their importance in the hinge region they are indicated for every protein separately. [ ] symbolizes deletions to maximize the homology; ( ) indicates uncertainty in amino acid sequence.

[b] IgG1 NIE, Ponstingl and Hilschmann (1972, 1976), *Hoppe-Seyler's Z. Physiol. Chem.* **353,** 1369; **357,** 1571.

IgG2 SA: Milstein and Frangione (1971); *J. Biochem.* **121,** 217.

IgG3: Michaelson et al. (1977), *J. Biol. Chem.* **252,** 883.

IgG4 VIN: Pink et al. (1970), *J. Biochem.* **117,** 33.

IgA1 TRO: Kratzin et al. (1975), *Hoppe-Seyler's Z. Physiol. Chem.* **356,** 1337.

IgA2 BUT: Torano and Putmann (1978), *Proc. Natl. Acad. Sci. U. S. A.* **75,** 966.

IgM GAL: Watanabe et al. (1973), *Hoppe-Seyler's Z. Physiol. Chem.* **354,** 1505.

IgE ND: Bennich and Bahr-Lindstrom (1974), *Prog. Immunol. II* **1,** 49.

[c] The number in the insertion accounts for the 47 additional residues found in the hinge region of IgG3 proteins at the position indicated, comprising a 65 residue long hinge region which contains a total of 22 polines.

[d] IgA1 proteins have a duplicated octapeptide in the hinge 1 + 2. The one letter code of symbolizing amino acid residues, *Biochem. J.* **113,** 1 (1969): A, ALA; B, Asx; C, Cys; D, Asp; E, Glu; F, Phe; G, Gly; H, His; I, Ile; K, Lys; L, Leu; M, Met; N, Asn; P, Pro; Q, Gln; R, Arg; S, Ser; T, Thr; V, Val; W, Trp; X, unknown; Y, Tyr; Z, Glx.

It is further interesting that μ and α chains contain an extra "tail piece" of 19 amino acid residues following the $C_{\mu}4$ and $C_{\alpha}3$ domain, respectively. The structural homology of these tail pieces is remarkable (Table 11.3). Interchain disulfide bonds within these pieces can be formed by the cysteine residue; this residue may also attach the J chain by a disulfide bond (3, 4, 6, 7, 12).

For the elucidation of the immunoglobulin structure enzymatic fragmentation procedures were widely used (26). The IgG molecule has originally been cleaved in the hinge region by papain yielding monovalent Fab and Fc fragments (14) as well as by pepsin yielding the bivalent $(Fab')_2$ and smaller, dialyzable fragments of the Fc region (27) (see Figure 11.1). The latter can be further isolated. The Fc fragment has been used for further cleavage into separate $C_{\gamma}2$ and $C_{\gamma}3$ domains (28). Intact subfragments, called Fv, have been isolated from the Fab by cleavage at the junction between V_L and C_L and V_H and $C_{\gamma}1$ (29—30). The fragment containing only the V_L and V_H domains retained specificity for hapten (DNP) and contained the idiotypic sites (29, 31).

Under conditions similar to those for IgG fragmentation human IgM was cleaved into Fab and $(Fab')_2$ fragments. These procedures usually destroy the Fc fragment. An IgM $(Fc)_5$ has been isolated by tryptic cleavage at

Table 11.3 Amino Acid Sequences of Extra "Tail Pieces" in Human Immunoglobulin (One Letter Code) at C-terminus[a]

IgM GAL...	RVTERTVDKSTGKPTLYNVSLVMSDTAGTCY—C(=O)OH
IgA1 TRO...	AF—QK—I—RLA——HV——V——AEVD———C(=O)OH
IgA2 BUT...	AF—QK—I—RLA——HI——V——AEAB———C(=O)OH

[a] Adopted in part from Beale and Feinstein (1976), *Q. Rev. Biophys.* **9,** 2, 135–180. With permission of Cambridge University Press.
IgM: Watanabe et al. (1973), *Hoppe-Seyler's Z. Physiol. Chem.* **354,** 1505.
IgA1: Kratzin et al. (1975), *Hoppe-Seyler's Z. Physiol. Chem.* **356,** 1337.
IgA2: Torano and Putnam (1978), *Proc. Natl. Acad. Sci. U. S. A.* **75,** 966.
A continuous line indicates sequence identity with IgM GAL. One letter code of amino acid designation as in Table 2.

60°C for 30 min. (32). Considerable species and subclass related differences in cleavage behavior have also been reported and must be taken into account if specific fragmentation is aimed at (33, 34). All of these procedures show, however, that fragmentation under suitable conditions occurs at loose or extended structures that are situated between compact domain folds (11, 35, 36).

11.3 THE PRINCIPAL FOLDING PATTERN OF THE IMMUNOGLOBULIN

The four domains of IgG molecules (V_L and V_H, C_L and C_H1, C_H2, C_H3) upon high resolution x-ray diffraction studies have been sufficiently analyzed to deduce a principal pattern of polypeptide chain folding which has been called the *immunoglobulin fold* (24, 37). When analyzed for a dinitrophenyl-(DNP-) binding light chain dimer both variable and constant parts consisted of straight segments of polypeptide chain forming two layers of antiparallel segments. One of these layers contains three chain segments and the other four chain segments; both layers are held together by disulfide bonds, and they surround a tightly packed interior of hydrophobic amino acid side chains (37, 38). In this light chain model the variable and the constant domains differ in the relative position of these layers such that the three strand layer faces the outside in the variable domains and the inside in the constant domains (Figure 11.2).

Generally in Fab fragments or complete immunoglobulin molecules variable region domains interact laterally through the three-strand layers to establish variable region modules. Variable region modules bring together the hypervariable region residues to line a well defined cavity forming the antigen combining site. In the slightly open structure of the variable region module transinteractions between the three strand β-pleated sheet are strong enough to hold together V_L and V_H functional units (24, 37, 39—41). Constant region domains contact each other through the four strand layers, largely by hydrophobic interactions that are difficult to disrupt, as shown for the $(C_H3)_2$ module of an IgG1 molecule (42—44). Interactions between C_H2 domains are somewhat similar to those between variable region domains; for example, they face each other by the three strand layers. In addition, the attached oligosaccharides on the interfaces may block stable transinteractions. As a result the C_H2 domains do not strongly interact. There is evidence even for repulsion between them. In the complete IgG molecule the C_H2 domains are disulfide-bonded.

Interpretations of longitudinal interactions between domains are more difficult. Clearly, there is segmental flexibility between domains, the most obvious one displayed by the hinge region. Flexibility between the variable

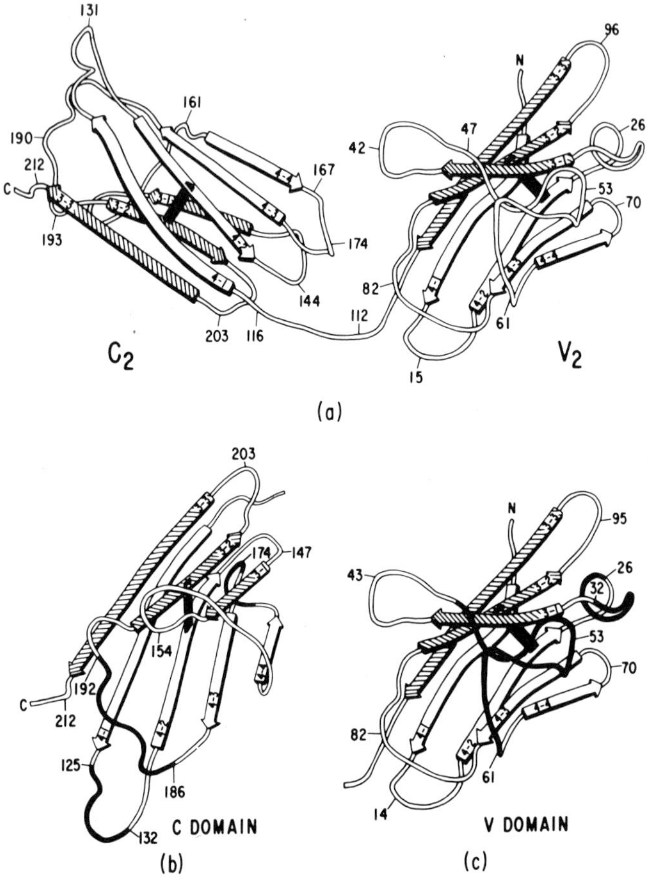

Fig. 11.2 The polypeptide chain of monomer 2 of the Bence Jones protein Mcg with directional arrows superimposed to indicate strands participating in antiparallel β-pleated sheets. Three chain layers are represented by numbered striated arrows, and four chain layers by white arrows. Black bars are the disulfide bonds. At the right of V_2 are the hypervariable regions. Residue positions are numbered to aid in the correlation of sequence and structure (*b*) and (*c*): For comparison of their structures, the C domain is rotated and placed in a position approximating that of the V domain. Chain segments present in only one domain or having different structures in the two domains are shown as solid black lines. Since the V and C domains have different sequences, but similar structures after rotation, we refer to them as rotational allomers (adopted from Edmundson et al., in E. Haber and R. M. Krause, eds., *Antibodies In Human Diagnosis and Therapy*, p. 144. Copyright 1977 by Raven Press. Reprinted with permission.

and the constant module can also be deduced by the variation in angles (113°—170°) observed (40—42, 44—45). Further there are only few contact points, if any, between the variable and the constant modules. In IgG_{Kol} no electron density map was obtained for the Fc region, indicating a disordered state of this region in the crystal, whereas the Fab fragment could be fully mapped in the crystal. A "minimal disorder" model of Huber and associates infers no other contacts between (V_HC_H1), (V_LC_L), (C_LC_H1), or (C_H1C_H2) and between $(Fab')_2$ or Fc fragments of this molecule but their covalent linkage through the hinge region peptide; the latter is believed to have a rather extended conformation (41). Because of this difficulty with the IgG_{Kol} crystal the x-ray structure was studied with a Fc fragment crystal (42, 44). In the Fc crystal the molecule has the shape of a "mickey mouse" where the compact C_H3 domains form the head and the C_H2 domains the ears. Here the C_H2 and the C_H3 domains show the common immunoglobulin fold, they are linked by a conformationally not completely extended peptide. Further, the so-obtained model of the Fc fragment could be fitted with the Fab model of the IgG_{Kol} crystal to form a complete model of an IgG molecule (41). Silverton et al. (45) examined the low-resolution structure of a complete human IgG_1 molecule. In this whole molecule the Fc portion had a structure similar to the isolated Fc fragment (41).

11.4 FUNCTIONAL PROPERTIES OF THE ANTIBODY MOLECULE

In this part of the article we will not describe the numerous studies on antigen antibody binding kinetics and measurements of affinity constants and so forth; these have been the subject of several recent reviews (12, 26, 36, 46). We intend to review here studies aiming at an understanding of the effector functions of antibodies after antigen binding has occurred, further interactions with membranes, with lectins, and with protein A on *Staphylococcus aureus*. In this context the question arises at several points "In what respect are antibody effector functions the result of allosteric changes of the molecule?" In contrast to this view is the "associative model," which correlates effector functions only with cross-linking (lattice formation) of antibody by multideterminant antigens (26, 46).

Complement Fixation

Complement fixation assays typically describe a method in which erythrocytes, generally sheep red blood cells (SRBC), are used as target cells for the lytic complement action mediated by an antibody antigen interaction. The classical and the alternative pathways of activation of the complement cascade have recently been extensively reviewed (47, 48) and are not dealt with here. It is sufficient to say in this context that the C_H2 domain has been localized as the complement fixing site (the first step of complement activation involves the binding of component C1) in mouse IgG (49), rabbit IgG (28), and also in human IgG1 (52, 53). In principal an IgG molecule has two C1 binding sites/mole, however, only one of these is occupied at one time; for C_H2 monomers, C_H2 dimers, and IgG all have the same molar activity (54). Since chemical modification of tryptophanyl residues on both IgG and IgG Fc fragments destroys the ability of complement fixation it is inferred that the indole side chain of tryptophanyl residue 277 (the only one exposed to solvent in the Fc fragment) is required for binding to provide the necessary hydrophobic center (53, 55). In contrast, the complement fixing site of IgM molecules is not known. When the $(Fc)_5$ fragment of human IgM was cleaved with cyanogen bromide or by limited tryptic digestion a peptide with weak binding activity was isolated which was identified as part of the $C_\mu4$ domain (56).

For studies on complement fixation by antibody two types of ligands have been used, *univalent* and *multivalent* ones. To study in detail the functional aspects involved we regard these, in agreement with H. Metzger (26), as critical for the relevant interpretation of the mechanism. Monovalent ligands were reported to induce comlement fixation with IgM but not with IgG antibodies (57). Ribonuclease was diazotized with a phenyl-lactoside determinant under conditions where a single determinant (mono-Lac RNase) would be present per ribonuclease molecule as well as under conditions where multiple determinants were introduced into the carrier molecule. Anti-phenyl-lactoside antibodies were elicited in rabbits and purified by affinity chromatography on phenyl-lactoside Sepharose 4B columns. IgM antibodies were subsequently separated from IgG antibodies by Sephadex G-200 filtration with 95% purity. For assaying complement fixation the Wasserman-Levine microcomplement fixation test was used employing SRBC and guinea pig complement. The affinity of the IgM antibodies for the phenyl-lactoside was assayed by equilibrium dialysis. The binding affinity of mono-Lac RNase was estimated by ultracentrifugation analysis of antigen-IgM mixtures. The Wasserman-Levine test measures depletion of complement. Under the conditions of this test the amount of fixation was related to the determinants added but not to the degree of hapten substitution. The authors concluded from their data that site filling by antigen was sufficient to generate changes in the IgM Fc structure, and cross-linking by multifunctional antigen was not required (57). As Metzger has pointed out, a significant inconsistency exists between the measured binding constant of the antibody for mono-Lac RNase and the concentration of IgM and antigen at which complement fixation was observed (26).

Further, the authors interpret the complement fixing capacity of the mono-Lac RNase upon interaction with IgM antibody as opposed to the incapacity of the hapten analog to fix complement, as indication for an allosteric change of the IgM molecule, even though the affinity constant for the free hapten was about 3.5—fold better than for the mono-Lac RNase. Apparently supporting evidence for the above data came from a system that used IgM antibodies specific for the "loop" of hen egg lysozyme in the presence of lysozyme or the isolated loop peptide, a monovalent ligand (58). This effect was, however, dependent on a critical "loop" concentration. Higher concentrations of loop destroyed this effect. Such data rather support an associative model (46), where aggregation occurs with higher antigen concentrations, than the opposite, an allosteric model.

Complement fixation by rabbit anti-DNP IgG antibody when liganded to a univalent DNP-lys_{20} or DNP-lys_{35} has been reported (59, 60). The criticism that this work meets is based on the fact that "the ligand itself can interact with the complement" (26); further, the IgG may indeed undergo a conformational change by the interaction with the large positively charged hapten (12).

The critical requirements for studies mentioned above are twofold: first, the antibodies used should be homogeneous and, second, the antigen in the test should only present a single, unique determinant (univalent) for these antibodies such that there is no ambiguity on either side of the two partners. Such a system has recently become available to test the question of whether aggregation of antibody by antigen is required for complement dependent lysis of SRBC. The system is as follows: Hyperimmunization of rabbits with streptococcal group A-variant carbohydrate bearing vaccines induces two kinds of antibodies when tested with the isolated group A-variant carbohydrate (Av-CHO), those that precipitate the antigen and another kind of antibodies that do not precipitate the Av-CHO (61). A detailed description of the antibodies follows the characterization of the formamide extracted, chemically pure Av-CHO. This Av-CHO is a linear homopolymer of L-rhamnose (linked alternately αl-2, αl-3) of 5000 daltons, corresponding to approximately 30 L-rhamnose units (62; Schalch and Braun, unpublished). Accordingly, this sugar moiety is composed of end group specific determinants (the nonreducing and the reducing end) and linear internal determinants. The valence for antibody of the antigen is theoretically 1 for each end group specific antibody and 2 or greater for any internally specific antibody. This group polysaccharide when tested at high concentrations does not activate the alternate pathway of complement activation (C. Koch and D. G. Braun, unpublished observation).

A detailed analysis of the two kinds of Av-CHO specific IgG rabbit antibodies yielded the following information. It needs to be stressed that these antibodies were purified by preparative isoelectric focusing in granulated gels to single band purity (63). Therefore the above mentioned stringent requirements were fulfilled to undertake a study that would eliminate all ambiguities with regard to a nonaggregated versus an aggregated state of antibodies binding to antigen. The results are clear-cut: Nonprecipitating IgG antibody of single band purity is specific for the nonreducing end group, the valence of the antigen for this antibody is 1 per monovalent, papain derived Fab fragment. From this all other characteristics follow: the antibody is nonprecipitating, nonhemolytic in a complement dependent lytic assay performed with SRBC to which the O-stearoylester of the Av-CHO has been coupled by virtue of the hydrophobic membrane anchoring stearic acid ester tail (Av-CHO-SRBC). The antibody does not agglutinate Av-CHO-SRBC, its intrinsic affinity for the antigen varies for different preparations between 10^5—10^6 M^{-1}. The affinity and valency measurements were obtained by a variety of techniques including fluorescence quenching, equilibrium dialysis, and ultracentrifugation analysis (64, 65). Specific binding of this antibody to Av-CHO-SRBC but not to SRBC was furthermore shown by radioimmunoassays using isotope-labeled antibody and by fluorescein-conjugated sheep antirabbit IgG. Since there is no complement dependent lysis observable we concluded that the association of univalent antigens with clean preparations of the corresponding antibodies induces no allosteric change that would lead to complement fixation (65). In contrast to this are the data obtained with the second kind of IgG antibody even from the same rabbit (66). This antibody precipitates the 5000 dalton Av-CHO, it agglutinates Av-CHO-SRBC but not SRBC, and it induces complement dependent lysis upon interaction with Av-CHO on SRBC. The intrinsic affinity of monovalent, papain derived Fab fragments is in the order of 10^8—10^9 M^{-1}, the valence of the antigen for Fab is 3–4, and in the ultracentrifuge concentration dependent aggregates between antigen and antibody are formed until precipitation occurs (65). On the basis of this data we feel that the role of the ligand is essential for the phenomena observed. Its univalency or multivalency determines whether aggregation can or cannot occur; this in turn initiates complement binding for which aggregation *seemingly in an ordered form* is a prerequisite for the attachment of Clq. Therefore, complement activation as an allosteric phenomenon is pretty much ruled out; the data with the Av-CHO, being both a univalent and a multivalent antigen, and the two kinds of homogeneous rabbit IgG antibodies therefore support the *associative model*, and they rule out both the allosteric and the distortive model (46).

In support of the action of multivalent antigens on the initiation of both IgM and IgG induced complement activation is a recent study on complement depletion mediated by IgM rabbit antibodies to spin-labeled lipid haptens in using model membranes (67). The authors found that IgM molecules must be bound to a critical multiple of antigenic determinants at a membrane surface for the induction of a complement mediated attack. In this paper it was also proposed, and this is in aggreement with the above findings on the anti-Av-CHO antibody systems, that the spacing and accessibility of antigenic determinants may influence complement activation critically.

Cleavage Rates of Liganded and Unliganded Antibodies

Another set of data using antipneumococcal type III rabbit IgG antibodies of restricted heterogeneity, but not of homogeneity, also suggests that the allosteric immunoglobulin model (41) is unrealistic, describing that the flexible antibody molecule is stiffened upon interaction with antigen through the formation of longitudinal C_H1—C_H2 interdomain contacts (68). This work tested, admittedly in a gross way, the accessiblity of the hinge region to reduction and limited proteolysis in the liganded and unliganded state, since according to the allosteric model it would be expected that the interaction with antigen would induce a change in folding in this area and thus affect the attack by reducing or proteolytic reagents. It was found that the rate of proteolysis or reduction of immune complexes formed with various oligosaccharide and polysaccharide units was not significantly different in the liganded and in the unliganded state, suggesting no major conformational change in the hinge region to occur after antigen binding (68). Similarly, in the same system the authors failed to observe allosteric interactions between the Fab and Fc fragments after binding hapten or antigen to the Fab region on the one hand, and after binding of *Staphylococcus aureus* protein A on the other hand to the Fc region when using a spectroscope approach (69).

The picture that emerges from the studies reported here and others (26) not dealt with is that the bifunctional property of IgG antibodies, antigen binding by the Fab domains and complement activation by the Fc domain, requires no "nervelike" signal transfer (70, 71) from the aminoterminal to the carboxyterminal part of the molecule in the liganded versus the unliganded state, but solely depends on an aggregation pattern dictated in a homogeneous test situation by the multivalence of the antigen as recognized by the antibody.

11.5 IMMUNOGLOBULINS AS CELL MEMBRANE CONSTITUENTS OF LYMPHOCYTES

Immunoglobulins have been demonstrated functionally as cell surface components of lymphocytes by the ability of antiimmunoglobulin antisera to activate rabbit lymphocytes *in vitro* to form blast cells, to divide and thereby incorporate ^{3}H-labeled thymidine as evidence for DNA synthesis (72). Since this early demonstration of immunoglobulins by a functional assay, immunoglobulins have been identified by a variety of means as constituents of membranes of bone marrow derived (B) lymphocytes (73). Whereas it has been possible to demonstrate with specific antisera the presence of light chain determinants and various classes of heavy chains the positive identification of the Fc fragment has proven difficult (74). The reason for this failure is generally seen in the fact that the C_H3 domain is buried in the membrane and therefore not accessible for a specific antiserum directed to this part of the molecule.

The presence of various immunoglobulin classes has been demonstrated on lymphocyte membranes. Determinants of IgM, IgD, and IgG were identified by specific fluorescent antisera (75, 76). It came as a surprise that the rare class determinant identifying IgD molecules was shown to be a major one on the lymphocyte surface; this immunoglobulin accounts for only a minor portion in the serum (Table 11.1) (76). IgM and IgD as well as IgM and IgG have been shown to occur simultaneously as membrane immunoglobulins. However, in keeping with the clonal selection theory, analysis with the aid of antiidiotypic antisera as well as with binding specificities has so far failed to positively identify more than one variable region per lymphocyte, indicating functional monospecificity. Recently, analysis of various differentiation states of lymphocytes in the rabbit appear to suggest that early lymphocytes may express more than one heavy chain variable region as well as light chain allotypes (77, 78). If this proves to become a general finding Burnet's clonal theory (79) would only apply to functionally finite lymphocyte states. Such a finding would have profound implications for the understanding of how bone marrow derived (B) lymphocytes function in response to antigenic stimuli either in a positive (immune induction and antibody formation) or in a negative way (tolerance).

There is also some evidence that the other class of lymphocytes, the thymus derived or T cells, acquire specificity for antigen by immunoglobulin or immunoglobulinlike receptors (80). While the immunoglobulin is considered an accepted receptor for the bone marrow derived lymphocyte, the B cell, which after antigenic stimulation differentiates and matures into the antibody secreting plasma cell, the positive demonstration of the T cell receptor is not yet at hand. Recent data suggest that endogeneous specificity of T cells for antigen is also conferred by an immunoglobulinlike molecule (81, 82).

The presence of immunoglobulins as membrane moieties of B and probably also of T cells make these molecules the candidates responsible for specific triggering by antigen. However, this statement is not at all cogent because it has not yet been shown that the immunoglobulin receptors on lymphocyte membranes are the signal transmitter for the cell to differentiate and mature, in other words, to divide and finally build up immune reponsiveness and an immunological memory. There is a possibility that the antigen specificity conferring receptors are merely concentrating the antigen at the surface of lymphocytes like a glue, and the trigger for differentiation and maturation is given through (an) independent receptor(s). Such a signal may either be given by mitogen receptors (83) or in immune responses that require the cooperation between T and B cells through the T cell or a component of it, so-called T cell factors (73, 84, 85). It is very likely that aggregation of immunoglobulin receptors, and more importantly of the Fc fragments, is again a key event for activation (26). In this respect the early activation of rabbit lymphocytes by aggregation through antiimmunoglobulin sera would have already described the basic event; until recently such a documentation has proven difficult for mouse lymphocytes (86). If aggregation of immunoglobulin receptors should promote lymphocyte triggering, this would be built into the immunoglobulin structure, since only bivalent antibodies are effective. The signal that is delivered across the membrane Fc domains to cause division of the cell is completely unknown. A similar signal probably is delivered to mast cells upon cross-linking of IgE receptors (see below).

11.6 CYTOTROPIC REACTIONS OF IMMUNOGLOBULINS

Immunoglobins are not only integral endogeneous cell membrane moieties but also they undergo cytophilic attachment to cell membranes of lymphocytes, leukocytes, and macrophages (53, 54, 87). Attachment of this sort is mediated by the Fc receptor on these cells, recently isolated and further identified as a glycoprotein with a molecular weight of 45,000 daltons (87), which appears to recognize $C_\gamma 3$ domains but not $C_\mu 2$ domains specifically (89). As such the $C_\gamma 3$ domain was successfully used to inhibit passive cutaneous anaphylaxis (90) to interfere with the rosette formation of human monocytes and tanned red cells coated with IgG (91). This view is supported by the finding that Facb, a fragment obtained by plasmin digestion C-terminal of $C_\gamma 2$ lacking the $C_\gamma 3$ domain, did not activate phagocytosis of sensitized bacteria by neutrophilic leukocytes and did not activate cytotoxic killer or K cells to lyse sensitized target cells (89).

Circulating antistreptococcal Av-CHO rabbit IgG antibodies when injected intravenously into rabbits with immunological memory for this antigen are able to completely suppress an *in vitro* immune response of subsequently isolated lymphocytes specific for this antigen (92). This effect was shown to be reversible both upon removal of the adherent cell population (monocytes) and under capping conditions with anti-V_H allotype antisera, indicating that the suppressive (feedback) effect on lymphocyte precursor cells for the Av-CHO is specifically mediated by the injected cytotropic IgG antibody (92). Specific antibody may by this mechanism selectively regulate immune responsiveness. The effect depends on the intact IgG antibody molecule, for $(Fab')_2$ fragments derived by pepsin digestion were without effect.

IgG is not only able to interact by way of the Fc receptor with cells that have direct or indirect effects on the immune regulation, but it also interacts in the first stage of specific transplacental movement from mother to fetus in man with receptors on the syncitiotrophoblast membrane (93). This binding of IgG to placental receptors can be saturated, and the order of affinity is different for different IgG subclasses (IgGl = IgG3 > IgG4 > IgG2). For the Fc fragment the same binding characteristics were shown as for IgG, however, fragments resembling $C_\gamma 2$ and $C_\gamma 3$ domains had no binding properties.

Also gut of rats has been shown to carry IgG receptors on the intestinal epithelium (94) with the following order of affinity: IgG2a = IgG1 > IgG2b ≃ IgG3. This binding depends on the Fc region of IgG molecules, and again $C_\gamma 2$ and $C_\gamma 3$ fragments were unable to inhibit the transepithelial transport of intact IgG molecules.

These quite diverse membrane localizing properties of the IgG may all be mediated by structurally the same Fc receptor. No evidence for or against this presumption is at present available. The importance of an intact Fc fragment for several of these effector functions is documented by the requirement of unreduced fragments. The evolutionary conservation of the interheavy chain disulfide bonds may find its explanation in this obviously important prerequisite for mediating effector functions (53).

A discussion of the cytotropic reactions of immunoglobulins by virtue of their Fc domains would be incomplete without detailed mention of the action of reaginic antibodies. In this system which mediates allergies and hypersensitivities the IgE molecule is the central actor. First *monomeric* IgE molecules bind tightly to the surface of peripheral basophilic leukocytes and to tissue mast cells. This binding, however, is *not* followed by histamine release, the basic event causing anaphylaxis. Cross-linking of the IgE molecules on the surface of these cells by antigen, by anti-IgE sera, or by antilight chain antisera mediates histamine release by degranulation (95). Isolated IgG Fc fragments also bind to basophils, and their cross-linking by specific antibodies leads to degranulation as well. In contrast, univalent Fab fragments of IgE molecules are unable to trigger degranulation (96).

It has recently been shown that dimeric immunoglobulin E is the unit signal for mast cell degranulation (97). To prove this rat IgE was treated with dimethyl suberimidate, a cross-linking reagent, and subsequently fractionated by Sephadex G-200 and Ultrogel AcA22 filtration into monomers, dimers, trimers, and higher polymers. The fractions retained their mast cell binding properties, and about one third of the cell bound dimers bound bivalently. Passive cutaneous anaphylaxis in rats mediated by the fractions was tested *in vivo*, and histamine and serotonin release was assayed on normal or tumor mouse mast cells. While monomeric fractions were without effect, dimers and higher oligomers caused good anaphylactic or release effects. The authors have thus shown that the minimal degree of cross-linking by dimeric IgE receptors on mast cells and basophils is sufficient to trigger exocytosis. *Dimers in this system, therefore, represent the unit signal for triggering* (97). Immunoglobulin E molecules thus are the link that permits the antigen to dimerize the receptor. How dimerization then signals the cell is not known (98). But again *aggregation* of receptors is the basic signal.

11.7 INTERACTION OF STAPHYLOCOCCUS AUREUS PROTEIN A WITH HUMAN IMMUNOGLOBULIN Fc FRAGMENTS

Protein A is a cell wall constituent of most strains of pathogenic *Staphylococcus aureus*. It has a high binding affinity for the Fc fragments of certain immunoglobulin classes; the biological significance of this protein A binding specificity is not at all clear (99).

Molecularly, protein A is buried in the *S. aureus* cell wall through a C-terminal fragment, called region X, with a very elongated shape and unique structural features (100). The rest of the molecule consists of four strongly homologous regions D, A, B, and C, consecutively arranged along the peptide chain (101). These structurally independent domains of 58 amino acid residues each are functionally fully active after isolation as tryptic fragments binding to the Fc part of IgG molecules at a molar ratio of 2:1 (102).

Recently Deisenhofer et al. (103) were able to obtain crystals of the complex formed by human IgG derived Fc fragment and fragment B of protein A. It was found that protein A tryptic fragment B is a small globular protein built of three parallel helices that are arranged in a triangular array; its globular folding scheme creates a hydrophobic interior, and it is the simplest one so far observed. Further, the three dimensional structure of the tryptic fragment B shows flexible amino- and carboxy-termini.

Upon interaction with IgG Fc fragments, fragment B makes two different types of contact in the crystal, a large one comprising hydrophobic residues of fragment B and the IgG Fc fragment and a small one involving several polar amino acid residues (103). The large contact is made with the C_H2 and the C_H3 domain, while the small one is made to the C_H3 domain alone. The large contact is the contact in solution, the small one is that due to the crystal packing contact; for isolated C_H3 domain does not bind fragment B.

Although protein A fragment B binds to the C_H2 domain, it does not interfere with complement binding (104), which is also through this domain (103). Porter (105) ascribed binding of the large globular Clq component head pieces of complement to the region between the two C_H2 "ears" of the Fc fragment. Deisenhofer et al. argue that binding to this site is incompatible with x-ray crystallography data (103). Rather, the upper half of the C_H2 domain pointing away from the neighboring C_H2 domain is considered as a possible contact area (103). Hence the IgG Fc fragment may accommodate binding of proteins of very different evolutionary origin, one that is produced by the body as part of an effector pathway and a bacterial one whose function remains to be resolved.

11.8 MULTISPECIFICITY OF ANTIBODY MOLECULES

The phrase "multispecificity" should not be confused with multifunctionality of antibodies. The property multispecificity simply implies that antibodies were found in some specific cases to have the ability to react not only with the antigen that was used for triggering but also with other, structurally seemingly unrelated antigens (106–108). It is not clear to what extent this may be a general property of antibodies. If antibody molecules would generally react with more than one ligand, that is, the antibody combining site could accommodate several antigens, or different interacting sites within the combining site would account for this property, the enigma of antibody variability would be partly solved by this phenomenon: fewer variable region genes coded for in the germ line would be required (109).

Several systems have been reported where multispecificity of specific antibodies was recorded (106, 108). Much of such multispecific antigen binding may be accounted for by cross-reactivity and cross-stimulation due to some common structures of the antigens (109, 110); but it has also been shown that single myeloma proteins, and even the aforementioned Bence Jones protein dimer, were able to bind structurally diverse molecules (38, 111, 112).

Recently new evidence for the occurrence of multispecific antibodies in rabbits to an adenosine-5′-phosphate-gramicidin S conjugate [$(AMP)_2$-gramicidin S] has been given (108). These elicited antibodies, pure by isoelectric focusing, were recognizing AMP, menadione, caffeine, and —less well—DNP-glycine, guanine-monophosphate, hydralazine, and sodium barbital. It had not been ruled out, however, in this study that all of these competing ligands resembled each other sufficiently to explain the phenomenon by cross-reactivity as had been shown by others (110, 113). Therefore, the argument of simple cross-reactivity on a common or related structural basis still stands (109, 110, 113). If cross-reactions occur with an association constant higher than that of the ligand presented by the antigen these antibodies are called "heteroclytic" (114).

The biological significance of such cross-reactivity may finally only come from systems dealing with protective immunity. Interesting is the original observation where the occurrence of such antibodies was first described by the phrase "original antigenic sin" for anti-fluenza immunity (115, 116); a similar argument holds for a system working with antistreptococcal group polysaccharide antibodies (113). We disagree with Cameron and Erlanger that complement dependent hemolytic properties of multispecific antibodies would automatically imply properties of biological significance (108, 117) simply because of this fact. Such properties may only reflect appropriate ligand presentation for activation of the complement cascade (see above).

11.9 ANTIBODY POLYPEPTIDE CHAINS: SPLICED PRODUCTS FROM SMALLER GENE SEGMENTS?

Soon after the identification of immunoglobulin variable and constant regions (1) the concept was put forward that light and heavy polypeptide chains are combined from different pieces at the DNA level to make up a complete gene for each structurally encoded variant (118). This phenomenon has been referred to as translocation in the case of the heavy chain (2). Recently it was found for selected light chain probes of the mouse that in the germ line (embryonic tissue) the sequences coding for the variable region and those coding for the constant region occupy separate stretches along the DNA sequence, that is they are physically separated (119). It was further shown by R-loop mapping and by heteroduplex analyses with embryonic genes that variable and constant genes move from distant positions in embryonic DNA to less distant positions in the polypeptide secreting lymphocytic cells but that they do not become contiguous (separation distance 1250 bases). It was hypothesized that the two genetic regions form a transcription unit only when moving closer (rearrangment) (120).

Kabat et al. in their recent analysis go a step further (121). By a thorough comparison of the published amino acid sequence data on immunoglobulin polypeptide chains they showed that the four sets of framework sequences within the variable regions inferred the following genetic solution to the problem: Tracing of each variable region from one framework set to the next revealed that members of the same set in one framework associated region were combined to a functional polypeptide chain with sets different in the other framework associated areas. This pattern was interpreted to indicate assembly of light and heavy chain variable regions during embryonic differentiation from sets of minigenes for each framework associated region (121). Braun, Riesen, and Huser came to a similar conclusion when they analyzed a large panel of rabbit antibody light chain sequences (unpublished observation). All of these data in fact imply that as far as immunoglobulin polypeptide chains are concerned the "one gene-one polypeptide chain hypothesis" (122) is untenable. Rather, it appears that proteins assembled from physically separated coding regions along the DNA sequence of chromosomes do already, by virtue of this physical separation, carry the imprint of different functions at the protein level. Apparently, evolution has solved the problem of acquisition of different functions within a single polypeptide chain by duplicating smaller genes early on, separating them by "spacer" sequences with presumably special signals, diversifying these separated pieces independently, and then splicing these genetic units that have diverged enough to functional polypeptides to combine these different functions. Following recent progress in this field it becomes very obvious that gene splicing may be a widely used mechanism to arrive at products (proteins) consisting of different domains (123–126) each of which is encoded in discontinuous segments at the DNA level. The immunoglobulin molecule, one of the best investigated molecules at the protein level, does combine different functions. Multifunctionality would, therefore, so it appears, already find its expression in a similar coding compartmentalization established in evolutionary time. Such a compartmentalization of presumably multigenic DNA segments, in addition, would favor the acquisition of new products by presumably short range recombinations.

11.10 CONCLUSIONS

It appears that the immunoglobulin molecule is a typical case among the examples discussed in this book combining several (diverse) functional properties. Clearly, the amino-terminal domain comprising the antibody combining site is a signal unit and carries its own specificity. So is the anatomically opposite end of the molecule as clearly shown by the properties of the IgE molecules or those of antistreptococcal group A-variant IgG antibodies as feedback unit in the immune response; both kinds of molecules first attach to membranes through specific receptors and subsequently mediate functions, for example histamine release or clearing free antigen by the adherent cell, respectively. In these two cases attachment of the immunoglobulin molecules to cell membranes occurred first. That is, binding of antigen by the antibody combining site is not necessarily the first step to trigger a function that involves cell surface (membrane) structures. There is reasonable evidence accumulating pointing to the basic importance of aggregation phenomena associated with effector functions of immunoglobulins: triggering of lymphocytes as result of specific immune responses, feedback phenomena in terminating immune responses, complement activation and subsequent lytic events, clearing of antigen, and anaphylactic reactions. Such mechanisms are all conceivable without the recently introduced network concept which grants importance only to the variable (recognition) regions (127). The very recent finding that 80 percent of the plaque forming cells occurring in the spleen during an immune or mitogenic (lipopolysaccharide) response secrete IgM antibody (rheumatoid factor) with specificity only for the Fc region of IgG antibodies (128) may invalidate the network concept (127) based only on a single functional domain of the immunoglobulin molecule. This report (128) demonstrates once more the multifunctional unit molecule "immunoglobulin."

REFERENCES

1 N. Hilschmann and L. C. Craig, (1965), *Proc. Natl. Acad. Sci. USA* **53,** 1403–1409.

2 J. A. Gally and G. M. Edelman, (1972), *Ann. Rev. Genet.* **6,** 1–46.

3 S. Watanabe, H. U. Barnikol, J. Horn, J. Bertram, and N. Hilschmann (1973), *Hoppe-Seyler's Z. Physiol. Chem.* **354,** 1505–1509.

4 F. W. Putnam, G. Florent, C. Paul, T. Shinoda, and A. Shimizu (1973), *Science* **182,** 287–291.

5 H. Bennich and H. von Bahr-Lindstrom (1974), *Progr. Immunol. II* **1,** 49–58.

6 H. Kratzin, P. Altevogt, E. Ruban, A. Kortt, K. Staroscik, and N. Hilschmann (1975), *Hoppe-Seyler's Z. Physiol. Chem.* **356,** 1337–1342.

7 A. Torano and F. W. Putnam (1978), *Proc. Natl. Acad. Sci. USA* **75,** 966–969.

8 G. M. Edelman (1970), *Biochemistry* **9,** 3197–3205.

9 R. R. Porter (1971), *Harvey Lectures* **65,** 157–174.

10 J. B. Fleischman, R. R. Porter, and E. M. Press, (1963), *Biochem. J.* **88,** 220–228.

11 J. B. Natvig and H. G. Kunkel (1973), *Adv. Immunol.* **16,** 1–59.

12 D. Beale and A. Feinstein (1976), Q. Rev. Biophys. **9,** 135–180.

13 M. E. Koshland (1975), *Adv. Immunol.* **20,** 41–69.

14 R. R. Porter (1959), *Biochem. J.* **73,** 119–126.

15 E. A. Kabat (1978), *Adv. Protein Chem.*, **32,** 1–69.

16 L. Hood, J. H. Campbell and S. C. R. Elgin (1975), *Ann. Rev. Genet.* **9,** 305–353.

17 A. Nisonoff, J. E. Hopper and S. B. Spring (1975), *The Antibody Molecule*, Academic Press, New York.

18 N. Hilschmann, H. Ponstingl, S. Watanabe, H. U. Barnikol, K. Baczko, M. Braun and W. Leibold (1972), *FEBS Symp.* **26,** 31–48.

19 H. Huser and D. G. Braun (1978), *Hoppe Seyler's Z. Physiol. Chem.* **359,** 1473–1480.

20 C. Milstein and J. Svasti (1971), *Progr. Immunol. I*, 33–45.

21 M. Cohn (1974), *Progr. Immunol. II* **2,** 261–284.

22 D. S. Secher, C. Milstein and K. Adetugbo (1977), *Immunol. Rev.* **36,** 51–72.

23 S. Tonegawa, C. Brack, N. Hozumi, G. Matthyssens, and R. Schuller (1977), *Immunol. Rev.* **36,** 73–94.

24 E. Haber and R. M. Krause, eds. 1977, *Antibodies in Human Diagnosis and Therapy*, Raven Press, New York.

25 J. Baenziger and S. Kornfeld (1974) *J. Biol. Chem.* **249,** 7270–7281.

26 H. Metzger (1978), *Adv. Immunol.*, in press.

27 A. Nisonoff, F. C. Wissler and L. N. Lipman (1960), *Science* **132,** 1770–1771.

28 G. E. Connell and R. R. Porter (1971), *Biochem. J.* **124,** 53P.

29 D. Inbar, J. Hochman and D. Givol (1972), *Proc. Natl. Acad. Sci. USA* **69,** 2659–2662.

30 K. Kakinoto and K. Onoue (1974), *J. Immunol.* **112,** 1373–1382.

31 J. Sharon and D. Givol (1976), *Biochemistry* **15,** 1591–1594.

32 A. G. Plaut and T. B. Tomasi, Jr. (1970), *Proc. Natl. Acad. Sci. USA* **65,** 318–322.

33 D. Beale (1974), *Biochim. Biophys. Acta* **351,** 13–20.

34 J. Zikan and I. Miler (1974), *Immunochemistry* **11,** 115–118.

35 H. Metzger (1970), *Adv. Immunol.* **12,** 57–116.

36 R. E. Cathou and K. J. Dorrington (1975), in S. N. Timasheff and G. D. Fasman, eds., *Biological Macromolecules—Subunits in Biological Systems*, Vol. 7 Part C, Marcel Dekker, New York, p. 91–224.

37 J. D. Capra and A. B. Edmundson (1977), *Sci. Am.* **236** (1), 50–59.

38 A. B. Edmundson, K. R. Ely, R. L. Girling, E. E. Abola, M. Schiffer, and F. A. Westholm (1974), *Progr. Immunol. II* **1,** 103–113.

39 D. R. Davies, E. A. Padlan, and D. M. Segal (1975), *Ann. Rev. Biochem.* **44,** 639–667.

40 R. J. Poljak (1975), *Adv. Immunol.* **21,** 1–33.

41 R. Huber, J. Deisenhofer, P. M. Colman, M. Matsushima, and W. Palm (1976), *Nature* **264,** 415–420.

42 J. Deisenhofer, P. M. Colman, R. Huber, H. Haupt and G. Schwick (1976), *Hoppe-Seyler's Z. Physiol. Chem.* **357,** 435–445.

43 P. A. Charlwood and S. Utsumi (1969), *Biochem J.* **112,** 357–365.

44 J. Deisenhofer, P. M. Colman, O. Epp, and R. Huber (1976), *Hoppe-Seyler's Z. Phys. Chem.* **357,** 1421–1434.

45 E. W. Silverton, M. A. Navia, and D. R. Davies (1977), *Proc. Natl. Acad. Sci. USA* **74,** 5140–5144.

46 H. Metzger (1974), *Adv. Immunol.* **18,** 169–207.

47 K. B. M. Reid and R. R. Porter (1975), *Contemp. Top. Mol. Immunol.* **4,** 1–22.

48 O. Götze and H. J. Müller-Eberhard (1976), *Adv. Immunol.* **24,** 1–35.

49 J. M. Kehoe and M. Fougereau (1969), *Nature* **224,** 1212–1213.

50 K. B. M. Reid (1971), *Immunology* **20,** 649–658.

51 M. Colomb and R. R. Porter (1975), *Biochem. J.* **145,** 177–183.

52 J. R. Ellerson, D. Yasmeen, R. H. Painter, and K. J. Dorrington (1972), *FEBS Lett.* **24,** 318–322.

53 K. J. Dorrington and R. H. Painter (1977), *Progr. Immunol. III,* 298–304.

54 D. Yasmeen, J. R. Ellerson, K. J. Dorrington, and R. H. Painter (1976), *J. Immunol.* **116,** 518–526.

55 D. E. Isenman, J. R. Ellerson, R. H. Painter, and K. J. Dorrington (1977), *Biochemistry* **16,** 233–240.

56 M. M. Hurst, J. E. Volanakis, R. B. Hester, R. M. Strond, and J. C. Bennett (1974), *J. Exp. Med.* **140,** 1117–1121.

57 J. C. Brown and M. E. Koshland (1975) *Proc. Natl. Acad. Sci. USA* **72,** 5111–5115.

58 I. Pecht (1976) in F. Melchers and K. Rajewsky, eds., *The Immune System,* Springer-Verlag, Berlin, Heidelberg, New York, p. 41–54.

59 J. W. Goers, V. N. Schumaker, M. M. Glovsky, J. Rebek, and H. J. Müller-Eberhard (1975), *J. Biol. Chem.* **250,** 4918–4925.

60 J. W. Goers, R. J. Ziccardi, V. N. Schumaker, and M. M. Glovsky (1977), *J. Immunol.* **118,** 2182–2191.

61 D. G. Braun, K. Eichmann, and R. M. Krause (1969), *J. Exp. Med.* **129,** 809–830.

62 J. E. Coligan, W. C. Schnute, Jr., and T. J. Kindt (1975), *J. Immunol.* **114,** 1654–1658.

63 W. Schalch and D. G. Braun (1978), in I. Lefkovits and B. Pernis, eds., *Immunochemical Methods,* Academic Press, New York, p. 123–130.

64 W. Schalch, J. K. Wright, L. S. Rodkey, and D. G. Braun (1979), *Eur. J. Immunol.* **9,** 145–149.

65 W. Schalch, J. K. Wright, L. S. Rodkey, and D. G. Braun (1979), *J. Exp. Med.* **149,** 923–937.

66 J. K. Wright, W. Schalch, L. S. Rodkey, and D. G. Braun (1978), *FEBS Letters* **93,** 317–319.

67 G. M. K. Humphries and H. M. McConnell (1977), *Proc. Natl. Acad. Sci. USA* **74,** 3537–3541.

68 J. K. Wright, J. Engel, and J.-C. Jaton (1978), *Eur. J. Immunol.* **8,** 309–314.

69 J. K. Wright, J. Engel, D. Ch. Brandt, and J.-C. Jaton (1978), *FEBS Lett.* **90,** 79–83.

70 J.-C. Jaton, H. Huser, D. G. Braun, D. Givol, I. Pecht, and J. Schlessinger (1975), *Biochemistry* **14,** 5312–5315.

71 D. Givol (1976), in P. Cuatrecasas and M. F. Greaves, eds., *Receptors and Recognition,* Series A, Vol. **2,** Chapman and Hall, London, p. 3–42.

72 S. Sell and P. G. H. Gell (1965), *J. Exp. Med.* **122,** 423–429.

73 F. Loor and G. E. Roelants, eds. (1977), *B and T Cells in Immune Recognition,* John Wiley New York.

74 S. M. Fu and H. G. Kunkel (1974), *J. Exp. Med.* **140,** 895–903.

75 B. Pernis, L. Forni, and L. Amante (1971), *Ann. N.Y. Acad. Sci.* **190,** 420–429.

76 D. S. Rowe, K. Hug, L. Forni, and B. Pernis, (1973), *J. Exp. Med.* **138,** 965–972.

77 *Conference on Immunogenetics of the Rabbit*, National Institute of Allergy and Infectious Diseases, National Institutes of Health, Bethesda, Md., May 3–5, 1978.

78 F. Loor and A. S. Kelus (1978), *Eur. J. Immunol.* **8,** 315–324.

79 F. M. Burnet (1959), *The Clonal Selection Theory of Acquired Immunity*, Cambridge University Press.

80 H. O. McDevitt, ed. (1978), *Ir Genes and Ia Antigens*, Academic Press, New York.

81 K. Eichmann, C. Berek, G. Hämmerling, S. Black, and K. Rajewsky (1976), in F. Melchers and K. Rajewsky, eds., *The Immune System*, Springer-Verlag, Berlin, p. 65–74.

82 H. Frischknecht, H. Binz, and H. Wigzell (1978), *J. Exp. Med.* **147,** 500–514.

83 A. Coutinho and G. Möller (1975), *Adv. Immunol.* **21,** 113–236.

84 J. Watson and R. Riblet (1974), *J. Exp. Med.* **140,** 1147–1161.

85 A. Schimpl, E. Wecker, L. Hübner, Th. Hüning, and G. Müller (1977), *Progr. Immunol. III*, 397–404.

86 C. L. Sidman and E. R. Unanue (1978), *Proc. Natl. Acad. Sci. USA* **75,** 2401–2405.

87 D. Yasmeen, J. R. Ellerson, K. J. Dorrington and R. H. Painter (1973), *J. Immunol.* **110,** 1706–1709.

88 B. Takacz, personal communication.

89 I. C. M. MacLennan, G. E. Connell, and F. M. Gotch (1974) *Immunology* **26,** 303–310.

90 J. O. Minta and R. H. Painter (1972), *Immunochemistry* **9,** 1041–1048.

91 G. O. Okafor, M. W. Turner, and F. C. Hay (1974), *Nature* **248,** 228–230.

92 D. G. Braun, W. Schalch, and I. Schmid (1978), in M. McCarty and J. B. Zabriskie eds., *Streptococcal Diseases and the Immune Response*, in press.

93 T. McNabb, T. Y. Koh, K. J. Dorrington, and R. H. Painter (1976), *J. Immunol.* **117,** 882–888.

94 R. L. Guyer, M. E. Koshland, and P. M. Knopf (1976), *J. Immunol.* **117,** 587–593.

95 K. Ishizaka and T. Ishizaka (1975), *Ann. N.Y. Acad. Sci.* **254,** 462–475.

96 K. E. Becker, T. Ishizaka, H. Metzger, K. Ishizaka, and P. M. Grimley (1973), *J. Exp. Med.* **138,** 394–409.

97 D. M. Segal, J. D. Taurog, and H. Metzger (1977), *Proc. Natl. Acad. Sci. USA* **74,** 2993–2997.

98 H. Metzger (1977), *Progr. Immunol. III*, 427–429.

99 J. Sjöquist, J. Movitz, I.-B. Johansson, and H. Helm (1972), *Eur. J. Biochem.* **30,** 190–194.

100 J. Sjödahl (1977), *Eur. J. Biochem.* **73,** 343–351.

101 H. Hjelm, J. Sjödahl, and J. Sjöquist (1975), *Eur. J. Biochem.* **57,** 395–403.

102 J. Sjödahl (1977), *Eur. J. Biochem.* **78,** 471–490.

103 J. Deisenhofer, T. A. Jones, R. Huber, J. Sjödahl, and J. Sjöquist (1978), *Hoppe-Seyler's Z. Physiol. Chem.* **359,** 975–985.

104 C. Wright, K. J. Willan, J. Sjödahl, D. R. Burton, and R. A. Dweck (1977), *Biochem. J.* 2429–2433.

105 R. R. Porter (1977), *Fed. Proceed.* **36,** 2191–2196.

106 F. F. Richards, W. H. Konigsberg, R. W. Rosenstein, and J. M. Varga (1975), *Science* **187,** 130–137.

107 D. S. Secher (1977), *Nature* **268,** 689–690.

108 D. J. Cameron and B. F. Erlanger (1977), *Nature* **268,** 763–765.

109 D. G. Braun, W. F. Riesen, and H. Huser (1976), in A. J. Cunningham, ed., *The Generation of Antibody Diversity. A New Look*, Academic Press, New York, p. 31–51.

110 M. C. Michaelides and H. N. Eisen (1974), *J. Exp. Med.* **140,** 687–702.

111 R. W. Rosenstein, R. A. Musson, M. Y. K. Armstrong, W. H. Konigsberg, and F. F. Richards (1972), *Proc. Natl. Acad. Sci. USA* **69,** 877–881.

112 A. B. Edmundson, K. R. Ely, R. L. Girling, E. E. Abola, M. Schiffer, F. A. Westholm, M. D. Fausch, and H. F. Deutsch (1974), *Biochemistry* **13,** 3816–3827.

113 M. Cramer and D. G. Braun (1973), *J. Exp. Med.* **138,** 1533–1544.

114 T. Imanishi and O. Mäkelä (1974), *J. Exp. Med.* **140,** 1498–1510.

115 T. Francis, Jr. (1953), *Ann. Intern. Med.* **39,** 203.

116 S. Fazekas de St. Groth and R. G. Webster (1966), *J. Exp. Med.* **124,** 331–345.

117 D. J. Cameron and B. F. Erlanger (1978), *Nature* **271,** 587–588.

118 L. E. Hood (1972), *Fed. Proceed.* **31,** 177–187.

119 N. Hozumi and S. Tonegawa (1976), *Proc. Natl. Acad. Sci. USA* **73,** 3628–3632.

120 C. Brack and S. Tonegawa (1977), *Proc. Natl. Acad. Sci. USA* **74,** 5652–5656.

121 E. A. Kabat, T. T. Wu, and H. Bilowsky (1978), *Proc. Natl. Acad. Sci. USA* **75,** 2429–2433.

122 N. Hilschmann (1969), *Naturwissenschaften* **56,** 195–205.

123. S. M. Tilghman, D. C. Tiemeier, J. G. Seidman, B. M. Peterlin, M. Sullivan, J. V. Maizel, and P. Leder (1978), *Proc. Natl. Acad. Sci. USA* **75,** 725–729.

124 M. Robertson (1977), *Nature* **269,** 648–650.

125 S. Tonegawa, A. M. Maxam, R. Tizard, O. Bernard, and W. Gilbert (1978), *Proc. Natl. Acad. Sci. USA* **75,** 1485–1489.

126 T. H. Rabbitts and A. Forster (1978), *Cell* **13,** 319–327.

127 N. K. Jerne (1976), *Harvey Lectures*, **70,** 93–110.

128 D. W. Dresser (1978), *Nature* **274,** 480–483.

INDEX